WINE GROWING
IN GREAT BRITAIN

A COMPLETE GUIDE TO GROWING GRAPES FOR WINE PRODUCTION IN COOL CLIMATES

STEPHEN SKELTON MW

Published by the Author
October 2020

E-mail: mail@stephenskelton.com
Telephone: 07768 583700

1st Edition published in paperback in Great Britain in 2014
2nd Edition published in paperback in Great Britain in 2020

A CIP catalogue record for this book is available from the British Library.

ISBN: 978-1-9163296-0-7

Proofreading: Sue Proof (www.sueproof.wales)
Index: Dr Laurence Errington (www.errington-index.demon.co.uk)
Design and typesetting: Prepare to Publish (www.preparetopublish.com)

Cover photograph is of Yotes Court Vineyard, West Peckham, Kent.

Other books by S P. Skelton:
Viticulture – An introduction to commercial grape growing for wine production
Vine varieties, clones and rootstocks for UK vineyards
UK Vineyards Guide
The Wines of Great Britain (Infinite Ideas Classic Wine Library)

Acknowledgements

My thanks go to the following for their help and assistance with some of the content of both editions of this book: Sam Barnes (machinery prices), John Buchan (vineyard nutrition), Chris Cooper (pest and disease control), Jim Newsome (trunk diseases), Mike Paul (wine sales), Rob Saunders (pest and disease control), N. P. Seymour (machinery photographs), Richard Smart (trunk diseases), Geoff Taylor (pesticide residue levels), Mike Trought (bird netting and frost prevention), Alex Valsecchi and Nick Wenman (organic and biodynamic viticulture), and Helen Waite (trunk diseases). Finally, I would like to thank the many clients and colleagues in the industry who have shared their experiences and their data with me which has been invaluable in writing this book.

Contents

Contents

Introduction to the 2nd Edition 2020

In 2014, when the 1st edition of *Wine Growing in Great Britain* was published, the area under vines in Great Britain stood at 1,884-ha (4,655-acres). This was double the area it had been ten years previously and we all wondered whether the upward trajectory of planting would continue. Well, today we know the answer to that question. The planted area of vines in 2020 has risen to around 3,500-ha (8,649-acres) – the exact figure is not known with any certainty owing to the lack of official records – an almost doubling of the area under vine in six years. Between 2013 and 2020, the number of commercial vineyards (0.10-ha or more) in Britain has increased from 448 to almost 800 and the average size from 4.21-ha (10.40-acres) to 4.51-ha (11.14-acres). The largest producer by volume, Chapel Down, produced 2.25 million bottles in 2018, over twice the total amount produced in Britain by all vineyards in 2012 (admittedly a very poor year) and the three largest growers, Nyetimber, MDCV, and Chapel Down, between them have 862-ha (2,130-acres) of vineyards. There are also almost twenty-five vineyards of 20-ha (49-acres) or more.

With this expansion of the planted area of course has come an increase in production and an increase in availability. English and Welsh wine, both still, but especially sparkling, is no longer the curiosity it once was and is to be found in many more outlets and on many more wine lists than ever before. Instead of being a rarity, wine merchants now say it's a 'must have' category as customers are actually asking for it! What a change from the dark days in the 1970s when I first started. Even the Champenois admit that the fall in sales of their wines since the financial crisis in 2008 is partly due to the rise in sales of 'ESW'. The quality of both still and sparkling wines has also risen, helped by some remarkable vintages and there are now several English sparkling wines priced at £100 and more, and even still wines, red and white, retailing at £20 or more. In 2020, the well-known wine writer, Mathew Jukes, saw fit to give one (still) wine, the 2018 Oxney Chardonnay, a score of twenty out of twenty commenting 'This is the most resonant and beautiful English wine I have ever tasted'. The industry is also developing in ways not foreseen a decade ago. Independent wineries have opened up in far flung places (Laneberg Wine in Gateshead for instance) and under railway arches and on trading estates in London and the south-east, buying grapes and making interesting wines. Charmat method and carbonated sparkling wines are here to stay, and English wines in cans and kegs are not unknown. Looking at it from the outside, the industry is on a roll.

However, with the expansion of plantings comes an expansion of production and the problems of an imbalance between demand and supply. The cropping area in 2014 stood at 1,506-ha (3,721-acres) with around 10 per cent of the planted area 'not in production'. For 2019 the cropping area was 2,438-ha (6,024-acres) – again, an estimated figure – but with around 33 per cent of the planted area 'not in production'. And 'not in production' means 'has not yet produced a single grape'. If you look at sparkling wine in isolation and take the area from which no grapes have been harvested and add on to that those vineyards who have harvested and who have made wine, but whose wines are too young to put on the market, then the amount of productive capacity which has not yet got wine on the market is around 50 per cent of the current planted area of sparkling wine varieties. And if this backlog of planted vineyards wasn't worry enough, yields in the last six years have also been above average.

Yields between 2008 and 2013 averaged 19.12 hl-ha a year; yields between 2014 and 2019 averaged 29.26 hl-ha a year; and yields in 2018 and 2019 averaged 39.11 hl-ha a year. To put it another way, the number of 75 cl bottles produced per year between 2008 and 2013 was 2.85 million bottles; between 2014 and 2017 it was 5.20 million bottles per year; and in 2018 and 2019 it averaged 11.79 million bottles a year. In around 3 years' time, when all the current planted area is cropping and if yields are the same as they were in 2018 and 2019, then we would be facing a harvest of over 18 million bottles. That's certainly food (or wine) for thought.

The writing of the 2nd edition of this book comes at a turbulent time in several different ways. The turbulence of Brexit has been overtaken by the turbulence of the Covid-19 pandemic, and with sales of all wines, still and sparkling, down and with growers in Champagne facing a limit to the amount of grapes they can harvest in 2020 which will mean leaving up to half their grapes unpicked, the 'elephant in room' for producers of English and Welsh wines – that of the unprecedented stocks of wine – seems to have taken a back seat. The actual level of sales of all English and Welsh wines is unknown and is the object of some debate and speculation amongst industry members. My best estimate is that in 2019, sales, including exports, were around 4.5 million bottles, and with the slowdown of sales due to Covid-19, are not expected to be any more in 2020. With stocks of wines estimated to be between six and eight times the level of sales (stock levels are also not known with any real precision) basic economic market theory would suggest that a stimulus to demand can only come from a drop in prices of wine, which in turn will have a knock-on effect on prices of grapes.

Of course, we have been here before as the reduction in planted area between 1993 and 2004 showed when 30 per cent of all vineyards were grubbed and production fell from an annual average of 2.53 million bottles between 1989 and 1996, down to 1.80 million bottles a year between 1997 and 2008. However, this time I feel that it's going to be different. Many of today's producers are bigger, better and better financed, especially sparkling wine producers, who make up 70 per cent and more of producers by volume. To be in the sparkling wine business you have to bottle last year's crop and once it's in a bottle and in a store, the urge to cut your losses, lower your price and sell is far less than still wine producers facing full tanks and another harvest round the corner. They are the ones most likely to sell at any price just to create space, or to leave grapes unpicked, thus distorting the market for grapes.

But these are temporary blips in what is otherwise a vibrant industry and they will be overcome, even if for a time the public gets to enjoy some cheaper, but good quality, wines. The two Champagne houses that have set up shop in Britain have yet to produce wine from their plantings, and when these wines arrive, there will be another round of interest in the sector, and these two producers may well be followed by others. With the quality of our wines rising with every vintage, with more well-trained vineyard managers and winemakers working in the industry than ever before, and with inventive marketeers looking for new ways and new outlets, both at home and abroad, this industry is here to stay.

Stephen Skelton MW
Fulham September 2020

Wine Growing in Great Britain has been written not only for anyone interested in growing grapes for winemaking in the British Isles, but also for those attempting to do the same in other ultra-cool climates. It has been written for those who know very little about the subject, but are keen to know more, together with those who have perhaps already whetted their appetites by growing vines on a small scale, but who now wish to move on to bigger things. It has also been written, as far as is possible, in non-scientific terms so that anyone can understand the subject and to relate to the very practical issues surrounding the growing of grapes in challenging climates. It has also been written from a commercial point-of-view and relates the costs of establishing and looking after vineyards to the potential income, something which I do not believe has ever been covered in a book of this type before.

I am not quite sure when I saw my first English vineyard, but it was probably Biddenden Vineyards, the nearest to Headcorn in Kent where we then lived. This would have been in about 1973-4. My interest was sufficiently aroused for me to seek out, and join, the English Vineyards Association and to find out about the small world of English and Welsh wine and we went to see vineyards at Nettlestead and Lamberhurst in Kent and New Hall in Essex. I also attended a one day conference at the National Agricultural Centre at Stoneleigh organised by the British Farmer and Stockbreeder at which several of the then-luminaries in the world of home-grown wine spoke and at which Robin Don MW, then owner of Elmham Park Vineyard in Norfolk, said that in order to get a return, you needed to average 2 tonnes-acre and sell wine for between £1.30 and £1.70 per bottle!

By degrees I learnt a bit more about grape growing and wine making and eventually decided that it was something I would like to pursue in a professional capacity. I was aged twenty-six. A visit to Germany in early 1975 took things a stage further and in June 1975 I found myself on a twelve-month placement in a vineyard in the Rheingau, to be followed by eight months studying at Geisenheim. All of this was in preparation to taking over a mixed fruit and arable farm which my late father-in-law had bought and generously made available to us for our new enterprise. This farm, then called Spots Farm, and subsequently Tenterden Vineyards (and today Chapel Down) became my base for the next twenty-plus years of grape growing and winemaking.

When I started growing grapes – I planted my first vine on Easter Monday in 1977 – my knowledge was probably better than most, but was still woefully lacking. How I planted, established and managed the vineyard was a combination of my experiences in Germany, what I had learnt at Geisenheim, plus what I could learn from fellow vine growers and winemakers in Britain and from the very few books on the subject that were written with British conditions in mind. Nick Poulter's *Wines from Your Vines* published in 1974 was about the best there was. Gillian Pearkes' 1982 book *Vinegrowing in Britain* and Jack Ward's 1984 book *Vine Growing in the British Isles* had yet to be written. Since those early days, I have been a viticulturalist, winemaker both at Tenterden and Lamberhurst and, since 2002, a consultant working with growers to establish vineyards all over the south of England (and one near Calais!) and this book is the culmination of my forty years of growing grapes and making wine in our sometimes exasperating climate.

Sadly, the history of English and Welsh winegrowing is littered with abandoned and grubbed vineyards. Poor site selection, the wrong choice of varieties and a lack of appreciation of the task ahead and of the skills required to manage a successful vineyard in a challenging climate have all played their part. This book has been written with the hope that it will help those who wish to become part of the small, but expanding, vinegrowing and winemaking industry in Britain to avoid some of the pitfalls which wait at every turn.

When I look back at the last sixty-plus years since the first commercial vineyard was planted in Britain and see what was achieved by a very small band of pioneers, the vineyards they created and the great wines made, and all achieved with very little in the way of official help and guidance, I am filled with a degree of pride and wonder. However, when I look forward to the next forty years (and more), and see the children and grandchildren of those pioneers, and look at the scale of some of the enterprises now planted and yet to be planted and at the skill, professionalism and, above everything else, the enthusiasm of today's winegrowers, I see a future of even better vineyards, better grapes and, of course, better wines.

Stephen Skelton MW
Fulham, May 2014.

Chapter 1

A short history of wine growing in Great Britain

The early days

Although vines have been grown in the British Isles since before the Romans arrived, and at various times ever since, the revival of viticulture in the 1950s and 1960s was based upon very different circumstances. Vineyards prior to then had, almost without exception, been grown to produce wine for consumption in monasteries, palaces and country houses and, with perhaps the exception of the third Marquess of Bute's vineyards at Castell Coch in South Wales, never for the production of wine for commercial sale.[1] The experiments carried out by Ray Brock (Raymond Barrington Brock) at his Oxted Viticultural Research Station, which started in 1945, showed that grapes for wine production could be grown in our climate and could be made into wine of something approaching a marketable quality.

It was these experiments that led directly to the planting, in 1952, of 0.40 ha (1 acre) of vines at Hambledon in Hampshire by Major-General Sir Guy Salisbury-Jones. This was to be the first modern vineyard, one based upon disease-resistant varieties, and would produce the first harvestable crop of ripe winemaking grapes in the autumn of 1955. What was also changing in these post-war years was the approach to food and wine by the public. Foreign travel and books such as Elisabeth David's Mediterranean Cooking (1950) and French Country Cooking (1951) made people more adventurous in their diets and the adoption by new wine drinkers of wines such as light, off-dry, German white wines, helped open the way to English and Welsh wines of a similar style.

Since those early days, British viticulture has undergone enormous changes. After Hambledon came Jack Ward's vineyards at Horam Manor in 1954 and the Gore-Brownes'

Ray Brock standing in front of his HRG 'Scientific' which he built and raced at Spa in Belgium in 1948

at Beaulieu in 1958, but together these three amounted to less than 5 ha (12.36 acres). However, their trials and tribulations and the actual production and sale of wines fuelled considerable interest in viticulture and the 1960s and 1970s saw a steady expansion in the number of growers, and in both the total area and geographical spread of vineyards. By the end of 1975, the total area under vine, as recorded by MAFF[2] in a voluntary survey, was 196 ha (484 acres) and what had been a minute cottage industry, peopled mainly by retired army and navy officers and a few gentleman farmers, was turning into something approaching a mini-industry.

The expansion of the vineyard area continued throughout the '80s and early '90s until it reached a total planted area of 1,065 ha (2,632 acres) with 479 separate vineyards in 1993, after which it started to decline. The

---

1 The history of almost 2,000 years of grape growing in the British Isles can be found in the opening chapter of my UK Vineyards Guide, as well as in books by Edward Hyams, Ray Brock, George Ordish and Hugh Barty-King. See bibliography for a full list.

2 MAFF, the Ministry of Agriculture Fisheries and Food, was the forerunner of today's DEFRA, the Department for Environment, Food and Rural Affairs. Until 1989, all MAFF vineyard surveys were voluntary and therefore certainly inaccurate. In 1988 the total planted area was recorded by the last voluntary survey as being 546 ha (1,349 acres). The first compulsory survey a year later (1989) recorded a total of 876 ha (2,165acres).

UK planted hectares of vineyards 1989 - 2020
Source: Wine Standards & UK Vineyards Guide

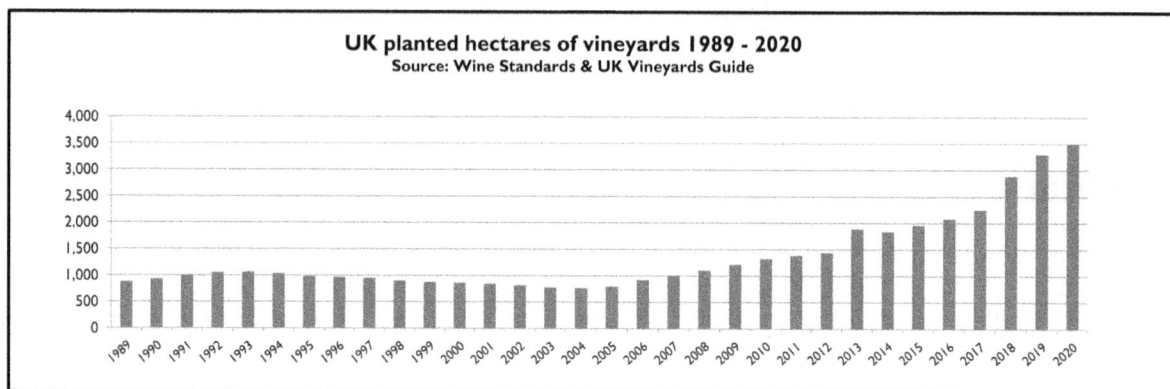

*Table 1 **Note:** The 1989–2017 figures in the graph above are the result of Wine Standards (WS) mandatory surveys and can therefore be assumed to be accurate. The figures for 2018–20 are industry estimates.*

main varieties at this time were the German cross-bred varieties (listed in alphabetical order): Bacchus, Huxelrebe, Madeleine Angevine 7672, Müller-Thurgau, Ortega, Reichensteiner and Schönburger, together with the French–American hybrid, Seyval blanc. Together these eight varieties accounted for around 75 per cent of Britain's vineyard area. Although these varieties could produce interesting and enjoyable wines (this was in the days when wines such as Blue Nun, Golden Oktober and Black Tower were the most popular wines in Britain), their acceptability by both the wine trade and the wine-buying public was limited. Although the total planted area started to fall after 1993, the area in production continued to rise (as young vineyards started to crop) rising to 842 ha (2,081 acres) in 1998, after which the area in production started to fall. By 2004, the total area under vine had fallen to 761 ha (1,880 acres), a drop of 304 ha (751 acres) – almost 30 per cent of the planted area – from its peak, with the area in production falling to 697 ha (1,722 acres) by 2007. How much money (let alone the time and effort) went into those grubbed 304 ha (751 acres) one can only wonder at, but it must have been at least £6–7 million in vineyards alone, to say nothing of wineries and other costs. Then, following the über-warm year of 2003, things started to change. Since the low point in the planted area of 761 ha (1,880 acres) in 2004, the area in Britain under vine has risen to where, in 2020, it will be approaching 3,500 ha[3] (8,649 acres). What has also increased is the average size of vineyards: in 1993 there were 479 vineyards with an average size of 2.22 ha (5.49 acres); in 2020 it is estimated that there are almost 800 commercial vineyards (of 0.10 ha (0.25 acres) or more)

Nyetimbers' Tillington Vineyard

with an average size of 4.51 ha (11.14 acres), an increase in average size of over 100 per cent.

The Nyetimber effect

This change in the fortunes of British vineyards has been dubbed (by me) the 'Nyetimber effect', although in reality it perhaps ought to be called the 'climate change effect'. In 1988, just as the plug was being pulled on Britain's own Liebfraumilch lake, two Americans, Stuart and Sandy Moss, started planting a substantial (for the time) area of Chardonnay, Pinot noir and Meunier solely for the production of bottle-fermented sparkling wines. Most vineyard owners, who were then happily growing and making still wine, thought they were mad (I certainly did) but when, in 1997, they released their first wine – the 100 per cent Chardonnay *1992 Nyetimber Première Cuvée Blanc de Blancs* – which won an IWSC[4] Gold Medal and that competition's English Wine Trophy, views started

---

3 Wine Standards (formerly the Wine Standards Board, and then the Wine Standards Branch) of the Food Standards Agency is the body responsible for implementing and policing the rules and regulations to do with wine marketing of all wines, home grown or foreign and wine making in Britain. It also collects vineyard data for the Vineyard Register and collects production data for the annual wine production statistics. Unfortunately, it has fallen behind in its collection duties, and the last official figure released was in 2018, based upon 2017 data, which showed there to be 2,382 ha. The industry estimate, based upon surveys and an estimate of vine plantings between 2017 and 2020, is 3,500-ha.

4 International Wine and Spirit Competition.

to change. Then, in 1998, their second release, the *1993 Nyetimber Classic Cuvée* (a true Champagne-variety blend), went one better, winning not only an IWSC Gold Medal and the English Wine Trophy, but also the Bottle Fermented Sparkling Wine Trophy. Some of us then realised that the game was up for German-variety-based still wines.

Whilst the bravery of the Mosses cannot be denied, it was the change in the climate that made it possible to start growing the Champagne varieties. Chardonnay, Pinot noir and Meunier[5] had already been tried in several different vineyards in Britain and found wanting.[6] With the summer temperatures between 1952 and 1988 seldom rising above 30°C (1976 being a notable exception) Britain fell well inside the accepted definition of 'cool' (and therefore marginal) for viticulture. However, in the eighteen years between 1989 and 2006, there were only four years when the temperature didn't rise above this magic figure. Met Office figures show that between 1960 and 1991, the mean night-time July temperature in the south-east of England was 11.56°C, whereas between 1981 and 2010 this had risen to 12.24°C. Likewise, the mean day-time temperature for the same month and region rose from 21.19°C to 22.01°C. These rises, although seemingly small, in fact take the temperature to the tipping-point where growing these varieties for sparkling wine becomes feasible.

Whilst the higher day-time temperatures are obvious in their effect, it is not always appreciated what impact the higher night-time temperatures have had upon vineyards in Britain. The vine's leaves have to reach a certain temperature[7] before photosynthesis starts and carbohydrates are produced. The warmer nights (and if you are a non-believer just ask bedding retailers how many 15 tog duvets they sell these days) mean that a vine's leaf warms up more quickly in the morning, giving the it more time in the day in which it can produce sugar.[8] This produces – along with the rest of the vinegrowing world – higher natural alcohol levels. If you take natural sugar levels in vineyards in Britain for, say, vintages between 1979 and 1999 and compare them with the sugar levels from the same vine varieties for vintages this century, in many instances they have almost doubled. Whereas winemakers in Britain were used to harvesting grapes at 6–8 per cent natural alcohol and enriching (chaptalising) the juice by 4.0 per cent or even 4.5 per cent, since 2000 it has not been unheard of to harvest grapes at 11 per cent, 12 per cent and even 13 per cent and to make wines without any sugar additions at all.

Looking back at my winemaking records for Tenterden Vineyards (today's Chapel Down) I see that between the first vintage in 1979 and 1989, the average potential alcohol level for Müller-Thurgau was 7.1 per cent (56.8°OE) and for Seyval blanc 6.5 per cent (52.4°OE). There was an exception in 1990 with Müller-Thurgau rising to 11.1 per cent (82°OE) – the first time in my winemaking career that natural alcohol level rose to double figures – and Seyval blanc to 9.2 per cent (70°OE), but for the next eleven years, until I stopped winemaking there in 2000, natural sugar levels for these two varieties were very seldom over 10 per cent (75°OE). (And for the sake of putting these seemingly low sugar levels into a quality context, between 1979 and 1989 the wines I produced from these two varieties won the Gore-Browne Trophy twice, plus two gold and two silver medals in the UKVA[9] competition.)

In 2019, however, winemakers in Britain experienced some of the lowest natural sugar levels seen for many years. The average for all varieties recorded for the 2019 ICCWS-WineGB Yield Survey[10] was 70.6˚OE (9.3%), with the average for the four most widely planted varieties – Bacchus, Chardonnay, Meunier and Pinot noir – only being 68.0˚OE (8.90%). As a comparison, the figures in 2018, a much warmer year, were 75.2°OE (10.0%) for all varieties and 76.8°OE (10.30%) for the four named varieties. These low natural sugar levels in 2019 remind us (as if any such reminder were really required) that Britain is an unpredictable place in which to grow grapes and make wine. (It is perhaps also worth putting British alcohol levels into the context of those in Champagne where, up until the late 1980s, the minimum natural alcohol at which grapes could be picked was 7 per cent and even today it is only 9.5 per cent).

Since the early days of Britain's winegrowing industry, there have also been other major changes which have had a huge influence upon the quality and marketability of our wines: experience and vine varieties. Those pioneer grapegrowers and winemakers typically had no experience of growing vines (and often no experience of growing anything very much at all), their winemaking skills were usually gleaned from home winemaking books, and they knew nothing of the sterile bottling of wines with residual

---

5 When reference is made to 'Champagne varieties' from now on it will only include Chardonnay, Meunier and Pinot noir, even though Arbane and Petit Meslier, Pinot blanc, and Pinot gris are permitted. These last four varieties only account for 0.3 per cent of the planted area in Champagne.

6 I remember visiting the Paget brothers at Chilsdown Vineyard (in Singleton, West Sussex) at harvest-time in 1981 and seeing their small patch of Chardonnay, the grapes of which were bright green and still as hard as bullets and joking that they would be good cannon-fodder to fire at the birds then starting to attack their Müller-Thurgau and Reichensteiner.

7 There is no single temperature at which photosynthesis starts as it is a progressive process which increases as temperatures rise. It is also affected by other factors – water availability, light intensity and wavelength – and is a hugely complex process. However, the warmer it is (within reason) the sooner it starts.

8 Since writing this, the Met Office has introduced the concept of 'tropical nights', nights where the temperature doesn't fall below 20˚C. Between 1961 and 1990 there were 44 such nights, around 1.5 per year (although most of them in 1976 and 1983). Between 1991 and mid-August 2020, there were 84 such nights, almost 3 per year.

9 United Kingdom Vineyards Association. This was the predecessor to WineGB which is now the national body representing Britain's vineyard owners and winemakers.

10 The ICCWS-WineGB Yield Survey is a survey carried out with funds provided by the surplus generated by the International Cool Climate Wine Symposium, held in Brighton in 2016, plus additional financial support from WineGB. Surveys have been carried out in 2018 and 2019 and will hopefully be continued in years to come.

Wine Growing in Great Britain

Great Britain - Vine Varieties 1990-2020		
Variety	1990	2020
Chardonnay	19.9	975.0
Pinot noir	32.4	925.0
Bacchus	76.0	295.0
Meunier	5.5	275.0
Seyval blanc	122.7	145.0
Reichensteiner	113.9	75.0
Frühburgunder, Blauer	Not recorded	70.0
Solaris	Not recorded	70.0
Rondo (Gm 6494/5)	2.0	65.0
Ortega	29.5	55.0
Madeleine x Angevine 7672	54.6	50.0
Regent	Not recorded	35.0
Pinot blanc	Not recorded	35.0
Pinot gris (Ruländer)	2.6	35.0
Müller-Thurgau	184.5	30.0
Phoenix	0.4	25.0
Schönburger	75.3	20.0
Siegerrebe	9.3	20.0
Dornfelder	5.4	18.0
Huxelrebe	43.9	15.0
Orion	0.7	10.0
Triomphe	7.5	8.0
Acolon	Not recorded	7.0
Auxerrois	9.1	7.0
Sauvignon blanc	0.5	7.0
Cabernet Cortis	Not recorded	2.5
Würzer	14.3	2.0
Albãrino	Not recorded	2.0
Dunkelfelder	3.5	1.5
Chasselas	2.5	1.5
Miscellaneous	112.5	218.5
Totals	928.5	3,500.0
Note: 2020 figures are an industry estimate		

Table 2

sugar.[11] Today, there are viticulturalists and winemakers, in some instances the sons and daughters and even the grandsons and granddaughters of those who started their vineyards, who have been to colleges both overseas and in Britain and have degrees and experience in grapegrowing and winemaking. Whilst no one would pretend that we have got it one hundred per cent right yet, we have

travelled a long way in seventy years.

In our selection of vine varieties, we have also seen major changes. Hambledon was initially planted with three French–American hybrids: Seyval blanc (Seyve-Villard 5-276), Aurore (Seibel 5279) and Seibel 10-868. Jack Ward and the Gore-Brownes also planted Seyval blanc, but twinned it with what was then often called Riesling Sylvaner,[12] but which we know today as Müller-Thurgau (or occasionally Rivaner) and these two became the default varieties for almost all the early vineyards. Jack Ward, who spoke fluent German, also imported vines and introduced growers to Bacchus, Huxelrebe, Reichensteiner, and Schönburger and these, together with a Brock special, Madeleine x Angevine 7672 (of which cuttings were sent to him from Alzey, Germany in 1957), became the foundation of almost all vineyards. In 1984, these seven varieties accounted for 74 per cent of the planted area of 430 ha (1,063 acres). Pinot noir accounted for only 11 ha (27 acres) or just 2.6 per cent of the total and Chardonnay was not separately recorded as it was barely planted at all.

Over thirty-five years later, the spectrum of vine varieties grown in vineyards in Britain is very different. Chardonnay and Pinot noir – the top two varieties – and Meunier together account for just over 62 per cent of the planted area; another seven varieties, Bacchus, Blauer Frühburgunder (Pinot noir Précoce), Ortega, Reichensteiner, Rondo, Solaris and Seyval blanc (listed in alphabetical order), account for another 22 per cent. Together these top ten varieties account for just over 84 per cent of the recorded vine variety area of 3,500 ha (8,649 acres).

The ability to grow and ripen classic 'noble' varieties such as Chardonnay and Pinot noir and craft them into world-class – in some instances even world-beating – bottle-fermented sparkling wines has also had some interesting consequences. Previously, a wine retailer or sommelier might have to say, when asked what an English wine was like: 'well, it's made from Müller-Thurgau and Schönburger with a touch of Reichensteiner and its tastes a bit like German wine or maybe a wine from Alsace' and in doing so lose the customer's attention and interest before the second umlaut. Today, when faced with the same question about a bottle of English sparkling wine, the same retailer or sommelier might say: 'well, it's just like Champagne' adding, for extra effect, 'and it's made from Chardonnay and Pinot noir' and in this short explanation convey a message that every wine drinker understands.

This change I believe has made a huge difference to the industry and has given the wine trade in all its many guises the confidence to get behind, to support and to sell the home-grown product. In still wines, the practice by wine producers in Britain of varietal naming has somewhat diminished and many still wines are now sold under brand names, thus getting around the 'umlaut' problem.

11 When I won my first Gore-Browne Trophy in 1981 (I won two more subsequently) the then Chairman of the English Vineyards Association – Colin Gillespie – said to me (I suspect only half-jokingly): 'Stephen, you cheated really, you were trained'. What he meant, of course, was that I had spent some time in Germany and spent some time at Geisenheim whereas almost all the other growers and winemakers in Britain at the time were self-taught.

12 Sylvaner is the old spelling and still used in Switzerland. In the EU, Silvaner is the official name.

Sir Guy and Lady Hilda Salisbury-Jones standing outside the front door of Hambledon in 1953

The exception to this rule are wines made from Bacchus, which have actually increased in popularity (in line with the planted area) and are now seen as something of an English speciality. The name is so non-German and so redolent of all things vinous, that most consumers have no idea that it's a vine variety (at least not until they read the back label) and even many wine trade people think it some sort of British-derived brand name. Let's thank Peter Morio and Bernhard Husfeld, its breeders, for not giving it a name like Dionysüs-reben.

Other still wines, made from international varieties such as Pinot blanc, Pinot gris and Pinot noir certainly benefit from having recognisable names and the days of trying to persuade the public to like wines made from Müller-Thurgau, Reichensteiner and Huxelrebe are mostly behind us. Seyval blanc, always the odd one out, is often used for sparkling wine production (where varietal labelling has never been as common as for still wines) and the variety is sufficiently peculiar to Britain (at least in a European context) and has enough supporters (Oz Clarke consistently rates it highly) that I am sure it will survive, indeed flourish, especially given that it can produce fruit and ripen that fruit even in poor years and usually requires little in the way of chemical treatments to keep it free of disease. Having said that, with the rise in sugar levels affecting all varieties including Seyval blanc – it averaged 70°OE in 2018 – botrytis, which was rarely seen on this variety, has been seen far more widely than it used to be.

However, climate change and the better consumer acceptance of our wines haven't changed the basic facts of viticultural life in Britain. We have a marginal climate for profitable grape growing. Climate change has not made the task of finding a suitable site for a vineyard any easier and it is no easier today to plant and establish a vineyard that will produce the quantity and quality of grapes required to create a viable, self-supporting, sustainable[13] enterprise.

13 The word 'sustainable' is being used in its true sense i.e. to keep going as a commercial enterprise. Vineyards in Britain, if they are to survive into the future, have to learn to become self-supporting businesses, not reliant on the benevolence of their founders. Only once they have learnt to become sustainable in this sense can they get to grips with other matters lumped under the banner of 'sustainability'.

Chapter 2

Why plant a vineyard in Great Britain?

Why plant a vineyard in Great Britain?

When I visit new potential growers for the first time – vineyard virgins as I sometimes call them – I often ask the question: 'why do you want to plant a vineyard in Britain?' The answer I get is often along the lines of: 'we thought it would be fun' or 'well, we've got this field that's not doing very much and it's very sheltered'. Seldom is it: 'I have researched the subject and I believe it can give me a good return over the next thirty years' or 'I really like English wine and would love the chance to be involved with its production and marketing'. It's a pity because if there were a few more of the latter and a few less of the former, our grapegrowing industry would be in a better position than it currently is.

However, having said the above, I will own up straightaway to being in the first camp myself. I have often been asked: 'what made you want to start a vineyard in England' and I usually make some comments about being interested in wine (just about true); having seen some local vineyards (true, although I didn't like many of their wines); and having a father-in-law who was (a) a farmer and (b) keen to support his daughter and son-in-law in what he saw as an interesting enterprise (thankfully, very true) – none of which really answer the question in anything but a roundabout way. Many prospective vineyard owners in Britain have little idea of what their total outlay will be or how long it will take them to get to the point where the vineyard is self-supporting, and what sort of return on capital they are likely to see. In short, many have very little idea of what they are letting themselves in for!

There are several reasons for this. The first is that they often appear not to care. They believe they know roughly what it is going to cost to plant and establish a vineyard; they know that it takes approximately four to five years to get vines to crop; and they live in hope that it will 'be alright on the night'. They also take the view that 'if others are doing it, then I can do it' and plough on hopefully,

forgetting (or perhaps ignoring) the fact that other people's pockets are sometimes both broader and deeper and other people's approach to a return on capital is not their bank manager's. I have often been told by those contemplating spending hundreds of thousands of pounds, even several millions, on vineyards, a winery, stock and a workforce, that: 'return on capital is not important. I've made the money and might as well spend it. Land isn't being made any more and it's a sound investment. If I create a viable business which I can sell at some stage, then it will all have been worth it'.

The truth is that to make money out of growing grapes, making wine and selling that wine for a profit is a hard task in ANY country, let alone in a country where the forces of nature are ranged against you. When I say 'hard', I don't mean that its physically demanding (which it is) or that it takes a lot of perseverance (which it does). What I mean is that it is difficult to know, in advance, before a single sod has been turned, before a single vine planted or grape harvested or bottle of wine made, what the following will be:

• The yield from your vineyard
• The quality of your grapes
• The quality of your wine
• The price of your grapes and/or wine.

Without knowing these factors, who can say what the returns will be?

The economics of grapegrowing in Britain

Note: References to a 'bottle' of wine refer to a 75 cl bottle unless otherwise stated. References to a 'case' refer to a nine-litre case of wine, i.e. twelve x 75 cl bottles, even though many wines, especially sparkling wines, are packed in six-bottle cases.

Vineyard economics worldwide are bedevilled by inconsistencies, and finding the true cost of production of a tonne of grapes or a bottle of wine is almost an impossible task. Over what time period do you write off the investment in vines and trellising which could last forty or more years? Do you look at expenditure on land in the same way as expenditure on other assets? Many do not. A winery consists of equipment, much of which is static, made of stainless steel which, with care and upkeep, will last for decades, even centuries. How does one put a true cost on this? Investment in stock is often regarded as money in the bank, and therefore not really part of the overall capital expenditure. When you consider, however, that someone making sparkling wine, which will only sell when it has been on the lees for between two and four (or five, or six, or even more years), might have to spend three to four times as much on stock as on the vineyard itself, then surely this has to be factored into the equation?

What follows is therefore an attempt to lay out, for the novice, the facts of life behind an investment in a wine producing enterprise in Great Britain.

Land prices

According to Strutt & Parker's Winter 2019–20 'English Estates and Farmland Market Review', the price of arable farmland in the South East of England for the top 25 per cent of sales averaged £26,563 ha (£10,750 acre) with the bottom 25 per cent of arable land sales averaging £18,038 ha (£7,300 acre). Pasture land for the top and bottom 25 per cent sold for between £20,386 ha (£8,250 acre) and £14,332 ha (£5,800 acre). The highest price achieved for arable land in this period was £43,243 ha (£17,500 acre), down from £49,420 ha (£20,000 acre) in 2016. Other regions such as East Anglia, the South West and Central and Midlands reported similar prices. Savills 'Farmland Values Survey' quote similar prices for the end of December 2019: 'Prime Arable' in the South East of England was £22,931-ha (£9,280-acre) with 'All Land Types' for the South East £20,119-ha (£8,142-acre). All the above prices are for bare land, i.e. with no accommodation or buildings of a substantial nature.

However, unlike sales of what might be termed 'general' agricultural land, land for vineyards is bought and sold infrequently and often in fairly small parcels. This makes average prices impossible to state. Added to this is the fact that many people looking to set up vineyards want such things as south-facing, sloping land, below a certain height above sea level, often with soil of a certain type and with good roadside access, and buildings to convert into wineries and tasting rooms, all of which will naturally increase the price. Any agent acting for a vendor of land which ticks all these boxes would surely be licking their lips at the commission they would be earning from such a sale. In 2009, Nyetimber paid £40,000-ha (£16,000-acre) for around 30-ha of prime chalk 'bare' farmland near Stockbridge and whilst this might seem excessive, similar land in Gloucestershire sold for £37,500-ha (£15,000-acre)

Painshill Park Vineyard, originally established betweeen 1738 and 1773 and re-planted on the same site in 1992

in 2013. Today (2020) really good land for vineyards, ideally situated and with all the right requirements, might well fetch at least £50,000-ha (£20,000-acre), possibly more, and I know of some sales taking place at prices substantially above this figure. It is also worth remembering that when one buys a hectare of land, typically only around 80 to 85 per cent of it is actually plantable, the rest being occupied by headlands, walkways, roadways, odd corners and other non-vineyard areas. This makes land costing, say, £15,000-ha, actually cost 15–20 per cent more for the planted area.

Someone wanting to set up a vineyard on a reasonable amount of land to plant an initial swathe of vines – say, 5-ha (12.5-acres), plus room for expansion – say, another 20-ha (50-acres), plus some amenity land – say, another 20-ha (50-acres), together with a modest house and some farm buildings and maybe a small cottage for agricultural staff, could be looking at an investment of £3–4 million and possibly more. Add in the specific requirements for a successful vineyard enterprise and the 'vineyard premium' (as I have heard several land agents call it) could add significantly to the bill.

Value of established vineyards

As with all land suitable for vineyards, already planted and established vineyards in Britain which come up for sale are few and far between. In the 1980s and '90s, I used to get calls from estate agents saying: 'I am selling a house with a vineyard and wonder what additional value it might have'. To which I would usually respond: 'minus £1,000 an acre since this is the cost of taking the vineyard out and reinstating the land to good grazing or arable land'. Whilst this might have been a slight exaggeration, the fact was that very few people wanted vineyards and they were usually only of value to the people who planted them. Today, the situation is very different. Whilst it would be wrong to suggest that every vineyard is worth more a fortune, there have been a small number of sales of mature vineyards over the last few years where good prices have been achieved.

In October 2012, the 4.17-ha (10.3-acre) Halnaker vineyard, near Chichester, planted in 2006 with Champagne varieties, with a patchy, but not hopeless, cropping record over four vintages, was sold to Shellproof PLC (majority owned by Lord Ashcroft and now owners of Gusbourne Estate) for a total of £360,000. If you factor out the 1.13-ha (2.8-acres) of unplanted land at a value of £20,000-ha (£8,000-acre) and another £4,500 for some machinery and equipment, the price paid for the planted land was £79,912-ha (£32,340-acre). This was despite the vineyard requiring quite a lot of remedial work on the trelliswork – many of the end-posts and intermediates needed replacing – and being situated on a busy road with a difficult access. Although the site was viticulturally sound, it was certainly not a good location to establish a roadside sales outlet or a winery and this price included a large premium for scarcity value. Had this vineyard had more additional land to plant, a house and perhaps buildings to turn into a winery and sales outlet, one wonders what the price might have been. When the same company bought Gusbourne Estate in September 2013, the total purchase price was £7 million for the property, business, stocks and goodwill. The property element of the purchase price was £5.287 million which was for 20.5-ha (50.7-acres) of established vineyards, plus around 100-ha (250-acres) of arable land, woodland and farm buildings. The offer documents do not assign a value to the planted land, but by making some educated deductions, the price paid for the vineyards appears to be around £86,000-ha (£35,000-acre) which I considered high, but then how often do vineyards of this size, age and quality come on the market? Very rarely. Given that the value of grapes from these vineyards, net of costs, would appear to be around £11,000-ha (£4,450-acre) per year, this represents a multiple of eight times yield, which is not unreasonable.

My view in 2020 is that a well-sited and well-established vineyard, planted with a spread of decent varieties with a fair cropping record, ought to fetch at least £80,000 per hectare (£32,376 per acre) and it wouldn't take many additional features to get a premium over that price. In 2018 I did a valuation for a bare vineyard, i.e. no buildings, where the price was over £100,000-ha (£40,000-acre). I have recently (2019) valued a ten-year-old 9-ha (22-acre) vineyard, planted with Chardonnay, Pinot noir and Meunier and with a very good cropping record of around 8 t-ha (3.2 t-acre) and a reputation for the quality of its wine, at £86,000-ha (£35,000-acre). This vineyard, together with other land, a house and buildings, sold in 2020.

Leased land for vineyards

Some vineyards in Britain have been established on leased land and this may be an option if a site to buy cannot be found or if the capital is not available, and leasing is certainly going to cost less than buying land. The length of the lease would need to be at least thirty years in order to recover the cost of the vineyard establishment and to get the best years of the cropping life of the vines. As for rents, they can either be straightforward annual amounts tied into some escalator like the Retail Prices Index (in one of its many different variants) or a combination of rent plus a percentage of crop value. In the 2014 edition of this book, I stated that 'I would not expect the rental to exceed £650-ha (£250-acre)'. However, since then we have seen Chapel Down take a lease on 157-ha (388-acres) on the Kentish North Downs at Boxley, near Maidstone and, in the face of a challenging bid from Mark Dixon's MDCV Ltd, pay around £865-ha (£350-acre) per year for a 30-year lease initially, and I would expect there to be an escalator clause in the lease which will raise this rent over the years. Whether this is now the benchmark remains to be seen. The land is very close to Chapel Down's other north Kent vineyards and is of an area big enough to make a significant difference to their overall production. Both of these factors would have been a contributory in achieving this level of rent.

I have helped several growers negotiate leases for land for vineyards, and whilst such things as the lease period, the rental and the basic tenure terms can be agreed, the question of what happens when the lease ends is always a tricky one. If the vineyard establishes itself well and you harvest good crops of grapes which produce award-winning wines, you, as the tenant, may feel that your thirty years or more of effort and energy ought to be recognised when you surrender the lease and leave the landlord with a mature, well managed vineyard. How this is to be calculated is tough, especially if your wine brand bears the name of the farm or estate where the land is as this doesn't allow you to recreate the wine on a different site. Most landlords are unwilling to agree to any terms that might commit them to paying out substantial sums when they cannot see what the winegrowing scene in Britain might be in a generation's time. How would you have worked out the future brand value of Nyetimber when it was planted in 1988? In March 2006, when the current owner bought it, the brand value element of the £7.4 million purchase price was around £2 million. Had the 16-ha (40-acres) of vineyards been leased, how much of that brand value would have been the landlord's? This aspect of a lease is often left 'to be negotiated'. Of course, if the vineyard, or the enterprise based upon it, fails and the tenant wants to hand the land back (or becomes insolvent and the landlord claims it back) the landlord may well want compensation to turn the land back into an arable field. Given the cost of taking out anchors, posts and wires, grubbing the vines and returning it to arable will be around £1500-ha (£600-acre), the landlord may well want this written into the lease. Either way, leasing land for a vineyard is never straightforward.

Establishing a vineyard

The cost of setting up a vineyard can vary hugely. My standard costings, based upon the spacings and equipment set out below, show that it costs a minimum of £30,000-ha (£12,000-acre) to prepare, plant, trellis and nurture 1-ha of vines for the first two years and quite easily up to £37,000-ha (£15,000-acre). The cost will depend upon a large number of factors: the size of the vineyard, the density of vines, the types of material used, and whether contractors and contract labour are used for the work, or whether it is done by employed staff. The conversion rate of the £ to the € will also affect prices of vines and most of the materials used as they all come from the eurozone. The prices above are for a reasonably sized vineyard – say, 4-ha (10-acres) – planted at 2.00 m row width x 1.20 m inter-vine distance (4,167 vines-ha or 1,686 vines-acre), GPS machine-planted, galvanised steel end and intermediate posts and using contractors for the trellising, i.e. a top-of-the-range, long-life, vineyard. To the prices above must be added extras such as access roads, water to the site, land preparation, fertilisers, partial or complete land drainage and rabbit and deer fencing. All of these could add several thousands to the bill.

The price of planting and trellising also depends very much on vine density, which can vary between 7,000 and 2,700 vines-ha[1] (2,833 and 1,093 vines-acre) and row width: a 1.75 m row width gives you 5,714 running metres per hectare (2,312 running metres per acre); a 2.50 m row width gives you 4,000 running metres per hectare 1,619 running metres per acre). The establishment cost will also vary according to who does the work. Newcomers to the vineyard business might well feel safer employing a vineyard consultant (a wise decision) and getting much of the work done using contractors who have experience in planting and installing vineyard trelliswork. This will, of course, be a more expensive way to establish a vineyard than, say, if you manage much of the work yourself and use farm staff or directly employed local labour. If you were prepared to plant at a lower density, plant by hand, use less-durable posts and do much of the work yourself, this price could be substantially reduced.

Equipment for the vineyard

A professionally run vineyard requires a basic level of agricultural equipment in order to farm it efficiently.[2] A suitable tractor, a mower, a sprayer, some weed control equipment are an absolute minimum. Other items such as prunings pulverisers, shoot-trimmers, and deleafers can be either done without (i.e. much of the work can be done by hand or with other machinery) or can be hired. The bill for the minimum equipment outlined above (bought new) would be unlikely to be less than £60,000 and quite possibly considerably more for top of the range equipment. For the smaller grower, and for equipment that is only used occasionally, the second-hand market is well worth investigating and for specialist vineyard (and winery) equipment, the French website Agriaffaires (which carries advertisements from many European countries including Britain) is a good place to start looking. See www.agriaffaires.co.uk.

Running costs of a vineyard

The annual costs of running a vineyard (excluding picking) will depend upon unchangeable factors such as vine density – 7,000 vines-ha takes a lot longer to prune than 3,500 vines-ha – and row width – the narrower the rows, the more running metres there are in a hectare of vines. It will also depend – hugely – on who does the work. Owner-operators, who do much of the pruning, weed control, pest and disease spraying and mowing themselves will see much lower costs per hectare than someone running a large vineyard and winery who uses farm staff, contractors and/or casual labour to do much – often all – of the manual and tractor work. One person might, with some additional contract and casual labour to help with canopy management, be able to look after 5-ha (12-acres) of vines, but any more would probably be too much for one person to prune on their own. However, if contractors and casuals were used for all the pruning, pulling out and tying down, plus the annual vine work and canopy management, then one tractor driver could probably spray and mow around 15-ha (37-acres) in a season.

The annual materials used in a vineyard consist of fuel for tractors, chemicals used in controlling weeds and pests and diseases,[3] annual replacements to vines and posts and small items such as vine ties. The cost of chemicals will vary according to several factors – a vineyard planted with Seyval blanc would cost a lot less to look after (in this respect) than one planted with, say, Chardonnay – but in most vineyards the annual chemicals bill will be significant. A sum of £900-ha (£364-acre) would be a minimum and, in many vineyards, it would be nearer £1,250-ha (£506-acre). In total, therefore, a reasonably sized vineyard, planted with a 2.00 m row width and with a vine density of 4,167 vines-ha (2.00 m x 1.20 m), will cost at least £7,000-ha (£2,833-acre) to run and quite possibly nearer £8,000-ha (£3,238-acre). There will also be maintenance costs for tractors and the other equipment, plus the depreciation of equipment, vines and trelliswork. Additions to the vineyard in the way

---

1 This is for cane pruned vines. Trellising systems such as GDC or wide-rowed Sylvoz might lower the density even further. For GDC at 3.66 m x 2.44 m you could get down to 1,120 vines-ha or 453 vines-acre (although I wouldn't recommend this system).

2 I am excluding what I would call 'hobby' vineyards where much of the work can be done by hand and/or with garden tractors. These would typically be less than 0.2-ha (0.5-acres). A vineyard above this size, which would hopefully produce at least 1 tonne of grapes, would be too big to mow and spray with garden equipment.

---

3 The pros and cons of using chemicals to control weeds and pests and diseases, plus the alternatives, are dealt with in Chapter 14: Weed control.

of fertiliser, compost and manure should also be allowed for but as vineyards vary hugely in this respect, the cost of these items is hard to estimate. A vineyard might spend £500-ha (£200-acre) every two to three years on fertiliser, but this is very much a ballpark figure. See *Appendix III* for additional information on vineyard running costs.

Picking costs

Picking costs are not normally included in the annual cost of running the vineyard as they tend to be directly crop-size related, i.e. picking 10 tonnes of grapes costs one price and picking 20 tonnes costs twice that price – all other things being equal. They are therefore seen as a deduction from crop value and not an addition to cost of production. Picking costs will, like almost all costs in a vineyard and winery enterprise, vary according to several factors, not all of which are the same each year and not all of which are the same for all growers. Apart from such obvious factors such as vine density and row width, plus whether you are using friends and family, permanent staff or contract labour, one of the most important factors is level of yield. In a heavy-yielding year where vines might be cropping at between 12 tonnes-ha and 25 tonnes-ha (5 tonnes and 10 tonnes-acre), the per tonne picking costs will be significantly lower than where a crop is small to very small and where pickers have to (a) hunt for the bunches and (b) select them for quality and (c) trim them of any diseased or rotten sections. The colour of the grapes will also make a difference – red grapes are easier to see than white (green) grapes – and some varieties are easier to pick than others – there are certain varieties that have a habit of tangling themselves up in the wires, making prising the grapes away from the vine that much harder.

The work done in the vineyard prior to harvest will also affect picking costs. A well-maintained vineyard, where pests and diseases have been controlled, weeds have been kept down, the grass mown and, most importantly, the vines have been deleafed in the 3-4 weeks prior to harvest, will be significantly easier (and therefore cheaper) to pick than a badly maintained vineyard. The method of picking – into small boxes or trays or into 350 kg bulk bins – and how the non-picking staff organise the distribution of empty picking boxes and bins and the collection and assembling of full picking boxes and bins, will all contribute towards speed of picking and therefore the cost per tonne. However, in a well-managed vineyard with an average yield of, say, 8–10 tonnes-ha (3.24–4.05 tonnes-acre), a motivated picker ought to be able to pick 500 kg in an eight-hour day, giving you a per tonne cost, including the non-picking staff, of around £275–£300 per tonne. In the exceptionally heavy-yielding 2018 vintage, some growers reported pickers achieving 1,000 kg a day, an almost unheard-of quantity. However, in low-yielding years, with a poorly managed vineyard and poor organisation, the cost might well be much more than that. If the vineyard site is remote from the winery, then haulage would be an additional cost

and, if the grower is selling grapes, then haulage to the purchaser would usually (although not always) have to be paid by the seller.

Following the huge 2018 harvest, when picking teams were overwhelmed by the size of the crop, at least two modern picking machines are now working in Britain. One of the German planting contractors came over with his self-propelled Ero Grapeliner 7000, the most up-to-date version from this well-respected viti-equipment company, and Sam Barnes, a British-based contractor, purchased a trailed Pellenc, both with on-board sorting capability. The costs of using these machines depend very much on the total area the machine can harvest in one place (travelling and downtime being expensive) and their picking rate is up to 0.75-ha (1.85-acres) per hour. Both contractors are quoting budget prices of around £1,100–£1,200 per hectare (£445–£486-acre). Of course, price per tonne picked will depend entirely on yield, but given a modest yield of, say, 7.5 t-ha (3.0 t-acre) this is around £150-tonne. With larger yields (such as were seen in 2018 and 2019), the cost comes down to under £100-tonne and Barnes said that in 2019 one grower ended up paying £60 per tonne. Cost, of course, is not the only benefit of using a picking machine. The speed of picking means that you harvest in one hour what one picker would take 10 days to pick, more if the yield is heavy. There is more on picking in *Chapter 12: Management – cropping years* in the section on *Picking*.

Yields

Note 1: There are several sources of yield data, some more reliable, some less reliable, and some that require interpretation. The 'official' yields come from WS, which (when we were members of the EU) was required to send an annual return to Brussels. It did this by collating the data from all wineries (who must be registered with both WS and HM Revenue & Customs (HMRC)) shortly after the harvest. The raw data was the total tonnage of grapes received into the winery, the hectares from which those grapes were harvested, the hectolitres (100 litres) of wine produced from them and the type of wine, white or red/rosé, and quality of wine – quality wine or table wine. This data then produced a figure for the hectares in production, the total tonnage of grapes and the hectolitres per hectare (hl-ha) of wine produced.[4] The problem with these figures is that the returns from very young vineyards, where yields were naturally low, or even very low, and from vineyards where yields were low for a number of different reasons (frost, disease, mismanagement), no allowance was made for these underperforming vineyards. Thus, a recently

---

4 The conversion of hectolitres per hectare into tonnes-ha or acre depends on the amount of juice pressed from each tonne of grapes. Depending on variety and pressing practices, a still wine producer might take 7.00–7.50 hl per tonne; a sparkling wine producer might take as little as 5.00 hl per tonne and may (or may not) press the grapes further to produce additional juice for other wines. I work on an average pressing percentage of 62.50 per cent (6.25 hl-ha). Thus 50 hl-ha = 8 tonnes-ha or 3.24 tonnes per acre.

Year	Total Planted Ha	Ha in production	Ha not in production	% of total not in prod	Total Yield in Hl.	No. of 75 cl bottles in millions	Yield Hl-Ha	No. of Commercial Vineyards	Av. Size of Commercial Vineyards	No. of Wineries
1989	876	652	224	25.57%	21,447	2.86	32.89	442	1.98	147
1990	929	629	300	32.29%	14,442	1.93	22.96	445	2.09	147
1991	992	650	342	34.48%	15,429	2.06	23.74	454	2.19	150
1992	1054	701	353	33.49%	26,428	3.52	37.70	457	2.31	157
1993	1065	767	298	27.98%	17,504	2.33	22.82	479	2.22	148
1994	1035	733	302	29.18%	17,693	2.36	24.14	435	2.38	123
1995	984	745	239	24.29%	12,651	1.69	16.98	413	2.38	115
1996	965	775	190	19.69%	26,080	3.48	33.65	408	2.37	123
1997	949	791	158	16.65%	6,460	0.86	8.17	386	2.46	114
1998	901	842	59	6.55%	11,202	1.49	13.30	382	2.36	108
1999	872	835	37	4.29%	13,272	1.77	15.90	373	2.34	106
2000	857	822	35	4.08%	14,215	1.90	17.29	363	2.36	106
2001	836	801	35	4.15%	15,817	2.11	19.75	350	2.39	105
2002	812	789	23	2.83%	9,385	1.25	11.89	333	2.44	114
2003	773	756	18	2.26%	14,503	1.93	19.20	333	2.32	109
2004	761	722	39	5.12%	19,071	2.54	26.41	339	2.24	106
2005	793	722	71	8.95%	12,806	1.71	17.74	353	2.25	90
2006	923	747	176	19.07%	25,267	3.37	33.82	362	2.55	102
2007	992	697	295	29.74%	9,948	1.33	14.27	383	2.59	98
2008	1,106	785	321	29.04%	10,087	1.34	12.85	416	2.66	116
2009	1,215	946	269	22.14%	23,835	3.18	25.20	381	3.19	109
2010	1,324	1,095	229	17.30%	30,346	4.05	27.71	404	3.28	114
2011	1,384	1,208	176	12.72%	22,659	3.02	18.76	419	3.30	124
2012	1,438	1,297	141	9.81%	7,750	1.03	5.98	432	3.33	128
2013	1,884	1,375	363	19.27%	33,385	4.45	24.28	448	4.21	131
2014	1,840	1,506	185	10.05%	47,434	6.32	31.50	473	3.89	131
2015	1,956	1,655	184	9.41%	37,977	5.06	22.95	502	3.90	133
2016	2,077	1,612	345	16.61%	31,116	4.15	19.30	525	3.96	135
2017	2,245	1,677	448	19.96%	39,574	5.28	23.60	523	4.29	151
2018	2,889	2,138	631	21.84%	98,289	13.11	45.97	674	4.29	164
2019	3,300	2,438	742	22.48%	78,607	10.48	32.24	725	4.55	171
2020	3,500	2,738	642	18.34%						

Source: Wine Standards & UK Vineyards Guide

Table 3

planted 20-ha vineyard I know of, which picked 2 tonnes of grapes in its second year and made around 1,200 litres of wine, recorded a yield of 60 litres per hectare or 0.60 hl-ha. Where you have an expanding region like ours, where in some years as much as 20–30 per cent (and even more) of the area in production is in its second or third year, thereby having very low yields, the annual yields reported by WS are distorted. My rule of thumb is that a well-sited, and well-managed vineyard in full maturity (i.e. from year four onwards) ought to achieve at least two and a half to three times the reported national yields.

The second source of data is from four annual yield surveys: two which I undertook privately in 2016 and 2017 and which came from vineyards I knew of and trusted; and two which were undertaken independently and anonymously via the ICCWS Yield Surveys undertaken in 2018 and 2019. The results of these four surveys have been used widely in this book.

The third source of data is from the many clients, friends and colleagues in the world of English and Welsh wine, whose data I have gleaned over the years and from which I have created a substantial database of yields. I believe, therefore, that the yields I discuss in the next few pages are as accurate as they can be. As ever, though, 'junk

in, junk out' and some of the varietal yield data is based upon small samples, and is sadly therefore not always that reliable.

Note 2: At the time of writing, the WS has issued the yields for 2019, but is unable to confirm the hectares from which the grapes were harvested, so the figures used in Table 3 are an estimate.

Ignoring such peripheral income streams as vineyard tours, catering and rent-a-vine schemes in the calculation of vineyard returns (although I would certainly not ignore them in the overall picture of total enterprise income), the two central factors in a vineyard's income are yield and the value of that yield. Yields in all agricultural and horticultural crops vary from year to year, but with vines, the yield variation is much greater than in most other crops. In regions with ideal growing conditions, it has been shown that annual yields in vineyards can vary by over 30 per cent and in regions with less than ideal growing conditions, into which Britain has to fall, factors such as spring frosts, poor flowering and difficult ripening and harvesting conditions can contribute to very significant variations in yield. In 1997, one night's frost in May reduced the average annual British yield to 8.17 hl-ha; in 2012 exceptionally poor

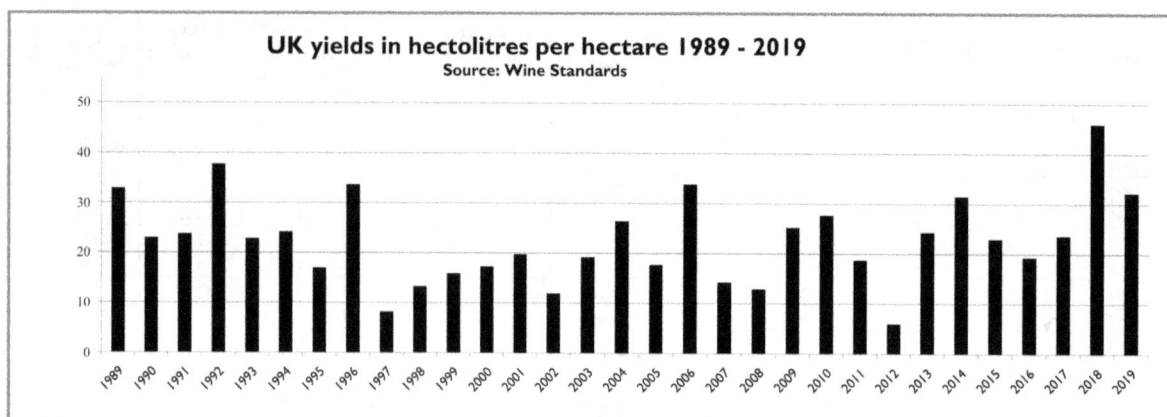

UK yields in hectolitres per hectare 1989 - 2019
Source: Wine Standards

Table 4

flowering weather lowered the annual yield to 5.98 hl-ha (around 0.8 tonnes-ha or 0.35 tonnes-acre) which was the lowest national average yield ever recorded. These two yields (1997 and 2012) are between 25 per cent and 35 per cent of the ten-year average yield (2010–19) of 25.23 hl-ha. In 2013, a very late spring and a cool and wet October, led to the latest harvest on record. Many grapes were harvested at very high acid levels and very low sugar levels and some growers left grapes unpicked as they were judged to be too low in quality. In 2017, 'inversion layer' frosts of down

to -6.0 °C and -7.0 °C on the 26th and 27th April hit the south-east of England and did substantial damage, made worse by the fact that a warm spring had seen the earliest bud-burst on record. However, not all varieties were equally affected. Chardonnay, an early budding variety, recorded average yields of 5.05 t-ha (2.04 t-acre) in non-frosted vineyards, whereas in frosted vineyards it only managed 2.54 t-ha (1.03 t-acre), a 50 per cent loss of crop. Meunier and Seyval blanc, however, both late-budding varieties, despite being frost-affected only saw reductions of 32 per cent and 20 per cent respectively. It might not sound like a huge difference, but it all adds up. Low yields, of course, are by no means a new problem and Jack Ward, the first Chairman of the English Vineyards Association (EVA),[5] in his 1984 book *Vine Growing in the British Isles*, wrote: 'the survival of English viticulture will depend almost entirely on the ability to maintain an adequate yield per acre' and that 'the average [yield] at present achieved is scarcely sufficient, if the cultivation of vines is to be regarded as a profitable undertaking'. More than a generation of vintages later, Ward's words are as true as ever.

The major problem in Britain is that the heat and light levels required for more consistent yields are right on the edge. If you take the seven highest yielding years since 1989, the year that accurate records were first produced, which were 1989, 1992, 1996, 2006, 2014, 2018 and 2019, the average national yield was 36.05 hl-ha or around 2.12 tonnes-ha. These years were in the main frost-free, early budding with good flowering conditions, with dry, disease-free ripening weeks. However, they were only seven years out of thirty-one. Take the seven lowest yielding years, 1997, 1998, 1999, 2002, 2007, 2008 and 2012 and you can see the results of indifferent weather. The average annual yield for these seven years was only 11.77 hl-ha, less than one-third of the seven best years. In fact, if you take the average of all the other years, twenty-four in total, it is only 18.97 hl-ha, around half of the average of

Year	Yield Hl-Ha	Yield T-ha	Yield T-acre
2000	17.29	2.47	1.00
2001	19.75	2.82	1.14
2002	11.89	1.70	0.69
2003	19.20	2.74	1.11
2004	26.41	3.77	1.53
2005	17.74	2.53	1.03
2006	33.82	4.83	1.96
2007	14.27	2.04	0.83
2008	12.85	1.84	0.74
2009	25.20	3.60	1.46
2010	27.71	3.96	1.60
2011	18.76	2.68	1.08
2012	5.98	0.85	0.35
2013	24.28	3.47	1.40
2014	31.50	4.50	1.82
2015	22.95	3.28	1.33
2016	19.30	2.76	1.12
2017	23.60	3.37	1.36
2018	45.97	6.57	2.66
2019	32.24	4.61	1.86
Average	**22.54**	**3.22**	**1.30**

Note: Hl per ha converted to tonnes per ha @ 700 litres per tonne

Table 5

5 The EVA was founded in 1965 and was the national vineyard association. It was wound up and its assets transferred to the UKVA in 1996. The UKVA has now been superseded by WineGB.

Yields Tonnes-ha	2016	2017	2018	2019	2016-19 T-ha	2016-19 T-acre
Top 25% of vineyards - all varieties	8.84	9.57	10.52	9.63	9.64	3.90
Middle 50% of vineyards - all varieties	4.41	4.12	6.16	5.35	5.01	2.03
Bottom 25% of vineyards - all varieties	1.36	1.31	1.63	2.00	1.58	0.64
All varieties - all vineyards	4.54	4.68	7.12	5.93	5.57	2.25

Table 6

Yields in a Kent vineyard 2004-2019
Planted in 2002 with Chardonnay, Pinot Noir & Meunier.
Average yield 10.40 t-ha (4.21 t-acre)

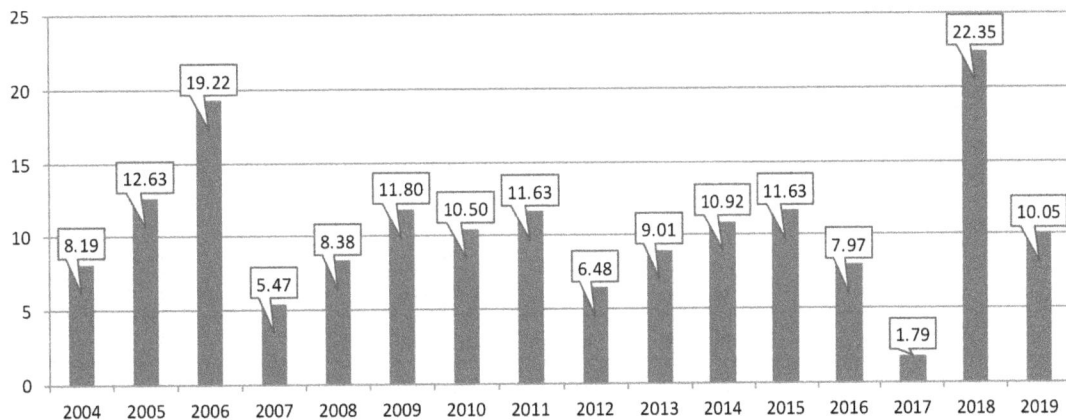

Table 7

those seven good years. Of course, 2018 with the highest yield ever (by a significant amount) and 2019 with a yield in the top four best ever, have skewed the recent figures. The combination of very much higher yields and a much larger area of vineyards in production mean that the total volume of wine produced in these two years (23.59 million x 75 cl bottles) is not far off the total amount produced in the previous five years (25.26 million bottles).

As you can see from Table 5, looking at WS yields since 2000, the average of all vineyards is only 3.22 t-ha (1.30 t-acre). However, as I have explained in the note to this section, these yields need some interpretation, and the yields in well-sited, well-managed vineyards can be two to three times these national averages.

Over the last four years, 2016 to 2019, from the data derived from my private surveys, plus the two ICCWS-WineGB Yield Surveys, I have ranked the vineyards by yield into three categories: the top twenty-five per cent, the middle fifty per cent and the bottom twenty-five per cent. The results are in Table 6. Of course, this data covers all varieties and all regions, but lower yields in some regions (of which more later) and the better performance of some grape varieties (also more later) probably cancel each other out, so that the figures represent a fair average. The only other caveat is that the four years surveyed do include the highest cropping year (2018) since reliable records began to be collected in 1989.

As can be seen, the top twenty-five per cent are streets ahead of the middle fifty per cent, by a factor of almost

two. A yield of 9.64 t-ha (3.90 t-acre) is very respectable and a yield that is certainly sustainable in an economic sense. A yield of 5.01 t-ha (2.03 t-acre) is not, let alone anything less than that. As for what makes a 'top twenty-five per cent' vineyard, the list is quite a short one: a good site, high vine density, good management, and good pest and disease control. Two other very basic factors are also important in getting good yields: vine variety and geography.

One of the best yielding vineyards I know off is Hush Heath Estate's first vineyard (which I planted). It was planted on the site of an old Bramley orchard which is very heavy Wealden clay and was pipe-drained at 10 m intervals. The majority of the site was planted in 2002 with 95 per cent Chardonnay and Pinot noir, with five per cent of Meunier planted in 2003. There are two clones of each variety, each on two different rootstocks, 41B and 3309C, and at 2.30 m row width and 1.30 intervine distance. Despite the relatively low density of 3,344 vines-ha (1,353 vines-acre) the site has performed extremely well and I do not know of a site planted with the same varieties that has a better yield over sixteen years. The highest yielding year was 2018 with an average yield of 22.35 t-ha (9.04 t-acre), closely followed by 2006 with 19.22 t-ha (7.78 t-acre). Low points include 2007 and 2012, both with bad flowering conditions, and 2017 which was very heavily frosted. These fluctuations in yields mirror quite closely the yields shown by the national figures, although not in 2012, which for Hush Heath was not the disaster it was for many other growers. This yielding record just shows the variability

13

GB Yields of different varieties Tonnes-ha	2016	2017	2018	2019	Average 2016-19	Av. 2016-19 T-acre
Reichensteiner	8.36	4.37	16.60	7.50	9.21	3.73
Seyval blanc	6.40	11.24	9.38	3.36	7.60	3.07
Regent	No data	5.70	11.09	4.12	6.97	2.82
Meunier	5.88	5.24	9.53	6.23	6.72	2.72
Ortega	No data	No data	No data	6.53	6.53	2.64
Rondo	No data	9.24	7.30	2.66	6.40	2.59
Pinot blanc	No data	4.86	7.85	No data	6.36	2.57
Chardonnay	5.58	4.42	8.67	6.61	6.32	2.56
Madeleine x Angevine 7672	6.05	5.64	6.70	6.44	6.21	2.51
Pinot noir	3.00	4.31	7.93	6.18	5.36	2.17
Bacchus	3.07	4.21	5.48	7.14	4.98	2.01
Blauer Frühburgunder (Pinot noir Précoce)	1.81	3.50	6.62	4.32	4.06	1.64
Other varieties	4.26	3.68	4.10	3.35	3.85	1.56
Dornfelder	No data	No data	No data	3.47	3.47	1.40
Solaris	No data	No data	No data	2.66	2.66	1.08
Pinot gris	No data	No data	No data	2.21	2.21	0.89
Average of all varieties	**4.93**	**5.53**	**8.44**	**4.85**	**5.56**	**2.25**

Table 8

Yields Tonnes-ha	2016	2017	2018	2019	2016-19 t-ha	2016-19 t-acre
East Anglia	6.89	4.35	6.92	6.09	6.06	2.45
South East	4.31	4.61	9.90	6.43	6.31	2.55
Wessex	2.08	3.44	9.93	5.66	5.28	2.14
South West	3.95	5.82	5.52	5.35	5.16	2.09
Thames and Chilterns	2.59	5.62	N/A	6.83	5.01	2.03
Mean of all regions	**4.31**	**4.56**	**8.07**	**5.88**	**5.70**	**2.31**

Table 9

of sites, showing that its weather conditions (principally flowering and frost) as much as anything that determines yields, and that you need a few over-performing years to compensate for the under-performing ones. See Table 7.

Vine variety in relation to yield

Note: Table 8 is only as good as the data it is built from, and whereas for the major varieties the data points were fairly numerous, for some of the less widely grown varieties, the data is less robust.

The selection of the right vine variety – right, that is, for your site, region, wine type and style, and your economic model – is an important factor in yield. As can be seen from Table 8, if you grow Reichensteiner, Seyval blanc and Regent, your average yields will be substantially higher than if you select Pinot gris and Solaris. However, yield in financial terms is composed of three factors: volume of product (grapes, wine), multiplied by the price of the product, less the costs of production. Therefore, selling nine 9 t-ha of Reichensteiner as grapes may be less profitable than selling five tonnes' worth of Bacchus per hectare as an award-winning still wine, or six tonnes' worth per hectare of a top Chardonnay, Pinot noir and Meunier-based sparkling wine.

Region in relation to yield

The second factor in yield, and one which will be dealt with more fully in *Chapter 3: Site selection*, is the location of the vineyard within Britain.

As can be seen from Tables 9 and 10, the further west you travel, the lower the average yields become. The reasons are fairly obvious: the further west you go in Britain, the wetter and windier it gets. If you were to map where dessert apples and hops are grown, you would see the same pattern. However, there are exceptions to this general trend which are more fully discussed later in this book.

Summary on yields

As can be seen from the above discussion on yields, getting a yield that in business terms is sustainable is possible in Britain, and such evidence as we have suggests that at least twenty-five per cent of vineyards currently planted and cropping manage this each year. They manage this by being in the right part of Britain and on the right sort of land; by planting the right varieties for their business model; by planting at an appropriate density; and by planting, establishing and managing their vineyards correctly.

Grape prices

So, if yields are unpredictable, what about the value of those yields? Once again – it depends. Let me deal first with prices achievable under long-term grape purchase (or grape supply) contracts.

The price of grapes has always fluctuated according to variety, sugar and acid level and, of course, supply and demand. In the 1980s, when Lamberhurst was under the ownership of Kenneth McAlpine and run by Karl-Heinz Johner (1976–88) and me (1988–91), grape purchase contracts were given to growers with prices based upon a combination of the then retail price of English wine, the variety being grown and the sugar and acid levels of those grapes. Grape prices were always tied to the Retail Prices Index (RPI) so that they increased with inflation.[6] Thus, really ripe Schönburger and Huxelrebe grapes, the basis of Lamberhurst's two highest priced wines, were valued more highly than mid-range Müller-Thurgau and significantly more highly than high-yielding, low-sugar Seyval blanc. In the big harvest of 1989, I recall the price of Seyval blanc grapes was around £350-tonne. These were always delivered prices, in Lamberhurst's bins, with growers bearing the costs of both bin and grape delivery. Since those days, things have changed quite significantly.

Whilst the main driver of grape prices remains the price that winemakers can command for their wines, there are a lot of factors which get in the way of a straight equation

Median yield tonnes-ha by variety by region 2018-19	South East	Wessex	South West
Chardonnay	8.30	8.16	2.94
Pinot Noir	7.59	7.68	3.33
Median Yields	**7.95**	**7.92**	**3.14**

Table 10

between these two. In 2000, Lamberhurst came under the influence of Chapel Down, who took on most of its long-term grape purchase contracts. Chapel Down was then still operating under its original model, that of being a grape buyer, winemaker and wine-seller and not really a grower of its own grapes – that is apart from the 7.40-ha (18.29-acres) of vineyards around its winery at Tenterden. It had decided that it needed to secure supplies of grapes, especially of Champagne varieties, as it wanted to build up its production of sparkling wine, and continued to hand out grape purchase contracts. The prices in these contracts, as they had been under the Lamberhurst model, varied according to variety and sugar level, with deductions for excess acidity and poor quality (as defined by percentage of rot or frost damage) and were tied to the RPIX. Therefore, in the harvest of 2019, the price of good quality Chardonnay, Pinot noir and Meunier lay between £1,650-tonne and £2,200-tonne, depending on when the contract was signed (as prices have tended to come down in recent years) and the actual sugar and acidity levels. Deductions for rot or frost are very seldom invoked, but deductions for acidity are certainly made. Most contracts have a maximum acidity above which a deduction of 5 per cent of the price for each 0.5 g/l is taken and the maximum acid level is usually set between 13.5 g/l and 15.5 g/l, depending on vine variety. Thus, in high-acid years, grapes picked with acids of 15.5 g/l, when the maximum level in the contract was 14.5 g/l, would suffer a 10 per cent price reduction.

Prices of grapes in grape purchase contracts will always be negotiated directly between buyer and seller and there is quite a wide range of prices known to me. Contract prices will always vary according to demand and supply, not on an annual basis, as contracts are typically eight year rolling contracts with three-year break clauses. Some contracts are not as reliant on sugar levels as Chapel Down's are (or at

Camel Valley Vineyard in Cornwall

---

6 Prices are now tied to the Retail Prices Index excluding mortgage interest payments known as the RPIX.

Wine Growing in Great Britain

Costs of grape growing - per ha	Capital per ha	Interest rate*	Annual charge
Land per ha	£ 50,000		
Establishment per ha	£ 30,000		
Machinery & equipment per ha**	£ 10,000		
Total capital	**£ 90,000**	3.416%	£ 3,074
Growing costs per year per ha			£ 6,795
Machine picking per tonne			£ 125
*Interest rate based upon a 30 year capital & interest repayment loan from AMC (2020)			
** Based upon spending £200,000 over 20-ha			

Table 11

least used to be), which is sensible given that sugar level is not such a quality-decider in sparkling wines as it is in still wines. Wineries that are some way distant from their grape sources, say, a winery in Cornwall buying grapes from Suffolk (if such a thing could possibly exist!), might have to contribute towards haulage costs. Today (2020), if I were negotiating a grape purchase contract on behalf of a grower, I wouldn't want to accept less than £1,750-tonne for Chardonnay, Pinot noir and Meunier, and hopefully even a bit more. For Bacchus, which until, say, ten years ago was always less valuable than the Champagne varieties and typically commanded prices of less than £1,000-tonne, has now come into its own and I would expect at least £1,500-tonne and, for a grower with a proven track record of quality, possibly more. Unfortunately, whilst I know of (and have negotiated) contracts where the prices have risen with RPIX additions to nearer £3,500-tonne, for the time being I fear that these prices are over. The grape growing industry in Britain now, in 2020, finds itself in a very different supply and demand situation than was the case when the first edition of this book was written in 2013-14. As has been pointed out previously, the 2018 and 2019 harvests produced almost as much wine as in the previous five years added together and it has been estimated that the stocks of English and Welsh wine might now amount to around 30-32 million 75 cl bottles' worth. This equates to six- or eight-times sales![7] With this level of stock, many producers, who might previously have bought in grapes to top up the supplies from their own vineyards, might understandably prefer to concentrate on selling the wine they already have in stock, rather than adding to it.

The other factor in the supply and demand situation for grapes is the massive amount of new planting that has taken place in recent years. The current planted area of vines in Britain is, depending on who you believe, around 3,500-ha (my estimate), or anywhere 200-300-ha either side of those figures. This includes such 2020 plantings as I am aware of, but there are undoubtedly some that I haven't recorded. This means that, say, by the end of planting season 2020, the area under vine will have almost doubled in five years. This will in time have a massive impact upon the quantity of grapes available, especially if the 2018-19 yield levels are the new normal. One sobering fact is that if you just concentrate on vineyards for sparkling wines, which typically take at least five to six years before any wine is ready for sale (three years to crop and two to three years to age the wine), then the amount of sparkling wine available for sale is likely to double in the next two to three years. This will surely put pressure on retail prices which in turn will have a downward effect upon grape prices and I can foresee the price of Chardonnay and Pinot noir on the open spot market falling to below £1,000 a tonne.

On a slightly more positive note, the last two to three years have seen more stand-alone wineries buying in grapes to make both still and sparkling wines, plus the emergence of a small number of producers using non-traditional methods to make sparkling wine, namely the Charmat (tank) method and carbonation. Whilst these are cheaper ways to make sparkling wines and some established producers think they will lower both retail prices and the perceived quality of traditional method sparkling wines, there is no doubt that, in the short term, these wines can be brought to market quickly and have the ability to consume large quantities of grapes. Given what many people see as a looming oversupply situation in the grape market, these non-traditional producers may actually help reduce the problem of too many grapes. However, what price they can afford to buy grapes at is open to debate. With retail prices as low as £10 a bottle (from which VAT, Excise Duty, production, packaging and transport costs, plus a notional profit have to be deducted) how much is there left for the grapes? Could it be as low as £500 a tonne (or around 50p a bottle)? There is no way a grower could afford to plant and run a vineyard if grapes have to sell at those sorts of prices. There is more on both traditional and non-traditional sparkling wines in *Chapter 4: Vine varieties, clones and rootstocks*.

---

7 Reliable sales figures of English and Welsh wine are not known and the best estimate is around four million bottles a year. See discussion on sales in Chapter 4 in the section on Sales of English and Welsh wines.

Retuns from grape growing	Yield t/ha	Price per tonne	Income per ha	Yield t/ha	Price per tonne	Income per ha
Gross returns	8.00	£ 1,750	£ 14,000	9.50	£ 1,500	£ 14,250
Less growing costs			£ 6,795			£ 6,795
Less picking costs	8.00	£ 125	£ 1,000	9.50	£ 125	£ 1,188
Sub-total			£ 6,205			£ 6,268
Capital charges			£ 3,074			£ 3,074
Net return per ha			£ 3,131			£ 3,194
Net return per acre			£ 1,267			£ 1,292

Table 12

Growing grapes and making a profit

Whether you can grow grapes in Britain as a profitable stand-alone enterprise is an interesting question. People do and, in some cases, have done so for several decades. The key factors are, of course, yield and price, plus your ability to establish the vineyard at least cost, and your ability as a grape farmer to manage and run the enterprise efficiently. On yields, if you grew one or more of the four major (and most wanted) varieties – Chardonnay, Pinot noir, Bacchus and Meunier – your average yield, even with a superb site and the best management possible, would be highly unlikely to be higher than 8.00 t-ha (3.24 t-acre) over a ten-year average and it would be risky to base your budget on higher yields. If you grew Reichensteiner and Seyval blanc, you might push this up to 9.50 t-ha (3.84 t-acre) but no higher and, of course, the price of the grapes will be lower. Prices have been discussed above and, given the current situation of wine stocks and planted area, are likely to be in the £1,5000–£1,750-tonne region. On establishment, you could, by using direct-labour to trellis and establish the vineyard and using cheaper trellising materials, cut the cost of establishment by a few thousand per hectare (although given that you are writing the vineyard off over thirty-plus years, this makes little difference to the annual charge for capital). If the area being grown is large enough, you could mechanise as much of the growing side as possible (including picking) thus bringing the growing costs down. These savings, however useful, are fairly marginal in the overall financial picture. A final factor in the economics of grape production is, of course, the cost of the land and how it is accounted for. Whether land should be included in the calculation of the cost of grape production is also an interesting question. Many will take the view that as 'land isn't being made any more'; it is one of the soundest long-term investments one can make. In addition, land prices in the grape growing regions of Britain are likely to increase in value in the future and that, as there are also inheritance tax advantages in holding land as an asset, then the cost of the land should be excluded from grape production. It is a view with which I have some sympathy. However, for the purposes of illustration, the economics of growing grapes for sale might well look something like the costs in Table 11.

Table 12 illustrates an example of the finances of grape growing, using two yield and price scenarios, as described earlier. The land price I have used is relatively high (£50,000-ha) but that is the way I believe good vineyard land is going, and I have not skimped on establishment costs. However, as I have said above, in an economy where interest rates are low, an extra, say, £10,000 of capital only adds £575-ha (£233-acre) per year to your costs.

As you can see, the returns are slim but, at 3.50 per cent, cover inflation and leave something over to live off. You would also, of course, get the upside of land values over thirty-plus years, plus whatever increased value accrues to that land from being a successful vineyard. Of course, if you ignore the cost of the land, and do much of the work yourself and/or with family help, then the figures look dramatically different.

Note: The question of oversupply of sparkling wine in Britain is dealt with more fully in *Chapter 4: Vine varieties, clones and rootstocks* in the section on *Oversupply of English and Welsh sparkling wine?*

The costs of producing wine

Surprising as it might sound, I doubt that there are many people in the world of wine production who can tell you accurately what it really costs to produce a bottle of wine. The reasons are that wineries are often put together over many years, even decades and often generations, with much of the equipment made from stainless steel which lasts indefinitely and which wears out very slowly. In addition, many of the overhead costs of running a winery – heat, light and power, finance costs and essential staff – which together account for a fair percentage of the cost of production – are fixed and therefore the amount attributed to each bottle produced fluctuates according to the annual throughput. Say, for instance, that your fixed costs of running a winery are £100,000 and one year you have a throughput of 500 tonnes and the next year only 250 tonnes. Do you increase the price of your wine

Wine Growing in Great Britain

Breaky Bottom, East Sussex in the snow

to take account of this? No – of course not – you have to take the longer view. In any event, the price of one's wine is a function of many factors: its market position, the overall demand for the type of wine you are producing, the demand for your wines in particular, the depth of your wallet to overcome cash-flow issues (if you decide to keep your prices higher than your buyers really want to pay), and the willingness of your competitors to lower their prices when offered a marketing opportunity that you are not able, or prepared, to take up. These will all play a part in how you price your wine.

Some of the cost of a bottle of wine, of course, is predictable and relatively simple to arrive at. The value of the grapes in the bottle is either known because that is what you paid for them, or can be worked out from knowing the costs of running your own vineyards, divided by average yields. The costs of materials used in the production of the wine – sugar and yeasts, fining materials, barrels (if used), bottles, labels, capsules and other packaging – can be costed and a total arrived at. They will vary, of course, depending on the style of wine and the type and quality of your overall package. The materials used for a quick-turn-around still wine with a lightweight bottle, a screw-cap closure, a simple label and the minimum of packaging, might cost a quarter the cost of materials for a bottle-fermented sparkling wine in a fancy sparkling wine bottle, with a three part label, a printed capsule, a printed muzzle cap, a six-bottle printed sparkling wine case with fibre dividers and even a single-bottle printed box.

Owning and operating a winery, especially in a climate like Britain's, where yields and qualities fluctuate and are very unpredictable, is very much an act of faith in the long-term future of the enterprise. Most wineries in Britain have started out small and grown as the business has grown, increasing capacity and upgrading equipment as they progress from one vintage to another. Some wineries start out by doing elements of the processing themselves, the pressing, fermentation and filtering are the most typical, then have the bottling and, in the case of sparkling wine, the post-storage operations – riddling,

disgorging and *dosage* – carried out by another winery or a contractor. Therefore, actually arriving at a true cost per bottle is often extremely difficult.

Another way to look at production costs is to see what a contract winery might charge you to produce a bottle of wine. At the moment in Britain, there are around twenty-five wineries that undertake 'contract crushing' as it is known (in the USA) and will take in your grapes and produce a bottle of wine to an agreed specification. Prices will (naturally) vary according to a number of factors. A winery would tend to give a better price to a grower sending in a larger tonnage and prices will vary according to the style of wine and type of packaging, as outlined above. As a guide, though, a bottle of still wine would cost around £3.50-£4.50 to produce; a bottle of sparkling wine around £6.00-£7.00. In addition to these costs would be such things as label design and labels, which might come to several thousands of pounds for the design and between 20p and £1.00 for the labels, depending on their complexity.

There is also the cost of storage which has to be taken into account. Even if you have your wine processed by another winery, they will usually be unwilling to provide storage for more than a few weeks after bottling (or, if they do, it will be at a high storage rate cost) and therefore most producers making wine to sell themselves will need to find suitable storage space. Still wines are typically placed in bulk bins after bottling and are generally unlabelled. The reasons for this are that some corks will weep in storage; labels will deteriorate, making them unfit for sale; and bottles inevitably get dusty and dirty. Better to store them as 'cleanskins' (Australian for an unlabelled bottle of wine) and then wash, capsule, label and package them as required. Washing, labelling and packaging are therefore additional costs to the above costs of production. In the case of sparkling wines, after bottling and closing with crown caps, the wines have to go through the secondary fermentation and, once that is finished, are put away for the next two to five years whilst they undergo *sur latte* ageing.[8] It is only after this process is finished that sparkling wines can be riddled, disgorged, have their *dosage* added and then finally be made ready for sale by corking, muzzling, labelling and packaging. Some larger wineries will have the equipment for these post-ageing operations; smaller wineries may well send their wines to another winery to have this work done under contract.

Whether you have your own winery or have your wine made under contract, there is also the matter of cost of stock to be discussed. If you are producing still wines and your wines are bottled in the spring following the harvest, unless they are the type of wines that require bottle ageing (not generally produced in Britain) then you will probably be carrying around two to three years' stock. Assuming

8 Sur latte ageing (literally 'on the lath') is the term used to describe sparkling wines undergoing ageing in bottle. In many cellars, bottles are stacked up with thin wooden laths placed under the necks to keep the bottles horizontal in the stack.

yields of 7.5 tonnes-ha (3 tonnes-acre) and an output of still wine of around 7,125 bottles per hectare (based upon 712.5 litres-tonne) and production costs of £3.50 per bottle, then the cost of stock, ignoring the costs of producing the grapes, will amount to £24,938-ha (£10,092-acre) for each year of production, i.e. for two years, £49,876-ha (£20,185-acre), and for three years, £74,814-ha (£30,277-acre).

For sparkling wines, however, the sums are considerably higher. Let us assume that yields are the same – 7.5 tonnes-ha (3 tonnes-acre) – giving an output of around 5,625 bottles per hectare (based upon a lower juice output for sparkling wines of 562.5 litres-tonne) and that production costs up to and including secondary fermentation (but not riddling, disgorging, *dosage* etc.) are £5 per bottle. In this case the cost of stock will amount to £28,125-ha (£11,382-acre) for each year of full production, ignoring a small crop in the first cropping year (year three or four). The number of years' stock of sparkling wine carried will be a function of two things: how long you need to age your wine *sur latte* and how quickly you sell the wine once you start disgorging it and offering it for sale. Ageing times depend on the quality of the vintage, the acidity at bottling, the style of the wine, and the level of *dosage* you favour (the drier the wine, the longer the ageing tends to be). Once you start selling your wine, your rate of sale will depend upon a myriad of factors, but you are highly unlikely to sell all of the vintage you are offering within twelve months of starting to sell it, meaning that you will probably be selling it over two, three, four and even more years. One British producer I know well, and who has been selling sparkling wine for more than fifteen years, takes on average six years to go from the point where he starts to sell a new vintage, to the point where he is down to a few hundred bottles of 'library stock'. So, taking a modest ageing time of three years, plus a modest sales time of another three years, you might well be carrying stock equal to six years' production of sparkling wine. Therefore, taking the production costs above, the cost of your stock for the six years will be £168,750-ha (£68,292-acre) which at an interest rate of 5 per cent adds exactly £1.50 per bottle to the cost of production for sparkling wines up to this point – almost the same as the value of the grapes! Of course, you will have income to set against this stock cost, but, somewhere along the line, the stock has to be financed. Chapel Down, a PLC whose accounts are available for all to see, held stocks in the twelve months to 31 December 2019 'valued at cost' of £10,791,361 and had sales of 'wines and spirits' of £10,102,000. Their 'cost of production', of course, is based upon running their own winery and is therefore much less than the cost of having your wine made by a contract winemaker. Also, their sales and stocks are a mix of non-wine items, entry-level and mid-range still wines, plus both non-vintage and vintage sparkling wines. Even so, stock levels running at over 100 per cent of sales are an expense that has to be financed and built in to your cost of production. Taking two other vineyard-owning PLCs in Britain, Gusbourne and Hambledon, both

principally bottle-fermented sparkling wine producers, we see a slightly different situation. Both of these companies are less mature than Chapel Down, sell sparkling wines with more bottle age, and are building up stocks as they expand their businesses. The consequence of this is that they have much higher stock levels. Gusbourne, to the year ending 31 December 2018, had wine stocks valued at £5,282,000 against sales of £1,150,000 – a ratio of 4.59:1 – and Hambledon to the year ending 30 November 2018 had stocks of £2,587,501 against sales of £963,986 – a ratio of 2.68:1. Stock, therefore, is very much a significant cost element in any wine business, but especially a sparkling wine business.

Returns from wine sales

The returns from sales of wine will depend very much on where the wine is sold and by whom. Assuming you are VAT registered (the pros and cons of the Agricultural Flat Rate Scheme[9] are dealt with in *Appendix V*), the income from all sales, apart from duty-free sales,[10] will have both VAT and Excise Duty deducted. Wine sales direct to the public will bring you 100 per cent of the bottle price, although selling wine direct is certainly not a cost-free sale. If you want to sell direct to the public then you have to be available when the public want to buy, not when you feel like selling. If you propose to sell a reasonable quantity to the public, then you will have to have some sort of shop, somewhere the public can taste your wines, somewhere they can park without getting their cars stuck in the mud, and of course, lavatories. Your premises must also be licensed for sales of alcohol and at least one of the people running the business will require a personal licence to sell alcohol. In addition, unless you propose to stand behind the counter yourself for eight hours a day, seven days a week, you will have to employ staff. However, selling from the farm-gate engenders loyalty from your customers, they become part of your wine 'family' and a satisfied customer will be your best ambassador. Although individual customers may not visit you every week or every month, they may well visit you once or twice a year, taste your new wines and buy a case or two. Almost all of the successful vineyards and wineries in Britain (and indeed those in many other parts of the world) see direct sales as a very valuable part of their sales mix and a very valuable part of their overall PR and marketing exercise. In my opinion, it would be a very brave wine producer who decided not to open up to the public at all, although I accept that there might be circumstances where public access is impossible or undesirable. Of course, what percentage of your crop you sell direct will depend on many factors. If you are producing 5,000–10,000 bottles,

---

9 See HM Revenue & Customs Notice 700/46 Agricultural Flat Rate Scheme.
10 Duty-free sales would be where wine is sold into a tax warehouse ('duty-free bond') where someone else will be paying the duty; or to an exporter where the wine is immediately exported; or to a purchaser such as a cruise line; or to the Government for overseas use in embassies; or to the Ministry of Defence for use on an overseas armed-forces base.

then you might well sell most of your wine direct; if you are producing 50,000-plus, then you will struggle, at least to begin with. In reality, however much you sell direct, you will also sell to the off-trade (retail shops), the local on-trade (pubs, hotels and restaurants), plus other off-site sales.

Selling direct to the public away from the farm-gate at places such as farmers' markets, country shows, fairs and other sales 'opportunities' are often less rewarding than on the face of it they might appear. You have all the costs of getting to the venue with your stock, setting up a stall, and staffing it and usually the organisers of the event will want you to pay for the privilege of attending. You also have the licensing issues to deal with. You will usually have no control over which other vendors will be attending (you are unlikely to be the only seller of wine and/or other alcoholic drinks), will be expected to give away endless samples and many events will be very weather-dependent. In short, such events are not always a very cost-effective way of selling high volumes of wine, although there may be circumstances where you get valuable publicity or exposure for being there. There are also sales over the internet and if you start to win prizes and can get mentions in your local and national press, these sales build up.

Exports have been mentioned by some as the way to sell quantities of English and Welsh wine, and most of the larger, more established and better-known sparkling wine producers have started to export, albeit on a small scale. Under the WineGB umbrella, six producers have taken stand space at the world's largest trade wine show, Prowein in Dusseldorf, over the years, starting in 2013 and, apart from 2018, continuing until 2019. Prowein was cancelled in 2020, but WineGB is planning to take space in 2021. Of course, going to overseas shows is a very expensive operation, at least £10,000 per producer per show, taking stand fees, travelling, accommodation and staff costs into account. The rewards of attending shows are rarely quantifiable in direct terms and sales leads take time to follow up and bring to fruition. Most of the producers I have spoken to (not just those attending Prowein) say that exports are slow to get going, sales are quite limited in scale and the costs of keeping export markets happy are high. Having said that, Ridgeview claims that around '15 per cent of their sales' are exported, which is a very respectable amount, and Simpsons Wine Estate, which has experience selling wines from its French vineyard, Domaine Sainte Rose, into the USA, is also starting to export its English wines. The topic of exports is covered more fully in *Chapter 4: Vine varieties, clones and rootstocks* in the section on *Sales of English and Welsh wines.*

However, if you are not proposing to sell one hundred per cent of your wines to the public and, in reality, this is only an option for smaller growers, then you will have to put yourself in the hands of someone who can: the wine trade.

The wine trade in Britain is split into several sections, all of which have their own peculiarities, requirements and pricing mechanisms. There is a broad division between 'on' and 'off' trade, on-trade being those that are licensed to sell alcohol for consumption 'on the premises', i.e. public houses, wine bars, restaurants and hotels, and off-trade being those that sell alcohol for consumption 'off the premises', i.e. retail shops, wine warehouses and other sales where the public are not permitted to open the bottle and pour the wine into a glass. There are also trade resellers who might loosely be termed 'wholesalers'. These might be traders who sell to the whole range of both the on- and off-trade, as well as to the public, or they might be traders who target specific sectors of the trade, say, just hotels and restaurants. There are also sectors such as the catering trade, airlines and cruise ships – all of them large buyers of wines – who have their own peculiarities and timetables which as an outsider you will find difficult to penetrate. Whichever section of the trade you end up selling to, you will find that they are, in one respect, all the same: they will require a discount.

My experience of the trade has been forged over my forty-five-plus years of being involved with growing, making and selling wine, both as a producer, a consultant and, for four years, as the wine-buyer and manager of a retail wine shop in Putney. The first time I sold any wine was on 21 June 1980 (I can remember it because it was my father's birthday) when our first two wines, Spots Farm Dry and Spots Farm Medium Dry, from the 1979 vintage, were ready for sale. Those sales were made direct to the public – friends and family – who had gathered to taste the fruits of our years of research, study and labour that had gone into establishing what became Tenterden Vineyards (and is today Chapel Down). Those first sales were all full price sales – £3.00 and £3.75 a bottle – and the question of discounts didn't arise. However, it wasn't long before one of the local off-licences wanted to stock our wines and wanted to know a price. I remember distinctly being torn between the pleasures of selling wine myself at full price and my reluctance to give anyone a discount. What was fair? 10 per cent, 20 per cent? Surely not more? I forget what we eventually agreed, but I do remember that having agreed a price, based upon our retail price, the off-licence immediately marked it up by another 50p or so, thus giving them even more profit!

Since those early days, I have come to realise that everyone has to take a slice –from producer to consumer – and the more attractive you make your product to the wholesaler, the retailer or the restaurateur, the more likely they are to want to sell it. A typical specialist wine shop, i.e. excluding the thousands of licensed corner shops and small grocery stores that typically sell branded wines and own-label wines bought from a cash-and-carry or trade supplier, will have at least 500 wines, maybe even double that, and one thing is certain: they cannot afford to carry a product that doesn't contribute to paying their overheads and to their profit. You must also understand that a retailer will work, not on a mark-up, but on a margin. If the retailer buys a bottle of wine for £5 (net of VAT) and

Still wines	Rate	Bottle price	Bottle price	Bottle price	Bottle price	Bottle price
Retail Price		£ 10.00	£ 12.50	£ 14.00	£ 16.00	£ 20.00
VAT @ 20%	20%	£ 1.67	£ 2.08	£ 2.33	£ 2.67	£ 3.33
Sub-total		£ 8.33	£ 10.42	£ 11.67	£ 13.33	£ 16.67
Excise Duty - Still Wine	£2.23	£ 2.23	£ 2.23	£ 2.23	£ 2.23	£ 2.23
Sub-total		£ 6.10	£ 8.19	£ 9.44	£ 11.10	£ 14.44
Production costs		£ 5.00	£ 5.00	£ 5.00	£ 5.00	£ 5.00
Net to producer		£ 1.10	£ 3.19	£ 4.44	£ 6.10	£ 9.44

Sparkling wines	Rate	Bottle price	Bottle price	Bottle price	Bottle price	Bottle price
		£ 15.00	£ 20.00	£ 25.00	£ 30.00	£ 32.50
VAT @ 20%	20%	£ 2.50	£ 3.33	£ 4.17	£ 5.00	£ 5.42
Sub-total		£ 12.50	£ 16.67	£ 20.83	£ 25.00	£ 27.08
Excise Duty - Sparkling Wine	£2.86	£ 2.86	£ 2.86	£ 2.86	£ 2.86	£ 2.86
Sub-total		£ 9.64	£ 13.81	£ 17.97	£ 22.14	£ 24.22
Production costs		£ 7.50	£ 7.50	£ 7.50	£ 7.50	£ 7.50
Net to producer		£ 3.14	£ 6.31	£ 10.47	£ 14.64	£ 16.72
Note: 2019 Excise Duty rates						

Table 13

applies a <u>mark-up</u> of, say, 33.3 per cent (as an example), this bottle becomes £7.99 on the shelf: £5 + 33.3 per cent (£1.67) = £6.67 + 20 per cent VAT = £8.00, less 1p to make the price look attractive. But, the £1.67 only represents a 25 per cent of the takings – the gross profit or the 'margin'. Thus, at the end of the day when the retailer has taken £1,200 in a day gross, i.e. including VAT, he or she wants to know how much of that is gross profit, i.e. profit before expenses. If 20 per cent is deducted for the VAT, leaving £1,000, what is left? The answer, in the case of the wine above and the day's takings, is £250, which is a margin of 25 per cent.[11] If that same bottle is to sell at £7.99 and the retailer wants 33.3 per cent <u>margin</u>, then he or she will only want to pay £4.45 per bottle 'DPD' (duty paid delivered) or will want to sell it at £7.50. Remember, too, that retailers will also probably have to give a discount for quantity sales – 5 per cent for six bottles is common and 10 per cent for twelve bottles is not unknown – and will also perhaps have 1–2 per cent of the retail price (including VAT) taken by credit-card costs. Merchants with multiple outlets may also need to build into their pricing structure an amount to cover head-office and internal distribution costs. Therefore, to the producer, who has after all done ALL the work, margins of between 25 per cent and 40 per cent may seem huge and out of all proportion. But, to the retailer, who has rent, rates, heat and light, staff and – I almost forgot – profit to pay for, this level of margin is needed in order to be viable and remain in business.

Most on- and off-trade buyers will be buying wines duty-paid, so the duty will be part of the price and it is up to you, as the producer, to pay that over to HMRC. This may seem unfair as it means that their margin represents an even larger percentage of your net income from the sale. But, right or wrong, you need to get used to this, as this is how the wine trade in Britain works. Some larger buyers may want you to ship the wine directly from your duty-free premises to a 'bonded warehouse' – either their own or one belonging to a licensed warehousing company – in which case they will want an 'IBD' (in bond delivered) price, even though they will most probably still calculate their margin on the DPD price.

The level of margin you are likely to meet will depend on exactly who you are selling to and it is difficult to be prescriptive about levels. However, as a guide, you could expect the on-trade (pubs, wine bars, hotels and restaurants) to want a margin of 10–15 per cent, multiple off-licences 25–40 per cent, independents and small-chain retailers 30–35 per cent and wholesalers 50 per cent or even higher, given that they will be reselling the wine to on-trade and retailers at a discounted price. Wholesalers may also want

---

11 For a fuller explanation see: en.wikipedia.org/wiki/Gross_margin.

you to consider promotional sales or additional discounts in order to get your wines noticed by their customers. Such things as lowering the retail price for a certain period of time, offering 'stock' as an incentive, i.e. they pay for five bottles but get six, and offering staff incentives for sales are all part of the game that has to be played if you want your wine to be noticed. These inducements to buy will all be at your expense and the wholesaler will still expect to make their margin.

Table 13 has been calculated at a 35 per cent margin for still wines and a 30 per cent margin for sparkling wines as these are higher in price and, therefore, on a cash basis, more profitable for the retailer to sell. The chart illustrates the importance of pitching the price of the wine as high as you can, justifying this on quality grounds, on the grounds of exclusivity and rarity value and, above all, making the product look the price. If you are wanting the retailer or the sommelier to persuade his or her customers that your wine is worth paying for, then before it gets anywhere near their lips (that's the consumer's lips – one has to assume that its already passed the retailer's or the sommelier's taste test) the product has to look professional: scuffed labels; poor quality printing; low-grade capsules, bottles, corks, muzzles and other packaging materials just will not do. You cannot, except in exceptional circumstances, get away with the story: 'we are only small, we only produce a few thousand bottles a year, and we don't have very good equipment'. The difference in net income between, say, a bottle of sparkling wine selling at £27.99, compared to £29.99, taking the above illustration, is £1.17. Let's say you have a 10-ha (25-acre) vineyard and sell 25,000 bottles a year, that's an extra £29,250! To get the higher price, of course, you have to work at it: grow good grapes, make good wine, enter competitions, win awards and publicise your successes.

In reality, of course, you are likely to be selling a proportion of your output directly to the public in one way or another and a proportion through the trade in all its many guises. Exactly what the proportions will be will depend upon many factors, size probably being the most important. Chapel Down, which has been selling wine from the farm-gate for forty-plus years, sells around 60,000 bottles a year direct to the public, but probably sells fifteen to twenty times that amount through the trade. One well-established, medium sized vineyard I know well sells around forty per cent by volume to the public at full prices and sixty per cent by volume via the trade at discounted prices. This would be fairly typical for this size of producer.

The one really important (and sobering) fact to always remember when selling wine is that a bottle of still wine sold direct to the consumer at, say, £12.50 is worth three times the margin compared to when you sell it to an off-licence. For sparkling wine sold at, say, £30 a bottle, because of higher Excise Duty rates and higher production costs, a bottle sold direct is worth at least twice the margin compared to when it is sold to the trade.

Note: There is a section on *Sales of English and Welsh wines* in *Chapter 4: Vine varieties, clones and rootstocks.*

Value of promotion in selling wine

Producers should also not underestimate the value of promotion in both selling their wines and of obtaining higher prices. Over time, producers can create a real demand for their wines by a combination of winning medals and awards, local (and even national) publicity and just good, old-fashioned self-promotion. I remember very well when, in 1981 we won the Gore-Browne Trophy with our 1980 Spots Farm Seyval blanc (very unexpectedly, as it was only our second vintage), the Sunday Times sent down their new wine correspondent, a pre-MW Jancis Robinson and a top photographer, Denis Waugh. The result of this was three full-colour pages in the Sunday Times Magazine which really helped us get on the English wine map and sell our wines. That wine, which pre-Gore-Browne had been retailing at £3.75 a bottle, immediately went up to £4.75 in order to (a) slow down sales (it didn't) and (b) increase our income (it did). If I were to single out one British vineyard which has created a solid market for its still wines by a combination of producing good wines, winning awards and medals and lots and lots of promotion, it would be Camel Valley. There, the Lindo family – Bob, Annie and Sam – have together got the message and proved that you can sell still wines at £12–£18, a price that is undoubtedly profitable and which many other producers can only dream about. Of course, it goes without saying that their wines are good, amongst the best in Britain, but the edge they get by good, solid promotion lifts them above other producers with similar quality wines. Of course, not everyone enjoys self-promotion like this and many find it difficult. If you are amongst this number, then perhaps a wine producer's life is not for you.

Buying a vineyard as a going concern

Over the past forty-five years, I have been involved one way or another with several instances of buying and selling existing vineyards, including selling, in 1986, what was then my own Tenterden Vineyards (and is today Chapel Down). What we sold then was the complete works: a 150-acre farm, 20-acres of vineyards, a winery, a farm-gate sales and visitor business, the stock, all the equipment and our house. It was a complex and, at times, long-winded process, given that it all had to take place in secrecy as we were determined not to let any of our staff or the public know that we had sold until contracts had been signed and (at least some of) the money was in the bank. Looking back on it I am glad that we did keep it hush-hush as it stopped the purchasers asking around too much!

Today, my advice to anyone wanting to buy an existing vineyard would be: tread carefully. Of course, each sale will be different. Is it just a bare vineyard, is there a business to be bought, and is there property involved? All of these will play their part in arriving at a value and a price. However, the central issues in all sales will be: what is the quality of the vineyard, what is its productive capacity and what are the quality of the grapes and wine like? Always ask to see the official WS production returns that every grower is required to make annually and, if they are selling wine, ask to see copies of their monthly HMRC Excise Duty returns. They are required by law to make both these returns and, as they form part of their accounts, ought to be kept for at least six years. Try and get them for as far back as they are available. If these documents are not forthcoming, then assume that whatever they tell you about the level of yields and sales is suspect.

Why plant a vineyard in Great Britain? – a summary

So – after reading this chapter, are you any the wiser? There is no doubt that owning a vineyard in Britain is not a quick or sure way to untold riches. But, given a good site and good management, money can be made. Whilst over the past forty-five years I have seen many vineyards disappear, I know some that have produced good wines, sold them well and prospered and in some cases these successful vineyards are in third-generation hands and expanding. Yes, they are in the minority, but they do exist and I am ever-hopeful that the numbers will grow, given the enthusiasm shown for good home-grown wines.

Prospective vineyard owners should realise that site selection is crucial, attention to detail is taken for granted and that creating a valuable wine brand is a slow, steady process, even more so when the wine is bottle-fermented and bottle-aged sparkling wine. Of course, the 'whether it is worth it' question is one that only you can answer. Different people have different views about the value and cost of their capital and the return they expect from it. Different people also place different values upon creating something from nothing and getting pleasure from joining the diverse community of people who make up the world of English and Welsh wine.

Now is the point of no return and the remainder of this book is about the A–Z of planning, planting, establishing and running a vineyard in Britain.

Chapter 3

Site selection

Where to plant?

Now that you have decided to go ahead and plant a vineyard, the first big question arises: where? The question, of course, is not a simple one and there will be many factors to be taken into account. Which areas of Britain do you know and like; are there family matters, work, schools, and relatives that have to be taken into account; are there budgetary constraints? All of these are important. However, the one thing you do know is that the better the site, the better the wine. Get this wrong and it will affect all aspects of your enterprise – for all time. However, we can ignore all these other non-viticultural factors for the time being and seek to answer one question: where in Britain will a vineyard produce the best grapes and therefore – all other things being equal – the best wine? This question of where the best grapes can be grown can only really be answered by looking at the problems facing British vineyards and seeing whether there are some regions that might help address these problems and some regions where these problems might be made worse.

Sunshine, light and heat

The main problem in Britain is not getting vines to grow: there are plenty of places where this can happen. It is not even really getting vines to fruit and getting that fruit to ripen: again, there are enough places in Britain for this to happen, especially if you select the right vine varieties. The main problem is to find a site where the vineyard can produce a high enough yield – in monetary value whether via grape sales or wine sales – of grapes of the right quality to turn into good wine. It is only on this basis that the substantial investment required will be repaid and the enterprise will become self-supporting and sustainable over the long term.

There will be those who will be able to fund vineyards from their back pockets – the wine world is littered with such vanity enterprises – but I have seen too many vineyards

in Britain like this and they don't always make the best wine. Yield, of course, is a moveable feast and the monetary value of a crop will be different to someone selling it as grapes at £2,000 a tonne, still wine at £10–£15 a bottle or sparkling wine at £25–£35 a bottle. However, one thing you do know is that the larger the crop, the more there is to sell, whether it be as grapes, still wine or sparkling wine. Yield, therefore, has to be a very important factor in the viability – the sustainability, if you want to use that word – of a vineyard, a winery and/or a wine sales enterprise. And it is because of low yields that the track record of many, many vineyards is poor and probably the main reason why so many vineyards have disappeared over the years.

In my 1989 book *The Vineyards of England*, I listed 338 vineyards. By the time I came to write *The Wines of Britain and Ireland* in 2001, only 118 of those still existed. Today, only fifty-seven of those original 338 vineyards remain, a loss of 281 vineyards covering 582-ha. Some of these were substantial vineyards, even by today's standards: Barnsgate Manor 20.0-ha, Barkham Manor 13.7-ha, Chapel Farm 8.0-ha, Chiddingstone 11.33-ha, Highwayman's 9.71-ha, Honeybee 11.37-ha, Horton Estate 3.6-ha, Isle of Wight 17.4-ha, Leeford 14.06-ha, Mapperton 9.8-ha, Meon Valley 6.4-ha, Pilton Manor 7.74-ha, St. George's 8-ha, Wellow 32.38-ha (and this is just a few of the larger ones – my database of grubbed-up vineyards runs to almost six hundred different vineyards). Where are they now? All gone or at least very much reduced. Many grubbed-up vineyards produced good wines, so it wasn't just a question of wine quality: Adgestone, Barton Manor, Brede, Chiltern Valley, Kelsale, Lamberhurst, Pilton Manor, Pulham, Valley, Wootton – all Gore-Browne Trophy winners and all gone or at least much reduced. If yields had been higher, might these vineyards have survived, been taken over by another generation or been sold on as going concerns?

And what drives yields? Sunshine, light and heat, and Britain often has too little of all three. As has been shown in the previous chapter, the tantalising thing is that, yes,

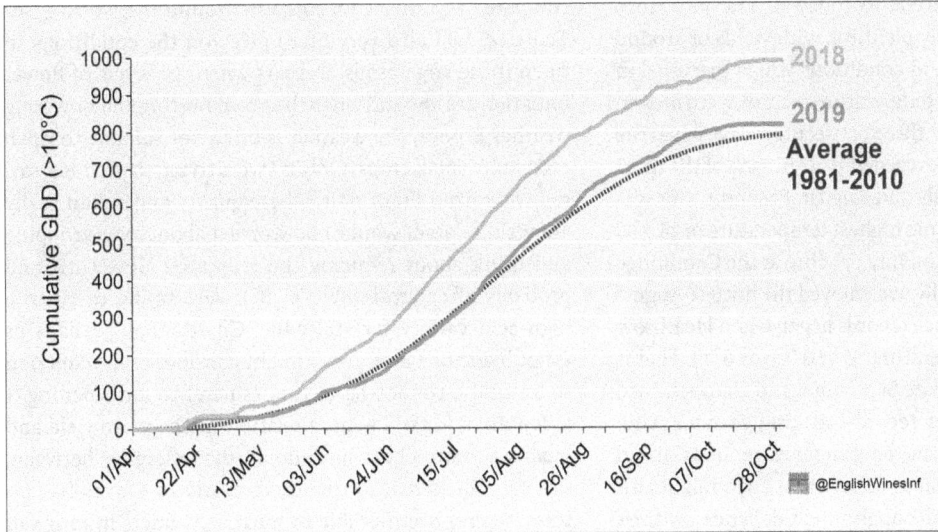

Figure 1

in some years everything is fine – 2018 being an extreme example. The vines get enough heat and the buds on the cane are filled with viable inflorescences (flower clusters) which in the following spring and early summer, given enough decent weather, will flower and produce a reasonable crop. But in years when we have too little, the potential is not even there in the buds and, given perhaps a cool spring and poor flowering weather, the yields are light or, as we saw in many vineyards in 2012, non-existent.

Many vinegrowing regions are categorised by the number of 'Growing Degree Days' (GDD). Originally a system developed for Californian growing regions by Amerine and Winkler in 1944; it is now sometimes accepted as a guide to how warm (or cool) any region is for viticulture. The figure of GDD is arrived at by taking the mean temperature on each day during April to October in the northern hemisphere (and usually October to April in the southern),[1] deducting 10°C from the figure, and arriving at a total. Thus, if the mean temperature on 1 April is 12°C, then this is two degree-days, if on 2 April it is 13°C, then this is three degree-days and so on until the end of October. In practice, the mean temperature of the month is often taken, 10°C deducted and the answer multiplied by the number of days in the month. This gives you a slightly different figure, but not massively so. In Britain, being a maritime climate, our GDD vary between the mid-700s in really cool years to the heady heights of 1,000+ in warm years. Champagne, in comparison, varies between 900 and 1,050, Chablis averages around 950 and the upper Loire 975. As you can see from Figure 1, the GDD have been climbing steadily from the 1950s, when the average stood at around 680, to today's average of 800. Still not high, but high enough for us to grow varieties today that were impossible seventy years ago. However, in my experience GDD is only of limited

use to British growers in deciding where to plant, as other factors (such as shelter and exposure) are more important on a site by site basis.

The effect of sunshine, light and heat is different for different vine varieties and one of the issues facing vinegrowers in Britain is that because of the warming of our climate, much of the area under vines has changed from using true cool-climate varieties such as Müller-Thurgau, Reichensteiner and Seyval blanc – varieties that had stood the test of time and which even in very poor years such as 2012, produced viable yields – to more marginal varieties such as the Champagne varieties. These are far more sensitive to both cool growing conditions in the preceding year, as well as poor conditions leading up to and during flowering in the cropping year. In 2012 this was very visibly demonstrated where I saw a block of Chardonnay with almost no harvestable crop right next door to a block of Seyval blanc with a 5.0 tonnes-ha crop: same site, same soil, same management, but different yield outcomes. The reason? Chardonnay requires more heat and light than we had at the critical times in 2011 and 2012 for producing viable flowers and for pollination. Study after study has shown that poor light and low temperature levels, both in the period when the bud-bearing cane is being produced and, in the period up to and during flowering, i.e. late May to mid-July, influence the production of cytokinins which are stored in the roots. Low levels of these hormones are one of the principal causes of low yields.[2]

In *The Grapevine*[3] (p. 68), the authors state that: 'maximal initiation [of inflorescences] for most varieties occurs at around 30–35°C. A pulse of only 4 hours in 24 hours of high temperatures, e.g. 30°C, during the critical period … is sufficient to induce maximal initiation'. Given that in Britain we only rarely see days when the temperature rises to 30°C, even in high summer (July–August) let alone in late May and June when initiation is occurring, it is not surprising that we suffer from low yields. When we do see excellent weather around flowering, as we did in 2017, the results show in the amazing, record-breaking 2018 crop when yields of almost 46 hl-ha surpassed the hitherto highest yield of 37.70 hl-ha set in 1992 (when the variety

---

1 I have sometimes seen a six-month October to March period taken for the southern hemisphere which makes comparisons between north and south difficult.

2 There are many sources of information on this subject. The paper Grape bud fruitfulness – what causes a bud to produce a bunch or not by Devin Carroll can be found at www.englishwine.com/wgigb.htm

3 The Grapevine – from the science to the practice of growing vines for wine by Iland, Dry, Proffitt, and Tyerman

spectrum was very different to today's). The year 2019 turned out to be extremely plentiful, with yields of around 37 hl-ha, set up by the good conditions which prevailed in the previous year. Several temperature records were broken in 2019. Firstly, there were three consecutive days when the temperature reached and exceeded 30°C in each of the three summer months: June, July and August. Secondly, this was followed by Britain's all-time highest temperature of 38.7°C (101.66°F) being recorded on July 25th 2019 at the Cambridge Botanic Garden. And finally, we enjoyed the hottest August Bank Holiday Monday since records began when Heathrow Airport recorded a temperature of 33.4°C (91.6°F). Had it not been for the very wet late September and October, 2019 might have been another record-breaking year for wine producers in Britain. Given the temperature 'firsts' listed above, there is every reason to believe that 2020 might also be a high-yielding year, although much depends on frost and the flowering weather.

Another factor in low yields in Britain is the lateness of our harvest and the strain it places upon the vines. In warmer climates, the harvest is often over well before leaf-fall, giving the vine a chance for a period of recuperation and reserve-building prior to the winter. Indeed, in many regions, vines are irrigated straight after harvest so that fresh shoots and leaves can grow and as many carbohydrates as possible are put back into the woody parts of the vine – the canes, trunks and roots – before winter. In Britain, with harvest taking place as late as the last week of October and even into early November, and with harvest-time frosts to contend with often, the vines seldom have time for any fresh shoot and leaf growth after harvest and before dormancy. Add to that the propensity of many young leaves to succumb to Downy Mildew in the period between when spraying has to stop and the harvest, and many vines will be entering dormancy underprovided with nutrients. In the following spring, if weather conditions around bud-burst are also poor – cool and wet – then the vine will struggle to produce good shoots in the spring, added to which bud development, i.e. the propensity to produce inflorescences on those shoots, will be impaired. A 2013 study carried out in Tasmania, which has a climate that, although better than Britain's for grape growing, could still be described as 'marginal', showed that: 'both reproductive development and shoot growth in the following season can be influenced by the amount of reserves stored over winter' and that 'cane starch was shown to be a significant predictor for inflorescence count and cane size was positively correlated with the mean number of inflorescences per node'.[4]

I accept that yield is not everything, and that there may be wine styles and there may be times where yield has to be restricted, but these are relatively rare styles and relatively rare times. And, even if yields are restricted, the grapes need to be ripe. In my experience, the conditions that give you large crops (large in a British context, not large when compared to other cool climates around the world – see Table 14) will also very often give you the conditions to ripen those large crops. In years when the levels of flower initiation are low and when the pre-flowering and flowering weather is poor, the weather is often not suitable to ripen even quite small crops. Only if I were trying to produce, say, red wine from Pinot noir, where quality and colour really are yield related, would I be worried about over-cropping and think about reducing the crop after flowering (and probably after *véraison*).[5] Or, if I were trying to ripen a high-acid variety for still wine, Chardonnay perhaps or Pinot blanc or Pinot gris, I might consider crop reduction to around 5 tonnes-ha (2 tonnes-acre) to aid ripening. I accept that, for still wine, a relationship between yield and quality is more likely, but I doubt the difference between, say, 7.5 tonnes-ha (3.0 tonnes-acre) and 10 tonnes-ha (4.0 tonnes-acre) is noticeable to most consumers in terms of quality in the glass for varieties such as Bacchus, Reichensteiner and Seyval blanc, although it certainly would be noticeable on the profit and loss account. At these yield levels, the degree of ripeness will be the deciding factor in quality and this is often a function of the quality of the summer and autumn, not the yield level. Harvest 2018 is a case in point: the largest yields per hectare ever, the largest bunches weights ever, and the highest natural sugar levels ever. There was not much evidence of vines struggling in 2018.

For sparkling wine, there is very little evidence in the context of vineyards in Britain that high yields are necessarily detrimental to final wine quality. There are so many stages in the production of a sparkling wine, from the *assemblage* of the base wines, via the *sur latte* ageing process, to the addition of *dosage* and post-corking ageing – all of which contribute in part to the style and quality of the final wine – that absolute grape quality at harvest time is less important, and certain imperfections can be overcome in the cellar. This is one of the reasons why yields in the Champagne region are so high and that the vast majority of Champagne itself is a multi-site, multi-variety and multi-vintage wine.

The figures in Table 15 are from an established vineyard in Kent, with 0.8 ha (1.98-acres) of both Chardonnay and Pinot noir (1.6 ha (3.95-acres) in total) planted on the same rootstocks (two for each variety) and on the same site. Over the three years, 2005, 2006 and 2007, for Chardonnay, the yields varied by a factor of four – from 6 tonnes-ha to 24 tonnes-ha (2.45 to 9.78 tonnes-acre) but sugar levels were much the same – between 9.4 per cent and 9.8 per cent potential alcohol – although acidities did vary some 5 g/l (expressed as grammes per litre of tartaric) but with the highest acidity being for the lowest yield, something common sense would tell you ought to be the other way round. It just so happened that the ripening conditions in

---

4 A statistical model to estimate bud fruitfulness in Pinot noir by Joanna E. Jones, Greg Lee and Stephen J. Wilson. AJEV.2013.12086.

5 Véraison is the stage of growth when grapes start softening, start turning red (in red varieties) and sugars start to accumulate and acids decrease. Véraison takes place at different times on different bunches and is a good guide to the order in which those bunches flowered.]

Region	Hectares	Tonnes per hectare	Tonnes per Acre
Luxembourg - all varieties	1,249	15.02	6.08
Champagne	34,300	14.70	5.95
Germany	103,000	14.30	5.79
New Zealand	38,000	11.10	4.49
Luxembourg - Chard & P. Noir	158	10.67	4.32
Tasmania	2,000	8.00	3.24
Oregon	36,600	6.70	2.71
Great Britain (2010-2019)	3,500	3.29	1.33

Table 14

Chardonnay					
Picking Date	Sugar °OE	Sugar % alc	Acid g/l as tartaric	Tonnes per ha	Tonnes per acre
24.10.05	74.00	9.80	10.50	11.11	4.50
23.10.06	72.00	9.50	10.30	24.17	9.78
22.10.07	71.00	9.40	15.50	6.04	2.45

Pinot Noir					
Picking Date	Sugar °OE	Sugar % alc	Acid g/l as tartaric	Tonnes per ha	Tonnes per acre
10.10.05	66.00	8.60	13.00	15.88	6.43
16.10.06	64.00	8.30	11.30	16.03	6.49
15.10.07	66.00	8.60	12.80	4.10	1.66

Table 15

Variety	Median sugar level 2018	Median sugar level 2019
Bacchus	76.0	68.0
Chardonnay	78.0	67.0
Früburgunder (PN Précoce)	73.5	79.5
Madeleine x Angevine 7672	75.5	70.0
Meunier	75.0	68.0
Pinot noir	78.0	69.0
Regent	78.0	63.5
Reichensteiner	75.0	76.0
Rondo	69.0	75.0
Seyval blanc	70.0	62.0
Mean of all Varieties	**74.8**	**69.8**

Variety	Median acidity level 2018 g/l tartaric	Median acidity level 2019 g/l tartaric
Bacchus	9.00	9.20
Chardonnay	12.00	12.90
Früburgunder (PN Précoce)	8.45	8.50
Madeleine x Angevine	10.10	9.50
Meunier	11.90	12.65
Pinot Noir	12.00	12.30
Regent	9.00	9.35
Reichensteiner	8.75	9.25
Rondo	9.15	10.70
Seyval blanc	11.25	11.25
Mean of all Varieties	**10.16**	**10.56**
Source: 2018 & 2019 ICCWS-WineGB Yield Surveys		

Table 16

2005 and 2006 were better than in 2007, although picking dates were within a day or two of each other.

For the Pinot noir the story was similar. Yields also varied by about a factor of four between around 16 tonnes-ha (6.5 tonnes-acre) in 2005 and 2006, and only 4 tonnes-ha (1.66 tonnes-acre) in 2007. But potential alcohol levels were almost the same – 8.3 per cent to 8.6 per cent potential alcohol – picking dates within a week of each other and acid levels not too far apart.

Looking at the above sugar and acid figures in isolation, it would be difficult to make any assumptions about yields and wine quality. What you would expect is that the higher sugars and the lower acids were in the lower-yielding years, but this is not the case. As for wine quality, the 2005 was clearly the best, doing very well in tastings. The 2006 wine was of lesser quality and, in terms of medals and awards, didn't win as many as the 2005 wine, yet sold for the same price. The lowest quality wine was quite certainly the 2007, the wine from the lowest yielding year! Looking at more recent vintages, we have the examples of 2018 and 2019 where the 'low yield – good quality' argument has been turned on its head. We saw record yields <u>and</u> record ripeness levels in 2018, whereas in 2019 we saw reasonable yields but lower natural sugar levels (lower than in 2018) owing to the cool and wet conditions in the 6–8 weeks prior to harvest. Taking the six most widely planted varieties, we can see in Table 16 that the sugar levels in 2019 were on average 7°OE (about 1.1% abv) lower than 2018, so quite a lot less ripe, and acids around 0.44 g/l higher. Given the amount of rain that fell up to and during harvest in 2019, the lower sugars and relatively low acids were the result of dilution of the grapes on the vine. However, my bet is that most of the sparkling wines of 2019 will be just as good and have as much staying power as the 2018s, despite the lower ripeness levels.

Site quality

So, back to the question of site selection. What we need for a site in Britain to establish a vineyard on is one that gives us the most sunshine hours, the highest light levels and the highest temperatures which will enable high-quality vine varieties to be grown, producing viable yields of adequately ripe grapes. In other words, the site with the highest site quality. So – where might these sites be found?

A windy site will never make the best vineyard. This photograph was taken at a vineyard on the Channel Isles

Location in Great Britain

In cool climates – and in mainland European terms this means much of Germany, the Loire from its source to its mouth, Champagne, Chablis, most of Burgundy and Bordeaux – the best sites are those which are the warmest and which will produce the ripest grapes, all other things being equal. This means sites that are well sheltered, not too high above sea level and typically on a south-facing slope. The reasons for this are well known and shouldn't come as a shock to anyone reading this book: sheltered, south-facing slopes under 100 m above sea level warm up more quickly, prompting the vines growing on them to break bud and come into leaf earlier than later sites, giving them longer to ripen their grapes to a higher degree of natural ripeness, with lower acid levels; riper fruit flavours; higher levels of dry extract; and more body, substance, length and concentration. Different soil types may help or hinder this process, but **site quality** is by far and away the most important determinant of fruit quality, all other things being equal. This is why traditional European winegrowing regions, especially cooler ones, put so much emphasis on *terroir*. Because, whatever else you may do to a site, you can never move it, change its altitude, its slope or its basic topography. And, along with better fruit quality,

Vineyards in Great Britain by Region - 2020 *				
Region	**Hectares**	**% of area**	**No. of Vyds**	**Average Size**
South East inc London	2,147.29	61.51%	222	9.67
Wessex	400.20	11.46%	101	3.96
East Anglia	353.78	10.13%	108	3.28
West	292.63	8.38%	176	1.66
Midlands & North	142.10	4.07%	85	1.67
Thames & Chilterns	97.37	2.79%	41	2.37
Wales	50.87	1.46%	31	1.64
Scotland	1.37	0.04%	2	0.69
Channel Isles	5.50	0.16%	3	1.83
	3,491.11	100.00%	**769**	**4.54**
*Vineyards of 0.10-ha and over. Regions as defined by WineGB Regional Associations			Source: UK Vineyards Guide 2020	

Table 17

brought about because the vines are growing in a better environment, comes the likelihood of better (and more) fruit in the following year as the vine is laying down its fruit buds for the year to come in the weeks leading up to flowering which, in Britain, would be late April, May and June.

The warmest regions in Britain are naturally in the south of the country and if you look at where the orchards for the best dessert apples are (or were) and where hops are (or most definitely, were) grown, you can see that they are: Suffolk; Essex and part of Hertfordshire; Kent; East Sussex; West Sussex and parts of Surrey; Hampshire and the Isle of Wight; the south coastal fringes of Dorset, Devon and Cornwall; the parts of Somerset, Gloucestershire, Wiltshire and South Wales that are influenced by the Severn; and up into Herefordshire and Worcestershire. This is a large area and of course not every hectare of land is suitable for vines. My personal choice would be Kent, followed by sites in East and West Sussex and Hampshire that are near to the coast, say at least 8 to 15 km away from it, but definitely not within sight of it. East Anglia is also worth considering and there have been some excellent crops (in terms of both yield and quality) coming from vineyards near to the Essex coast and in sheltered parts of Suffolk and Norfolk. In the early days of the revival (1960s–1980s) East Anglia had around 15 per cent of Britain vineyard area, with some notable producers. Today, despite some large plantings especially in the region between the Blackwater and Crouch estuaries, the whole of East Anglia only has just over 10 per cent of Britain's total. This compares to, say, the South East which in 1989 had 46 per cent of the planted area and today has 61.5 per cent. Climatically the Isle of Wight might also be considered suitable, but perhaps not from a pure business location point of view and most of the vineyards once established there have now disappeared. I am probably biased towards Kent because I have spent forty-five-plus years in and around the region, but it is notable how many trophies and gold medals have come out of this county.

If you look at where the vineyards in Britain are today (Table 17), the South East (Kent, East and West Sussex and Surrey) is by far the most widely planted region with just over 59 per cent of all vineyards. The next region, Wessex (Hampshire, the Isle of Wight, Dorset and Wiltshire) which is actually slightly larger in total physical area, only has just over 13 per cent of vineyards. One could argue that the proximity to London and the populous and wealthy Home Counties makes the South East a better place to have a vineyard offering wine for sale, although this hasn't helped East Anglia or the Thames and Chilterns regions, both of which have the same proximity to London and, if anything, even wealthier populations. The fact is that, climatically, the southern one-third of East Anglia and the majority of the South East are the most favourable places in Britain to grow a sun-loving and sugar-producing plant like the grapevine.

Taking the yield data derived from the sources already mentioned in *Chapter 2: Why plant a vineyard in Great Britain?* in the section of *Yields* we can see from the table that there is a definite east to west bias to the yields, with East Anglia and the South East the only two regions with yields over the four years surveyed to exceed 6.00 t-ha (2.43 t-acre), and with the South West yielding 1.15 tonnes-ha (0.47 t-acre) less than the South East. Another factor to be borne in mind when comparing these figures is that they are for 'all varieties' and therefore include a bigger proportion of the higher yielding varieties such as Reichensteiner and Seyval blanc.

If you look at the two most widely grown varieties, Chardonnay and Pinot noir, the regional yield differences are far greater, as can be seen from Tables 18 and 19.

I would also consider the influence of rainfall when selecting an area to establish a vineyard, as vines naturally prefer dry conditions over wet, not least because when it is raining the sun is usually not shining! However, although the east of Britain is quite a bit drier than the west in general terms, in my experience rainfall on a given spot tends to be quite variable and subject to influence by local factors. One has often heard of sites that are drier than sites only a few miles away, usually because there is an intervening hill which deflects rainfall-containing clouds. How easy it is to select a site based upon this level of micro-weather knowledge is another matter. Dr Alastair Nesbitt, in his

Yields Tonnes-ha	2016	2017	2018	2019	2016-19 t-ha	2016-19 t-acre
South East	4.31	4.61	9.90	6.43	6.31	2.55
East Anglia	6.89	4.35	6.92	6.09	6.06	2.45
Wessex	2.08	3.44	9.93	5.66	5.28	2.14
South West	3.95	5.82	5.52	5.35	5.16	2.09
Thames and Chilterns	2.59	5.62	N/A	6.83	5.01	2.03
Source: Private surveys 2016-17. ICCWS-WineGB Yield Surveys 2018-19						

Table 18

PhD[6] on climate change and the suitability of Great Britain for viticulture, singled out rainfall during June as the most significant factor in climate variability. He wrote:

Table 3: Median yield tonnes-ha by variety by region 2018-19	South East	Wessex	South West
Chardonnay	8.30	8.16	2.94
Pinot Noir	7.59	7.68	3.33
Median Yields	**7.95**	**7.92**	**3.14**
Source: ICCWS-WineGB Yield Surveys 2018-19			

Table 19

For the full 1989–2013 period ... June precipitation had a negative relationship with yield, that is the greater the precipitation the lower the yield. It was found to be the single statistically significant variable explaining 34.7 and 64.1% of the variability in yield, respectively.

In his subsequent 2018 paper[7] 'A suitability model for viticulture in England and Wales: opportunities for investment, sector growth and increased climate resilience' he concluded by saying:

As well as opportunities for expansion within the currently dominant regions of Kent and Sussex, new areas such as Essex and Suffolk in particular, where relatively few vineyards currently exist, have been shown to have highly suitable land, express high degrees of climatic suitability and greater levels of stability from season to season than other areas.

It should also be remembered that the Lizard in Cornwall, Britain's most southerly mainland point, is actually slightly further south than Geisenheim which is in the middle of the Rheingau, Germany's best winegrowing region, yet nobody would pretend that south Cornwall is a great place to grow grapes: two windy and too wet. Camel Valley Vineyard, situated in what must be the most sheltered part of Cornwall and almost equidistant from the north and south coasts, is on the same latitude as Germany's Ahr winegrowing region, famous for its Pinot noir wines, yet even the exuberant Bob Lindo wouldn't claim that his valley was Britain's red wine capital. Again, too wet and windy for some varieties and one of the reasons why two-thirds of their grapes come from Essex and East Sussex. Norwich, however, is some 300 km (187 miles) to the north of the Lizard (almost on the same latitude as Hanover in Germany – not well known for its vineyards) and the few vineyards in Norfolk are producing some good wines. People tend to forget that Britain tilts to the west!

As to the question of how far north to grow vines, all I can say is: *caveat emptor.* As I have pointed out above, growing vines is not the issue, you can probably do that in the Shetland Islands (given enough protection from the wind).[8] The problem is (a) growing vine varieties that give good quality wine, (b) getting them to fruit, (c) ensuring fruiting at the right level of yield to make it worthwhile, and (d) getting that fruit to ripen. Whilst I applaud the efforts of those growers who have planted vines north of Birmingham, I cannot say I would recommend it.

In early October 2013 I made a pilgrimage north to visit all of the major Yorkshire vineyards – Holmfirth, Leventhorpe, Ryedale, Summerhouse and Yorkshire Heart – plus Somerby and Three Sisters in Lincolnshire and was pleasantly surprised to see in most of them (sadly not all) good crops of grapes, in some cases ripe and already being picked, in sharp contrast to some vineyards in Kent and East and West Sussex, which were still waiting for their grapes to ripen! The vine varieties that appeared to be doing the best were Madeleine x Angevine 7672, Rondo, Seyval blanc and Solaris – some of which I would not rate in the first division of wine quality – but still, grapes are grapes and I wish these producers all the best in producing award-winning wines. Varieties such as Bacchus, Ortega, Phoenix, Pinot noir and Siegerrebe were faring less well. However, it does prove it is possible to grow grapes in some unlikely parts of these islands.

The proof of the pudding, of course, is in the eating (or in this case, the drinking), but I would point out that if you chart the location of trophies and gold and silver medals that have been awarded to wines entered into the national competition, they get sparser the further north you travel. Yorkshire's main producing vineyards have, to date, between them amassed a few silver and bronze medals, plus some highly commendeds, but nothing better: no trophies or golds. It may not be much evidence, but it's some. Of course, you can probably survive and maybe even prosper if you are a tourist-magnet, can charge high prices, sell lots of tours and teas and can purchase extra grapes when your own harvest is lacking, but that is not really the point of this book.

Altitude

The altitude of the site is important for two reasons: the higher a site is above sea level, the cooler it is and the more exposed it will be to the prevailing winds (which in the south of Britain are typically from the south-west). Both of these factors will make the site cooler. It is an often quoted fact that for each 90 m that land rises above sea level, the average annual temperature drops by 0.5°C and the growing season is shortened by two days – one in the spring and one in the

6 https://onlinelibrary.wiley.com/doi/10.1111/ajgw.12215

7 A. Nesbitt, S. Dorling & A. Lovett (2018) A suitability model for viticulture in England and Wales: opportunities for investment, sector growth and increased climate resilience. Journal of Land Use Science, 13:4, 414-438.

8 You can certainly grow raspberries and blackcurrants on the east coast of Scotland in the open (although they are mostly grown under plastic tunnels these days) but I would not advise growing vines. Despite plenty of light, there is just not enough heat to ripen grapes.

Height above sea level	Median yield 2018	Median yield 2019	Median yields 2018-19
Lower than 60 m	5.73	5.76	**5.75**
60 m-100 m	8.33	6.23	**7.28**
Higher than 100 m	8.40	4.00	**6.20**
Average yields	**7.49**	**5.33**	**6.41**
Source: ICCWS-WineGB Yield Surveys 2018, 2019.			

Table 20

autumn.[9] I would therefore start getting concerned about sites where the land is above 100 m, get very concerned about land over 125 m and would not consider land over 150 m. Of course, on a sloping site, some of the land is bound to be lower and some higher, so a site that had 50 per cent of the land below 100 m and the rest above, especially if the site had excellent shelter with rising land behind it and good cover to the south-west, might be better than one with all the land below 100 m but on the top of a hill with views of two counties. An additional factor might also be soil type (dealt with more fully later in this chapter) and one might be more forgiving about altitude on a site with a dry, stony, lean soil which retained heat compared to a fertile, loamy, damper soil which would be cooler.

As can be seen from Table 20, the ICCWS-WineGB Yield Surveys looked at attitude and yield and the data showed significant differences between heights above sea level, with the 'sweet spot' being 60-100 m. The poor performance in both years of the 'lower than 60 m' vineyards was attributed to losses due to frost – which tends to be worse at lower altitudes – and the poor performance of the 'higher than 100 m' vineyards due to cooler sites and increased wind exposure. The differences in 2018, which was a benign year in terms of weather and saw record yields in almost all vineyards, was less marked than in a more difficult year such as 2019.

Aspect

The next most important requirement for a top vineyard site is that it should slope towards the south, the south-west or the south-east. The vines in vineyards facing east of south will probably dry out faster in the morning and those facing west of south will get the benefit of evening heat. Any generally south-facing site will be warmer and will therefore give you a better quality and quantity of grapes than a flat site, or one facing north which would be around 1°C cooler. However, having said that, I would be prepared to sacrifice a slope facing due south if one facing south-east or south-west was better sheltered from the prevailing wind or was at a lower altitude. At Breaky Bottom vineyard,

9 Lockhart and Wiseman's Crop Husbandry, eighth edition 2002, p. 79.

a site I have known and visited regularly for almost forty-five years, a small block of Chardonnay on a north-facing slope, but a block very well sheltered by a large hedge and rising ground, is always earlier, and crops better than the Chardonnay growing on the south-facing slope in the valley bottom which is exposed to winds that blow up the valley off the sea. This, however, is the exception that proves the rule: vines like shelter.

A sloping site will also always help with air drainage which will allow frost to drain away, so long as there is lower land for the frost to drain to. On all accounts, avoid what is known as a 'frost pocket'. Sites lucky enough to be near a body of water – a large lake or river – will also be less prone to frost damage as the water will both hold heat and help create air circulation which will keep frost moving. Frost pockets and the problems of frost and how to avoid and/or overcome them are dealt with fully in *Chapter 13: Frost protection.*

Whilst gently sloping sites are ideal for vines, steep sites, sloping at, say, more than 30°, can also be very good, but they can also be problematical. Unless terraced, something that has never really been done on anything other than a very small scale in Britain, steep sites can only really be planted directly up and down the slope and will require more powerful four-wheel drive tractors in order to get up the slope towing and powering sprayers and mowers. In addition, turning at the top and bottom of the rows on a steeply sloping site, unless there is a flat area that has been created specifically to turn on, has its dangers and tractor drivers will have to take extra care if they are to avoid tipping over. However, steep slopes will be typically warmer than gently sloping sites and drainage will often (but not always) be better. As to how steep is too steep, this depends entirely on the equipment used. In many European regions, sites with slopes as steep as 45° are successfully planted with vines.

Shelter

A major factor in site quality is shelter and this cannot be emphasised enough. A sheltered site is one which has rising ground, or perhaps a block of woodland, or failing those two, some shelter belts and/or windbreaks planted on the windward side in order to slow down the prevailing wind, stop it moving the leaves and stop it cooling the site. Forget all about having a breezy site to blow away diseases – I have never known a site for vines in Britain that didn't have some degree of wind blowing through it – you are after a site with as little breeze as possible. The windier the site, the cooler it will be: the more a vine's leaves are moving, the less they will photosynthesise; the less they photosynthesise, the lower will be the sugars in the grapes, the reserves of carbohydrates in the trunk and the roots and the fruiting ability for the vine in the coming year. Shelter belts and windbreaks (which are dealt with in *Chapter 6: Site planning and preparation* in the section on *Windbreaks*) can be very valuable to a site, but they are never as effective as a naturally well-sheltered site. This also means that sites within the influence of sea breezes should be treated with caution, even though they may well be less troubled by frosts. I am very familiar with sites such as Adgestone on the Isle of Wight, 2.1 km from the coast, and Breaky Bottom, near Lewes in East Sussex, 3.9 km from the coast, where grapes grow happily and good wines have been made over 40+ years, but I know that they are sometimes negatively influenced by the cooling effect from their proximity to the sea and their yield track record shows it. My advice, therefore, would be to think carefully before attempting to establish a vineyard within 5 km of the coast, especially the south coast which catches the full brunt of the prevailing wind travelling up the Channel. Rathfinny Estate, with vines first planted in 2012 and now with 93.50-ha (231-acres) of vineyards, has suffered considerably from wind issues, being just over 3 km (less than 2 miles) from the south coast in East Sussex. As owner, Mark Driver, admits to in his blog:

Before we bought Rathfinny I thought long and hard about wind and I looked at historical weather statistics from the Met office. They seemed to be okay, the average wind speed during the summer growing months was 4.5 metres per second, which is less than 10 mph, and that is the average for the whole site ... the lower part of the slope at Rathfinny gives much greater protection from the south-westerly winds and we could plant windbreaks to slow it down further. How wrong could I be? Well as it turns out the average is the 24-hour average and the winds tend to be stronger during the afternoon, when the vines are meant to be growing. And the trees we planted as windbreaks are taking a lot longer to grow than I had expected.

The original windbreak trees at Rathfinny proved difficult to establish, so 4 m high 'Parafence' windbreaks were installed in order to protect the windbreak trees, so that they could protect the vines (at a cost, it is said, of £250,000). Wouldn't selecting a more sheltered site have been a better option? Of course, it is a beautiful part of the world with the rolling South Downs as a backdrop, and with the impressive visitor facilities on site, and the effort being out into attracting wine-buyers to pay a call, farm-gate sales should build up. However, on the basis of the site's yield alone, only time will tell whether this multi-million-pound venture will be financially sustainable in the long term.

I suspect that one of the reasons why so many vineyards in Cambridgeshire, Norfolk, Suffolk and the less sheltered parts of Essex did not prove successful in the past, is the cold east winds that hit this part of Britain. When I look back to the 'East Anglia' section in my 1989 book *The Vineyards of England*, the roll call of grubbed-up vineyards

is alarming: Boyton, Brightlingsea, Broadwater, Bruisyard, Cavendish Manor, Clees Hall, Coton, Deben Valley, Elmham Park, Essex Crown, Felsted, Finn Valley, Fenlandia, Fyfield, Gamlingay, Greens, Harling, Helions, Heywood, Highwayman, Isle of Ely, Ilsington Lodge, Langham, Lexham Hall, Nevards, Oak Hill, Priory, Roding Valley, Pulham, Sotterley Farms, St Mary Magdalene, Swafield House, Uggeshall, West Hanningfield, Willow Grange, Writtle. It would have been easier to list those that have survived: Chilford, Giffords, Mersea, New Hall, Staverton, Wissett (now Valley Farm) and Wyken. Of the 117-ha (290-acres) of East Anglian vineyards listed in my 1989 book, only 48-ha (119-acres) survive of which 35-ha (87-acres) are at New Hall,[10] situated in the southern, coastal part of Essex. However, global warming has brought benefits to East Anglia, as it has to other parts of Britain, and there are today several new plantings across Essex, Suffolk and Norfolk. East Anglia today (2020) has almost 381-ha (941-acres) of vineyards and is expanding fast. The CM3 postcode, whose southern end is basically all of the land on the northern side of the river Crouch estuary (and therefore south-facing) is – so growers there claim – with 192-ha (474-acres) the most be-vined postcode in the country.

Location, altitude, aspect and shelter are for me the key elements to a good site in Britain for growing vines. Get these wrong and the enterprise is already starting with a handicap. I am not alone in this view. Peter Hayes, the WineSkills[11] Viticulture Mentor, wrote in his first report, dated March 2010, following visits to twelve different vineyard operations, several with multiple sites:

Site selection and site amelioration

Some evidently superior sites were observed in the course of these visits. These were characterised by:

- *Relatively low elevations (30–100 m)*
- *Southerly aspects with considerable slope for sun exposure and soil drainage*
- *Protection from, in particular, S-W winds*

Several other high potential sites at greater elevations appeared to have adequate wind protection either in place or planned and some had clearly inadequate arrangements for moderating impacts of wind. Wind can severely inhibit vine function and vineyard establishment and beyond the establishment phase may also severely suppress vineyard performance. For example, wind speed greater than 3m/sec are reported as suppressing stomatal conductance and photosynthesis with effects potentially lasting days depending on variety.

Business consideration

The next consideration is the location of the site in relation to the type of business enterprise you are proposing to create. Are you growing grapes to sell to another winery? Or are you hoping to make wine and sell it? If the latter, then

---

10 New Hall have since increased their planted area to 44.61-ha (110-acres).
11 WineSkills was the Plumpton College based training scheme which was partly DEFRA funded.

the proximity of your site to your market and its suitability for establishing a wine production and wine sales enterprise will be critical. You might have found the best site in Britain, but if it is miles from anywhere, only accessible down a farm track and impossible to get to except with a four-by-four, then think twice. Fine if it's just for growing a crop and the only visitors will be you and the farm staff – but even then you will have to think about getting the crop out – but if you want to build a winery, a farm-shop and a visitor enterprise, maybe it's not the right place.

Remember also that by planting a vineyard, you are signing up to a thirty- to forty-year investment and whilst you might hate the idea of dealing with the public and having visitors crawling all over your property, one day you might well want to sell up and move on. The potential buyers of your well-established, high-production and therefore high-value vineyard might have other ideas. They may well want to get involved in farm-gate sales, vineyard tours and teas, rent-a-vine visitors – the full wine experience – and if there is nowhere to put the car park, the shop, the winery, the lavatories and all the other add-ons that come with the territory, then you might lose out. In addition, they might also want somewhere to live, so having nearby property or a suitable site where a house could be built might be a huge advantage.

Therefore, if you are looking for a site for a vineyard and are thinking about the prospect of having a winery and/or selling wine from it, remember such things as: sight-lines for access roads for the public in their cars; visitors in coaches, and lorries collecting or delivering grapes and/or wine; whether there is three-phase electricity somewhere nearby (and preferably on site); and whether there are any buildings that might be converted into a winery and wine store or added on to.

Although, under 'Permitted Development Rights', wineries up to a certain size and subject to certain other requirements do not generally require planning permission to make wine from fruit grown on the site where the winery is situated, it is so much easier to start with an existing building.[12] There is more about planning permission in *Chapter 19: Getting started.*

Soil

I am very sure that some people will be thinking that the question of soil ought to have come first in this long list of site requirements and, whilst I have sympathy with this view, I do not agree with it for sites in Britain. Soil is an extremely important factor in a vineyard and countless studies on the influence it has on wine style and quality have been written and absorbed by generations of viticulturalists and winemakers. Soil is, after all, the basis for the theory of *terroir.* It is where the vine has its roots and from where it draws almost all its water and nutrients. Soil comprises not only the physical material from which it is made – clay, silt,

sand and stones – it also contains chemical elements, both major and minor, plant material which we generally call 'humus' and an unimaginable range of organisms – yeasts, moulds, fungi, mycorrhizae,[13] bacteria, protozoa,[14] algae and worms – all of which play a part in the ecology of the soil and which in all probability influence the way in which the vine grows and fruits. Soil therefore certainly influences a vine and the grapes it bears. Studies at Stellenbosch University[15] have hypothesised that variations in tanks of wine made from grapes harvested from a vineyard which was in all respects identical, i.e. same variety, clone, rootstock, age and vineyard management, may be due to differences in the yeast composition within the vineyard. The authors state in their abstract:

Importantly, yeast species distribution is subject to significant intra-vineyard spatial fluctuations and the frequently reported heterogeneity of tank samples of grapes harvested from single vineyards at the same stage of ripeness might therefore, at least in part, be due to the differing microbiota in different sections of the vineyard.'

Soil microflora, of which yeasts of many different types (and not always winemaking types) are a part, are undoubtedly important in keeping soil fertile and vines happy, but their influence on the taste of the final product is unknown, especially if cultured yeasts are used for primary and/or secondary fermentation. In addition, soil microflora can be negatively or positively influenced by the way the soil is treated and managed and have little to do with the actual type of soil they reside in.

The question of whether soil actually 'flavours' a wine though, is a different matter and the notion that it does is rejected by many well-respected voices. Despite everything you read about soil and its relationship to the style and character of a wine (usually on back-labels or in publicity material) there is no scientific proof that any substance, any element or any chemical in the soil can jump the divide between the soil and the root, travel up the roots, trunk and cane and lodge itself in the grape, thus flavouring the wine. The 2018 paper 'Minerality in Wine: Towards the Reality Behind the Myths'[16] concludes that 'the popular notion that we are simply tasting inorganic minerals in the wine transmitted from the vineyard ground is not scientifically plausible'. Dr Jamie Goode, the respected wine writer (whose PhD was in plant biology) states in his book *Wine Science:*

Do chalk, flint or slate soils impart chalky, flinty or slate-like

---

13 This is the RHS (Royal Horticultural Society) website description: 'Mycorrhizae are beneficial fungi growing in association with plant roots, and exist by taking sugars from plants 'in exchange' for moisture and nutrients gathered from the soil by the fungal strands. The mycorrhizae greatly increase the absorptive area of a plant, acting as extensions to the root system.'

14 Protozoa are microscopic single-celled organisms that live in the soil. They are vital to the break-down of plant matter in the soil.

15 The Vineyard Yeast Microbiome, a Mixed Model Microbial Map by Mathabatha Evodia Setati, Daniel Jacobson, Ursula-Claire Andong and Florian Bauer. December 2012.

16 'Minerality in Wine: Towards the Reality Behind the Myths' by Wendy Parr, Alex Maltman, Sally Easton MW, and Jordi Ballester. Beverages Journal, October 2018

characters to wine? As a scientist who has a working knowledge of plant physiology, I find this notion, which I call the 'literalist' theory of terroir, implausible.

Goode also quotes Dr Richard Smart who was asked what he thought of popular notions of *terroir* which propose direct translocation of flavour molecules from the soil to the grapes, and hence the wine. Smart said, with his typical Antipodean honesty: 'This is an absolute nonsense'.

Professor Jean-Claude Davidian of the *Ecole Nationale Supérieure Agronomique* in Montpellier echoes these sentiments: 'Nobody has been objectively able to show any links between the soil mineral composition and the flavour or fragrance of the wines. Those who claim to have shown these links are not scientifically reliable'.[17] David Farmer, a geologist and graduate of the University of Tasmania, in a posting on the website www.glug.com.au makes the quite interesting point that even if soils vary greatly from one site to another and trace element levels also vary:

... why would a vine take up more than was required of any element just because it was more abundant? If say the trace element selenium varies from region to region why would vines take up more than needed, assuming of course that these trace elements are playing some active rôle and are not just sucked up and have no real end use?

Farmer continues to question why a vine would: 'want to concentrate more of an element than it normally needed just because more was available?'

Professor Alex Maltman, Emeritus Professor of Earth Sciences at Aberystwyth University, who has made a study of the 'influence of vineyard geology on wine', points out in his paper titled 'Minerality in wine: a geographical perspective'[18] that most soils which produce wines said to have 'minerality' are in fact those soils which are poorer and less nutrient-rich, whilst the soils which produce wines which tend towards the soft and fruit-driven, are likely to be higher in nutrients and minerals – the exact opposite of what pro-mineralists believe. In his discussion and conclusion Maltman states:

In fact, it may turn out with further research that the nutrient minerals of geological origin in vines and wines – miniscule in concentration and flavourless though they may be themselves – are pivotal in determining wine character and flavour. However, this would have to take place in complex and circuitous ways. Thus, perceiving minerality in wine would not involve tasting minerals but permutations of complex organic compounds whose production has been influenced by inorganic cations. Future research will no doubt evaluate this speculation. In any event, for all the reasons explained here, minerality in wine – whatever that perception is – cannot be in any literal, direct way, the flavour of minerals derived from vineyard rocks and soils.

Soil profile in New Zealand showing three distinct soil layers

In 2018 Maltman published *Vineyards, Rocks & Soils: The Wine Lover's Guide to Geology* which contains a mass of information about the relationship between soil and wine. There are a lot of useful insights into why so many claims (in his view, baseless claims) are made with regard to soil and the way it affects the taste of wine. One thought in the epilogue to the book I felt really made for interesting reading. Why is that grapes (and therefore the wine made from them) are said to be 'flavoured' by the soil their parent vine is growing in, but root vegetables such as carrots, which have much more contact with the soil and are often eaten raw, are never said to be flavoured by them? The same of course goes for many other root crops – potatoes, turnips, parsnips, beetroot etc. We know that grapes can be flavoured by wind-blown particles such as smoke from wild fires and particles from neighbouring trees, eucalyptus and conifers being examples, and these particles can survive the scouring process of fermentation. But minerals in the soil? If it was that simple what is there to stop vineyard owners working out what the right minerals are and adding them to their soils?

Chalk soils are held in particular esteem by many vineyard owners and winemakers and provide a perfect example of 'mineralist' thinking. Why do growers of Chardonnay in Champagne prefer chalky soils? Not because the chalk gets into the wine and causes 'steely', 'flinty', 'mineral' or even 'chalky' characteristics (although many of those involved in the business of making and selling Champagne often claim it). No, the Champenoise prefer chalk soils because, as Alex Maltman says, 'its restrained nutrient provision and high porosity providing excellent water storage coupled with good permeability'. This means that these soils retain moisture, even in drought years, and their vines, especially their Chardonnay, can root deeply and have plenty of access to water. This keeps the leaves green and active which in turn sustains high yields and keeps acid levels higher than on less chalk-rich soils. And acidity, especially in these days of climate change, where the acidity in Pinots is lower than it has ever been, are needed in all good sparkling wines, but especially in Chardonnay-based blends and *blanc de blancs*. I sometimes wonder if those British growers who

---

17 For a fuller discussion on this, see Dr Jamie Goode's paper on the mechanisms of terroir: www.wineanorak.com/mechanisms_terroir1.htm and in his: Wine Science – the application of science in winemaking in Chapter 2. Terroir: how do soils and climate shape wine?

18 Minerality in wine: a geographical perspective, Professor Alex Maltman, Journal of Wine Research, Volume 24, Issue 3, 2013. Available at www. englishwine.com/wgigb.htm

have actively sought out (and paid high prices for) chalk land to grow their vines on, have actually taken on board the reasons why Chardonnay growers in Champagne like chalk. After all, in Britain, we have no requirement to retain acidity in our grapes (or at least not in most years).

Origin – which includes the influence of soil – does of course have an effect upon the character, style, flavour and quality of wine, a fact which partially enables wine tasters to distinguish one wine from another. (The other factor in the party trick of wine identification is, of course, the memory of the flavour and taste of individual wines.)

However, what different soils certainly do is to cause the vine to grow in a certain way, perhaps promoting high yields, perhaps promoting low yields. The pH[19] of the soil, a topic which will be dealt with in later chapters in some detail, affects the way the plant's roots are able to access and absorb nutrients. Soils can vary in their pH value in adjacent plots and pH values can also vary at different layers of soil in the same plot. This factor alone may account for quite substantial differences in the growth habits of the vine. Therefore, some soils will cause the vine to be vigorous, leading to a crowded canopy and shaded grapes and this will make the grapes taste different to ones grown on a soil which reduces vigour, causing a vine's canopy to be very open and airy which would tend to produce riper, lower-acid grapes. Many of the differences in the way vines grow are due to the way in which a soil receives, stores and releases water, whether from rainfall or via irrigation. The way a vine deals with water is at the heart of how a vine grows, the quality of its canopy and the degree of air and light that the grapes are exposed to. Some soils are able to store water, others not. The porosity and permeability of a soil and how the vine's roots access the water falling on it, passing through it and stored in it are at the heart of how a vine grows. The soil in a vineyard must, therefore, not be taken lightly. The question, therefore, of what type of soil is best for a vineyard in Britain, a soil that will enable us to produce viable yields of good grapes, is another matter.

Britain is in a different situation to almost all other established European winegrowing regions. There are an estimated 3.3 million hectares of vines in Europe (OIV report April 2019[20] and if you go to France, Germany, Italy, Portugal or Spain, where they have been growing vines on the same sites for hundreds and in some cases, thousands, of years, they have been able, by experimentation and by measuring the results, to determine which section of which hillside or river valley or estuary or escarpment produces which quality of grapes and in doing so, have built up a picture of where the best sites and soils are for certain types of wine. Would that it were like this in Britain: but it is not. Firstly, we have only been growing vines commercially for wine production in the modern era for around seventy years and for almost half of those years the area under vine

was (a) extremely small and (b) spread over an extremely wide area with distinct differences between the east and west of the country in terms of rainfall and exposure to prevailing winds. Secondly, until around twenty years ago we were growing a selection of varieties that have, if not now totally gone, then greatly diminished. All of these make comparisons of sites, areas and regions almost impossible. The major varieties that we now grow – Chardonnay and Pinot noir, plus Meunier – have in reality only been producing wine in Britain since the early 1990s and the area under vine of these varieties is only around 2,175-ha (5,374-acres). Compare this to Champagne where there are 34,300-ha (84,755-acres) of these three varieties and where they have been growing vines for several centuries. Couple that to the fact that all vinegrowing in Champagne is entirely controlled and regulated by the CIVC, the *Comité Interprofessionnel du Vin de Champagne*, which carries out endless research and experimentation in all aspects of both viticulture and winemaking, and you can see that British vineyard owners are currently at a distinct disadvantage when it comes to deciding where to plant vines.

The truth is that in Britain, as far as viticulture is concerned, we have very limited data to work with. Much of what people believe about different sites for vines in Britain is circumstantial and not based upon evidence. Additionally, many wine producers (especially the larger ones) either have sites in several different locations, sometimes many miles apart and on completely different soils, or are buying in grapes either under contract or on the open market, again from sites in many different locations, so that trying to associate soil type with wine style and quality is almost impossible. However, the yield surveys I conducted in 2016 and 2017, plus the ICCWS-WineGB Yield Surveys (already mentioned) conducted in 2018 and 2019, have started to produce some interesting data, some of which is reproduced in this book. It's early days, but hopefully, as growers see the value of this research, more will be prompted to take part, thus making the results more statistically reliable.

Note: For general information about soil types see Cranfield Soil and Agrifood Institute (CASFI) National Soil Resources Institute's Soilscapes website: www.landis.org.British/soilscapes/

Drainage

For the same reasons that we know the best fruit-growing soils are well drained, we know that vines like well-drained soils. Good drainage can be natural drainage and of course that is the cheapest form of drainage. But, for generations, growers on heavier soils, clay soils and the like, have produced good crops of top-fruit – apples, pears, cherries and plums – on soils which have been so-called 'tile drained'. These days, the half-round clay drains, their tile covers and their straw covering may have disappeared and been replaced by efficient plastic piping and 300 mm of 'beach' over-fill, but the reasons for field drainage have

---

19 pH is a measure of the soil's acidity or alkalinity.

20 OIV – International Organisation of Vine and Wine. The world area of vines cultivated for all purposes, wine, juice, table grapes and drying grapes, is 7.4 million hectares.

not changed. A well-drained soil will warm up more quickly (because wet soils require more heat to dry them out before they warm up) and will provide better conditions for access by tractors and machinery at both ends of the season leading to less compaction, rutting and damage to the structure of the soil. A drier soil will also create less humid conditions and lessen the pressure of fungal disease. A vine growing in a well-drained soil will tend to have both a deeper and more extensive rooting system than one growing in soil where the topsoil is wetter (sometimes a problem in vineyards with poorly designed irrigation systems). Then, when the soil dries out in the upper reaches in the summer, especially post-*véraison*, the vine will have the ability to draw water from lower down. Claude Bourguignon, who, together with his wife, Lydia, runs probably the viticultural world's best known soil analysis laboratory (LAMS),[21] and is a self-confessed soil 'nut', believes fervently that, for vines, depth of rooting is a very significant quality marker and that, effectively, the deeper the roots, the better the wine. Whether this is totally true is another matter. Given that the deeper you go, the less nutrients there are in the soil, one might think that a deeper rooting vine would pick up less in the way of minerals, which runs counter to the *terroir* argument. However, it may be that because there are less minerals the deeper you go, a vine has to develop a much larger root system, which gives it two advantages: the larger root system is able to store more carbohydrates; and the deeper roots are able to access a steady supply of water less influenced by rainfall. The question of drainage is further addressed in *Chapter 6: Site planning and preparation* in the section on *Drainage*.

Soil type

The question of what type of soil suits vines growing in Britain is another open question and I would say that as long as the site conforms to the criteria laid down above – somewhere in the traditional fruit and hop growing regions of the south of England (and possibly Wales), below 100–125 m above sea level, well sheltered from the south-west, sloping to the south, the south-west or the south-east, well drained, or capable of being drained, and preferably well sited for the establishment of a winery and farm-shop, then I would say, in all honesty, forget the type of soil you are presented with and learn to live with it. Drain it if it needs draining, lime it if it needs liming, add humus if it needs humus (it will), subsoil it (always), plant your vines at the right density and on the right rootstock, treat it with respect and almost any soil type can be made to work. I know vineyards on heavy clay (New Hall, Hush Heath), on light sandy soils (Gusbourne, Tenterden), on brick earth (Nyetimber's original vineyards), on high-pH chalk-rich soils (Chapel Down Kit's Coty, Evremond, Hambledon, Leckford, Squerryes, Wiston and many others) and they all produce good to very good grapes. I have also known

vineyards on the same soils which have failed. It is NOT the soil type that ultimately makes the difference.

In a generation's time, when maybe the area under vine has risen from today's bare 3,500-ha (8,649-acres) to perhaps 20,000-ha (49,420-acres) and the whole of the Thames Valley within the M25 is planted with vines and all the available land between Pulborough and Petworth is planted with vines, and someone has bothered to do the research, then we might know a bit more about which soil types produce the best grapes in Britain. But, until that happens, don't worry about it.

My attitude towards the soil of a vineyard does not mean however, that I don't think it important – far from it. I was brought up in the countryside, was a member of my school's farming club, a member of my local Young Farmers' Club and started working for my father-in-law, on his large-scale arable and dairy farm, in 1971. I started my studies into vineyards in Britain in 1973–4 and spent almost two years in Germany, working in vineyards, studying at Geisenheim, visiting vineyards belonging to friends and fellow students, and since 1977 have been very actively involved with growing vines, producing grapes and/or making wine. My final paper in the Master of Wine Theory examination, which I took in 1997 and which helped me win the Mondavi Award for the highest marks in that section of the exam, was eight handwritten pages titled: 'The Secrets of Soil'. In short, I have been involved with, and thinking about, soil for a long while.

Soil is, after all, the medium into which the vine places its roots; it is the medium from which it almost entirely draws its moisture and its nutrients; and it is the factory floor which has to take the traffic of our feet when we work in the vineyard and of our machinery when we tend the vines. How could it not be important? But, what effect does the type of soil (note: the type of soil not the condition of the soil – that is a different matter) have upon our vine and the grapes it produces?

The type of soil is important because it has an effect upon the way the vine grows. The water-holding capacity of the soil will dictate how much water the vine has access to and will affect the vigour of the vine, how much shoot and leaf growth there is and how crowded the canopy will be. A vine growing in a soil that drains well in its top layers, but holds on to its water lower down, is likely to retain its foliage longer, grow longer shoots and more leaves and keep a fuller canopy than a vine in soil that is the opposite. Whether this is good for growing vines in Britain is the question and even if you have a soil which does promote growth, by having a high vine density, growing the vines on a deep-rooting, devigorating rootstock, using an appropriate pruning system, carrying out the correct canopy management and growing something in between the rows of vines (grass, weeds, green manure etc.) that uses up the moisture, you will almost certainly be able to counteract the effect of a soil with a higher water-holding capacity.

The humus content of the soil is a second major factor in a soil's qualities. Humus is a vital source of nutrients for

21 Laboratoire Analyses Microbiologiques Sols www.lams-21.com

the multitude of different organisms that are crucial to a healthy soil. The worm population of a soil is a very good indicator of its overall health and there can be as many as 4 million worms per hectare which together would weigh over 8tonnes. Each worm will produce around 4.5 kg of casts – earth, grit and vegetable matter – per year which will help improve the soil. Worms will be active in many soils down to a depth of 2 m and will take down vegetable matter from the surface as well as provide channels for water to drain into the soil and for carbon dioxide to escape. Appreciating what is happening beneath a vineyard's surface is key to understanding your *terroir*.

Plant a vine in a soil too low in humus and it will struggle to establish as it tries to grow new roots and settle into its environment. If it is also faced with other hazards: water and/or nutrient shortage, competition from weeds, damage from pests and diseases, and then it may fail altogether. Roots cannot of themselves take up moisture and nutrients, but require mycorrhizal fungi to act as intermediaries between the tip of the fine root hairs and the water and nutrients.[22] Much of the root activity in a vineyard in a region well supplied with summer rainfall (such as we enjoy in Britain) or in vineyards which are irrigated, takes place in the top 500 mm of soil.[23] Vines that are staked and trellised as soon as they are planted do not require their roots to act as anchors and roots will spread sideways, as well as down, taking the easy path to find water and nutrients. Once the roots have colonised the upper reaches of the soil, only then will they start diving deeper. Although typically in Britain we allow the vineyard alleyways to grass down after planting (and I much prefer to allow this to happen naturally from self-sown grass and weed seeds, rather than actually sowing grass seed), in Europe and in drier, un-irrigated regions, vineyard alleyways are often cultivated, at least for the first few years, if not for the entirety of the vineyard's life. The reason for this is that if any roots do grow near the surface they will be disturbed by the cultivations, thus forcing the vine to root deeper.

The humus in a soil before planting a vineyard is a combination of what has been left behind from previous crops plus what has been added. Even if the straw from an arable crop is baled and taken away, some will be left behind and this will be incorporated into the soil after harvest. If the farmer has no use for the straw, then all of the crop's straw will be left behind – especially now that straw burning has disappeared. Grassland used for grazing is generally rich in humus as the animals who graze it return much of the humus element of the pasture that they eat in the form of manure. Grassland which is used to make silage or hay, a so-called permanent pasture, will be less humus-rich, as large quantities of plant material are actually removed from the land each time the grass is cut, although a good farmer

will rotate crops so that perhaps after five years, a grass field is put into cereals for a few years to restore fertility before it is planted with grass again.

Land that has been used for fruit crops, especially hardwood fruits where little in the way of vegetation, apart from leaves and prunings, is left behind after harvest, is often in danger of being low in humus content which makes the grass alleyways between trees (or vines) especially important providers of this element. Vineyards are usually better in terms of their humus content as the grassed area covered by the alleyways is typically 70–80 per cent of the surface area of the vineyard, with the rest, i.e. the area under the vines, being kept free of vegetation. The weight of leaves, shoots and prunings, which are generally left behind in the vineyard, is high compared to the weight of crop taken out. A yield from an apple orchard might be as high as 50 tonnes-ha (20 tonnes-acre) compared to, say, 8.00 tonnes-ha (3.24 tonnes-acre) in a vineyard in Britain and the area of the alleyways might be as low as 30 per cent of the area of the orchard. This is because an apple tree has a much wider spread than a vine.[24]

In almost all situations, before planting a permanent crop such as vines, the land will benefit from an addition of humus and any addition is better than none. If there is time before planting, then a 'green manure' crop can be planted and either grazed off or ploughed in and incorporated into the soil. Otherwise additions of farm yard manure (FYM) will always help vines establish, although, if not well rotted, may aggravate the weed situation. FYM will also provide some nitrogen and other fertilisers and trace elements and help improve the overall soil fertility. Otherwise, compost such as commercial green waste (BSI PAS 100)[25] or spent (i.e. used) mushroom compost are useful sources of humus, although the latter will not improve the nutrient status that much. Green manuring is discussed in *Chapter 7: Vineyard nutrition*.

The third aspect of a soil is its physical character, i.e. how deep is the topsoil, how deep is the subsoil and what is the soil made of?

A deep topsoil, such as is found in many vineyards in Britain, is good for getting vines established and will usually be rich in humus, nutrients and soil organisms that promote healthy root growth. Soils such as this will often promote vegetative growth and therefore vine density and the choice of rootstock will be important. These soils will typically have a fairly neutral pH – 6.5 to 7.0 – and will require little in the way of pH adjustment at the pre-planting stage. Drainage in these soils may also be an issue that needs addressing. Sites with leaner topsoils – those with visible stones, flints and gravel, for instance – will certainly make fine vineyards, but will require good soil preparation, subsoiling and deep ploughing and may well require some additions of humus, both pre-planting and

22 Understanding Vineyard Soils, Robert E. White, p. 60-1.

23 Although you will often read about vine's roots growing down to five, ten, even twenty metres below the vineyard surface, very few of the writers who state this know this for a fact and it is highly unlikely that any more than a small percentage of a vine's roots in Britain's well-watered vineyards ever get below 1 m to 1.5 m deep.

24 Some modern systems of growing apples – the tall spindle system for instance – where the trees are trained and trellised much more like vines, are better in this respect, but the alleyway to non-alleyway area is still typically only 50 per cent of the total orchard area.

25 This specification, plus much other useful information about compost for use in agriculture, is available from www.wrap.org.uk.

at times throughout the life of the vineyard. Natural drainage on these soils may well be good, but there can always be areas and sections of any site (where there is a spring, perhaps, or compaction) where drainage should be installed. Sites with very shallow topsoils are probably best avoided and those with subsoil constraints certainly should be. If possible, it is always best to dig some soil pits of at least 1 m deep (and probably nearer 2 m deep) using a mechanical digger in order to see what's down there. How many pits and to what depth will depend upon the size of the site and the uniformity (or lack of it) of the soil. The MAFF publication *Bulletin 105: Soils and Manures for Fruit* (dated 1975 and now out of print, but still obtainable second-hand) has some very good advice about soil types and the digging of soil pits.

Soils with noticeable stones, flints and gravel (and also dark soils) may help retain heat in the vineyard into the early evening when the sun has gone down, thus keeping the vines dry for as long as possible and aiding ripening, although with vines in Britain typically having their fruiting wires at least 700 mm from ground level (and often quite a bit higher) and with the breezes that usually blow through our vineyards, this benefit may be more in the imagination than in reality. Soils with a high stone content may also prove problematic when installing posts and anchors, but this is usually a one-off problem and can be lived with. Also, if mechanical methods of either weed control or alleyway management are used on stony soils, then the vineyards will be more difficult and more costly to run.

The pH of a soil, which is dealt with more fully under the section on rootstocks in *Chapter 4: Vine varieties, clones and rootstocks*, should not be a determining factor in whether a site is suitable for vines or not. With the correct lime additions (to an acid soil) or with the correct rootstock (in an alkaline soil), most soils can be planted with vines. Having said that, soils below pH 5.0 and above pH 8.5 are best avoided unless they have some exceptionally positive attributes in their favour.[26]

Previous land use

Land that has been previously cropped with arable crops or grazed by animals will probably pose no problem when it comes to establishing a vineyard. However, land that has previously been planted with fruit trees or fruit bushes or was previously woodland – either managed or scrubland that has run to rough woodland – may pose a problem. On old fruit land, vines may suffer from something similar to Replant Disease which affects several species, especially apples and pears. Whilst this classically only occurs when plants are replanted on a site that was previously planted with the same species, I have known situations where vines following apples have failed to establish in certain small patches of the vineyard, possibly due to the presence of old roots which are rotting below the surface. On land that was previously

woodland there may be traces of *Armillaria spp.* (honey fungi) which might give rise to problems with establishing vines, as it does in other countries. If you are concerned about the land use prior to planting vines, the best thing to do is to make sure that as many as possible of the old roots of the previous species are removed and burnt off-site, make sure the site is well-subsoiled and cultivated, plus allow a crop (or two) of a cleansing green manure such as mustard[27] to grow and be incorporated back into the soil. Follow that with a good application of FYM or compost and also consider the use of dips based upon mycorrhizal and/or a *Trichoderma spp.* fungi dip on the vines at planting. See *Chapter 8: Planting your vineyard* for more discussion on these dips.

Note: The best book on the subject of vineyard soils that I have found is: *Understanding Vineyard Soils* by Robert E. White.

Other factors

Public footpaths (and other right of way) can often be a big problem when they are near, within or even run across sites for vineyards. Footpath users are very territorial about 'their' paths, which will probably have been in use for centuries, and seem to dislike new owners planting what they see as ugly vineyards. If a footpath actually crosses the land you propose to plant (and you cannot get it moved) then the only practical way you can accommodate it is to align your rows with the footpath (assuming it runs in a straight line) and you must allow at least 1.50 m of clear space for the footpath. Whilst footpaths can in theory be re-routed or diverted, in practice it can be a long and often unfruitful battle to get them moved.[28] In addition, footpath users will almost certainly complain about any weed spraying or pesticide spraying taking place and in addition, under the buffer zone rules you may have to keep your sprayer a considerable distance from any public access areas. The DEFRA (Department for Environment, Food & Rural Affairs) website has detailed information on public rights of way for England. For Wales, refer to the Natural Resources Wales website. There is more on buffer zones in *Chapter 16 - Pest and disease control* under *Buffer zones*.

The residents of houses close to vineyards can often prove problematical, especially with regards to such things as early morning, late night and weekend vineyard and winery operations, spraying, and noise made by bird scarers.[29] Apart from the legal aspects surrounding both

---

26 For a more complete discussion about vineyard soils, see Chapter 5: Soils for vineyards (p. 41-7) in my book: Viticulture – an introduction to commercial grape growing for wine production should also be consulted.

27 Mustard, when allowed to grow and then incorporated into the soil, releases compounds called isothiocyanates which act as a biofumigant and suppress certain pests and pathogens.

28 When I wrote this originally, in 2013, this was true. In 2016, however, not long before the Taittinger vineyard, Domaine Evremond, was planted, we managed to get a footpath re-routed in just 6 weeks following the sending of a begging letter, written by Pierre-Emmanuel Taittinger, to the council. Strutt and Parker (who were advising us on the purchase) said it was the quickest footpath diversion anybody had ever known!

29 See Chapter 16: Pest and disease control in the section on Bird damage for more information on audible bird scarers.

spraying and audible bird scarers, disgruntled neighbours can make life unpleasant for vineyard owners. There have been quite a few cases of vineyard vandalism – anchor wires cut, vines cut off at ground level, and vines sprayed with weedkillers – and it pays to make friends with any neighbouring owners and telling them of your plans, especially if you intend to welcome visitors, have farm-gate sales, start a winery etc. Disaffected neighbours can easily disrupt planning and licensing applications.

I would also guard against selecting a site too near a pheasant shoot and especially too near a pheasant release pen. How near is debatable, but in my experience a site within 1.6 km (1 mile) of a well-stocked pheasant shoot is to be avoided. Although the case law established under *Rylands vs Fletcher* in 1868 gives landowners protection against damage done by livestock that escapes from neighbouring land, the 1971 Animals Act specifically exempts 'pheasants, partridge and grouse' from this provision and when they escape (or are released) they instantly become 'wild' birds. It is also very difficult to prove that the birds actually eating your grapes have come from a particular source. I have been involved as an expert witness on two substantial claims for pheasant damage (both of which were successful) and in these instances the shoots were established after the vineyard was planted, so it was not a case of poor site selection.

If the site you have chosen for your vineyard is right next door to an arable farm, a golf course or horse paddocks, you should also be aware of the possibility of herbicide damage. Vines are easily damaged by hormone weedkillers which can vaporise in warm weather and then drift to adjoining fields. The chief culprits in this respect are farmers and horse-owners with grass fields spraying off nettles and thistles and other broad-leaved weeds early in the season, prior to cutting the field for hay, and golf course greenkeepers who are very keen on keeping their fairways and greens free of weeds. There is also the possibility of spray drift contaminating your grapes with a substance which might show up in the final wine. Whilst this is probably a more remote possibility, it is something to be borne in mind should drift occur. If in doubt, test for it. Of course, if your vines are damaged by drift from a neighbouring farm, you are entitled to compensation for any damage caused, but proving it and getting their insurers to pay is another matter. It is best to keep on good terms

with your neighbours (farming or otherwise) and tell them of your plans and remind them about their legal obligations not to let sprays drift. If you do suffer weedkiller damage, you will need to collect evidence straightaway and notify your neighbours, their insurance company, the Health and Safety Executive (HSE) and call in the loss adjusters. (There is a considerable amount of detail on the HSE's website about spray drift.)

Sites with telegraph poles, electricity pylons and other obstacles are best avoided if possible, not only because they give somewhere for birds to perch as they size up the ripening grapes, but also because they make laying out and planting a vineyard more difficult. However, if the site is perfect in every other respect then such obstacles can be accommodated. Sites which are long and thin which can only be planted with the rows in the short direction are of course plantable, but will result in higher establishment costs (more end-posts and anchors), higher working costs as tractors will spend more time turning round than driving in a straight line and more land will be lost to headlands.

Site selection – a summary

Choosing the right site for growing vines in Britain is without doubt one of the most difficult tasks facing a potential vineyard owner, yet is probably the most important. Ignoring the factors of availability and affordability, finding a site that fits all the major criteria – location, altitude, aspect and shelter – and is then in the right place for the business model that fits your plans, has a soil that is suitable and isn't blighted by any of the other considerations that need to be taken into account, is a major, major problem. Many of the pioneers of course, selected their sites on the 'because it was there' principal, planting vines in the field 'I can see from my front door' or in the field empty 'because my daughter has got fed up with her pony' or – and probably the worst reason I have yet come across – 'because it was the only land on the farm that wouldn't grow sugar beet'. All of those reasons were actually given to me as the criteria for site selection. Much as one might envy those pioneers the ease with which they chose their sites, all of the above examples are now ex-vineyards. Inevitably compromises have to be made when selecting a site, but my advice is not to compromise on the ability of the site to grow good grapes.

Chapter 4

Vine varieties, clones and rootstocks

What varieties to plant

After you have chosen your site, the next major decision is: which varieties do I plant? The answer, strangely, is actually quite easy: these should be the ones that (a) suit your site and (b) make the style of wine you will be able to sell! Of course, this is far easier said than done. Given that on a site with no previous history of growing vines – a common situation for many winegrowers in Britain – prior knowledge of what varieties 'suit' your site is, of course, difficult, but not impossible. And again, the question of what style of wine 'you will be able to sell' may be something you have not considered and is therefore also quite difficult to answer.

But, by asking these two questions it brings to the fore two essential points: the quality of your site will dictate the best varieties for you to establish on it; and you need to have a vision of the wine you would like to produce and how and where you propose to sell it before you plant. Do not fall into the trap of growing a variety that your site will not support, or attempting to make a wine for which there is a limited market. I often ask potential growers at my first meeting 'which English wines have you enjoyed recently' or 'tell me what styles and types of English wines you like' and it always amazes me how many look rather blank, shuffle around a bit and then mention one or two wines that they vaguely recall having tasted. In order to have a successful vineyard, especially if you are proposing to make and sell the wine, it is imperative that you should know and enjoy the product. The other major question of course is still or sparkling and this is dealt with at length later in this chapter.

The range of varieties available to plant is huge and, to the uninitiated, the question of which to plant is either a problem to be run away from or an opportunity to experiment. There are advantages and disadvantages of both views. For those who see the selection of varieties as a problem, the answer is to stick to the varieties that have proved themselves and which suit your site and the style of wine you wish to make. Don't try and re-invent the wheel. If you were a grower in Chablis or Sancerre, Burgundy or Alsace would you be worried about what variety to plant? No. The *appellation* has already determined which varieties suit your region and whilst you may have the choice of clone – although even this is not certain – your task is to grow the best grapes you can.

The selection of the right varieties for your site, your financial position and your marketing plans is the most important decision you will have to take, apart from that of site selection itself. In taking this decision, the best advice I can give is to learn from others and to look backwards to what has been successful, rather than to look forwards and hope the obscure varieties you have heard about will produce award-winning wines. When looking backwards, I am guided by the collective wisdom of the 500-plus growers in Britain, by the membership of the UKVA and WineGB, and of the judges who judge the wines entered into the annual 'Wine of the Year' competition. If you just take the trophies and gold medals for both still and sparkling, and white, red and rosé wines, the list of successful varieties is actually very small. The only real criterion is to choose those varieties that have proved themselves capable of making high-quality wines, because it is only by making wines of high quality that your enterprise will be successful in the long run. I accept that looking back has its problems, as growers can only make wines from the varieties they have in the ground and, therefore, some of today's (and yesterday's) wines are made from varieties that, given a fresh start, many growers would never plant today and will certainly not plant again.

Varietal change

For those that see the selection of a variety (and, more often, varieties) as an opportunity to experiment, let me issue a health warning straightaway. Although we have seen considerable changes in the spectrum of varieties

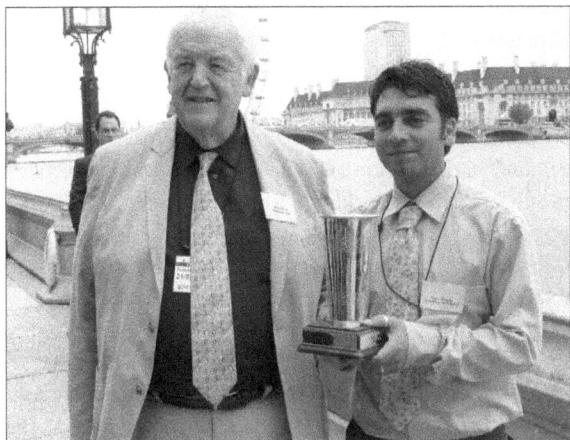

The late Mike Roberts and his son, Simon picking up some silverware at the House of Lords for their award winning Ridgeview wines

being grown in Britain since modern commercial vineyards started in the 1950s, some of the old favourites are still there. The top ten varieties following the first compulsory survey in 1989 when the total planted area was shown to be 928.51-ha (2,294-acres) were (in order of area planted): Müller-Thurgau, Reichensteiner, Seyval blanc, Bacchus, Schönburger, Huxelrebe, Madeleine x Angevine 7672, Pinot noir, Kerner and Ortega. Thirteen years later, in 2002, when the area under vine had shrunk to 812-ha (2,006-acres) the top eight varieties were, not surprisingly, almost exactly the same! Only Kerner had dropped out (to number 12) to be replaced by Chardonnay, creeping in at number 9. However, jump another eighteen years (to 2020), when the total area under vine had more than quadrupled, and the variety spectrum in the top ten looks a little different. The old favourites Bacchus, Ortega, Reichensteiner and Seyval blanc are still there, but the list is, of course, dominated by Chardonnay and Pinot noir, which together occupy almost 55 per cent of the planted area. The other 'old school' varieties have all dropped out of the top ten: Müller-Thurgau to number 15, Madeleine x Angevine 7672 to number 11 and Huxelrebe to number 20. They have been replaced by Meunier, Blauer Frühburgunder (Pinot noir Précoce), Solaris, and Rondo – three red varieties and one modern PIWI[1] hybrid.

Not only has the change in varieties over the last thirty years affected the type and style of wines produced in Britain, with sparkling wine now probably accounting for 70 per cent of total production (and likely to rise in the future), but it has also changed consumers' perception of our wines. Again, as I said in my opening chapter, in earlier times, when a wine retailer or sommelier was asked what an English wine was like, he or she might have said: 'well, it's made from Müller-Thurgau and Schönburger with a touch of Reichensteiner and its tastes a bit like German wine or maybe a wine from Alsace' and in doing so lose

the customer's attention and interest before the second umlaut. Today, when faced with the same question about a bottle of English or Welsh Sparkling Wine, the same retailer or sommelier might say: 'well, it's just like Champagne' and then perhaps as an afterthought, in order to make sure the customer was completely at ease, add: 'and it's made from Chardonnay and Pinot noir'. In saying this, in such simple terms, a message which every wine drinker understands has been easily and simply conveyed. I believe this has made a huge difference to the industry and has given the wine trade in all its many guises the confidence to get behind, to support and to sell, the home-grown product.

Those vineyards planted with German varieties in the 1970s, '80s and '90s weren't defeated by poor quality wine, and they weren't necessarily defeated by poor yields. What defeated them was the problem of marketing their (mainly still) wines in the time frame you have to sell still wine. Most new vineyards start cropping in their third or fourth year and, if you are producing still wine, six months after picking, your wines are usually bottled and ready to sell. Initially, with a small maiden vintage, and with some local interest in your wines, they may sell quite quickly, but come October and another vintage, this time hopefully a full-sized one, is ready for picking and by the next May, you have another vintage for sale. The pressure is on.

Most buyers, especially trade buyers who perceive our still wines as being mostly fresh, fruity, lively wines, at their best within twelve to eighteen months of picking, are not so interested in older vintages and would rather taste your more recent wines. If sales are not matching output, you cannot afford to store your wines in tank indefinitely – there is another harvest approaching – and the temptation (and the reality) is to cut your losses, lower the price and move the stock. In 1984, Kenneth McAlpine, then the owner of Lamberhurst Vineyards, the largest and (seemingly) most successful vineyard in the country,[2] found himself, after the larger than average crops in 1982 and 1983, with one million bottles in his cellars and sales lagging behind production. At the time, the vineyard's best-selling 'Lamberhurst White' sold for £2.99. Faced with the prospect of another large crop in 1984, McAlpine looked at the alternatives: take on more sales staff, advertise, promote, attend more shows and fairs and maintain the price; or slash the price and stimulate sales. He took the latter route, cut the price to £1.99 and within six months had cleared the excess stock. Short term, the problem was cured; long term, it set the benchmark price for entry-level English wine at a low – and probably unprofitable – point from which it would struggle to recover. Other large-scale producers, such as Carr Taylor (who would face bankruptcy a few years later),

---

1 PIWI stands for *pilzwiderstandfähig* or fungus resistant varieties.

2 I say 'seemingly' as in fact Lamberhurst Vineyards at the time was loss-making, something that didn't seem to trouble Kenneth McAlpine too much as he was probably able to write it off against other income. When I took over as winemaker and general manager in 1988, I was given audited accounts which showed annual losses running at around £350,000 pa, which equates to around £900,000 in 2020 prices, and this level of loss had been going on for several years.

Wine Growing in Great Britain

This is the sort of newspaper article that helps sell English Sparkling Wine and convinces customers that its worth buying

Barkham Manor, Penshurst, Three Choirs and Wellow all found it hard to lift their prices above those of their main competitor – Lamberhurst.[3] Sparkling wines, however, are different.

Still or sparkling?

You cannot get into the (bottle-fermented) sparkling wine business without putting the wine into a bottle, putting that bottle into a store (usually in a storage or riddling crate) and stacking the crates in an insulated, secure, temperature-controlled store. Once the wine is in the store, all you can do is wait for it to mature. Therefore, why sell it (a) before the wine is ready and (b) until you can find someone willing to pay what you perceive to be the right price? With sparkling wines, bottle age is seen as a plus: two years, good, four years, better, six years, even better. This puts the decision of when to sell firmly back in the producer's hands and although building up stocks creates havoc with the cash-flow, at least it means that the temptation to cut and run towards an easy sale at a low price is lessened. You can probably sell enough to keep the cash-flow positive, leaving you time to re-double your sales efforts, open that farm-gate outlet, build that website, visit more hotels and restaurants, send out more samples to merchants, attend more shows, investigate export opportunities. In short – find different ways of getting your wine to market. I believe this difference between still and sparkling wines has helped the prices of British-grown sparkling wines stay resiliently high even though both the number of different sparkling wines and the amount available for sale have risen in recent years. Whether this price premium for our sparkling wines will remain, in view of the volumes now coming to market,

is an interesting question and one which only time will answer.

However, the sparkling wine business requires large amounts of capital to be tied up in stock. As I have already pointed out in *Chapter 2: Why plant a vineyard in Great Britain?*, in the section on *The costs of producing wine*, many producers will be working with at least three, and maybe up to six, vintages in stock at a cost of around £28,125-ha (£11,382-acre) for each year of production, based upon a modest 7.5 tonnes-ha (3 tonnes-acre). Take four years as an average stock level and the total is a frightening £112,500-ha (£45,528-acre); six years' stock and the bill is an astronomical £168,750-ha (£68,292-acre), which adds over £1 a bottle to your production costs: almost as much as the grapes cost! Larger crops and longer-aged wines – Breaky Bottom and Coates & Seely are both still selling their 2009s in 2020 – and the overall cost of stock will be even greater. Your selection of varieties and wine styles must therefore be taken not only from a marketing perspective, but also from a financial one.

Over the past forty years, the viticultural landscape in Britain has changed dramatically from one that was producing almost entirely still wines to one that produces predominantly sparkling wines. Exactly how much of Britain's wine production is still and how much sparkling is not known reliably. Although WS collects overall production statistics, it does not differentiate between still and sparkling wines. Likewise, HMRC, which publishes statistics on the amount of Excise Duty paid on wines coming from 'UK registered premises' (which does not include 'British Made Wines'), does not split the amount between still and sparkling wines. In 2018–19, WineGB conducted a survey amongst its members and its commercial report showed that, of plantings, 69 per cent was for sparkling wine and 31 per cent for still. The same survey said that sales by volume were 72 per cent sparkling and 28 per cent still. Given that WineGB members tend to be the larger, more sparkling-orientated producers, we can take the often quoted two-thirds sparkling and one-third still as a fair working division.

The reasons for the switch from still to sparkling are not difficult to see. The quality of the best sparkling wines made in Britain is by any measure considerably ahead of our still wines. They have elegance and finesse, plus lasting power and have proved themselves in tasting after tasting to be the equal of Champagne and other fine sparkling wines. In addition, as I have already pointed out, they are much easier to explain to the wine-buying public and therefore much easier to sell, especially through the on-trade. The early successes of brands such as Nyetimber, winners of major national and international awards with their first wines (from the 1992 and 1993 vintages) in 1997–8, and Ridgeview, whose first wine (a 1996) was released in 1999 and who won the Gore-Browne Trophy in 2000 (with their 1996 Bloomsbury) and in 2002 (with their 1998 Cavendish), plus the high prices they achieved, gave credence to the idea that sparkling wine was the way forward. The idea

---

3 At the time I owned Tenterden Vineyards (today's Chapel Down) which was only a few miles from Lamberhurst, and I found that this price reduction didn't have quite the effect I feared, mainly because we hand-sold most of our output from the farm-gate. When people said: 'Why is your wine more expensive than Lamberhurst's?' we always said: 'Our wines are better, of course, and you get what you pay for!'

that 'because they can do it, so can I' – much in the way that the prominence of Lamberhurst Vineyards persuaded still winegrowers that they too could grow grapes, make wine and make money – persuaded many growers that sparkling wine was the product for the future. In addition, it wasn't very difficult to see that a bottle of Nyetimber or Ridgeview really was of an equivalent quality to a bottle of Moët, Lanson or Taittinger costing much the same, whereas a bottle of still wine from Three Choirs or Chapel Down was OK, but perhaps slightly (or even very) overpriced and therefore poor value for money. Couple this to the fact that bottle-fermented sparkling wine (aka Champagne) has the aura of prestige and quality about it and the average customer expects to pay more for it, and the appeal of sparkling wine to the producer from a financial standpoint is an obvious one.

Still wines hold their own

When I wrote the first edition of this book in 2013-14, English and Welsh Sparkling Wine was definitely on a roll and it looked like, given a decade or so, the switch from still to sparkling would accelerate to a point where still wine was perhaps only 10-15 per cent of production. However, that appears not to be the case and whilst all the larger producers are still predominantly sparkling, many of them have now started to produce still wines, and many of the smaller new producers, who rely much more on direct farm-gate sales, have a mixed offering of both still and sparkling. My best estimate now is that still wines account for 30-35 per cent of production, and probably higher than this in plentiful vintages such as 2018 and 2019.

The reasons are several and varied. Undoubtedly competition to sell English and Welsh sparkling wines through the trade, both on-trade and off-trade, has

One of London's less well-known vineyards. Six vines planted on the terrace of Decanter Magazine's offices on the 6th floor of a Southwark office block with the Shard in the background

increased and there is only so much shelf space and so many hotel and restaurant wine lists that can find room for home-grown sparkling wines. Likewise, as competition has increased, the need to get customers to visit the vineyard, taste the wines, enjoy the ambience and – before they leave – put a couple of cases in the boot of their cars, has grown. Almost all of the larger vineyards have invested in visitor facilities in the last few years, ranging from simple tasting rooms and retail shops, to very substantial offerings including catering facilities for large groups, and even, in some cases, hotels and other on-site accommodation. This realisation that vineyard tourism has definite advantages and that retail sales direct to the consumer are much more profitable than trade sales, has meant that the wine offerings, and also the varietal spectrum, have changed. Varieties such as Bacchus, Ortega, Phoenix, Pinot blanc, Pinot gris, Siegerrebe, and Solaris for the production of white wines, and Blauer Frühburgunder (Pinot noir Précoce), Regent, and Rondo, for the production of red and (more commonly) rosé wines have increased their percentage of the overall planted area. In addition to these 'still only' varieties, the percentage of still wines made from Chardonnay, Pinot noir and even Meunier, has risen considerably in the last decade. Pioneers such as Chapel Down and Gusbourne, who to my palate made the best still wines from Chardonnay and Pinot noir, have been joined by others and whilst there are still a few 'sparkling only' producers, many of the major producers now offer both styles.

Another important factor in the (modest) resurgence of still wines as a useful category in the sales mix is undoubtedly climate change. The last ten years – with vintages such as 2014, 2016, 2018 and 2019 – have seen significantly riper grapes, with natural alcohols often in double figures (and even Chardonnay with 12-13.5 per cent natural alcohol), which have helped increase the quality of still wines. Assuming no cataclysmic change to our climate which brings icebergs knocking into our shores (although writing this whilst under a Coronavirus lockdown, it easy to believe that anything is possible), I see the still wine category being given a new lease of life and the still wine varieties mentioned above continuing to be planted, plus increased production of still wines from all three Champagne varieties, with Meunier possibly becoming one of Britain's 'signature' still wine varieties. There are also now a number of stand-alone wineries who are buying grapes from established producers and making some very interesting wines, both still and sparkling. Although small-scale, these independent wineries are pushing the boundaries of what it is possible to make from home-grown grapes and introducing English and Welsh wines to a wider, often metropolitan, audience.

Oversupply of English and Welsh sparkling wine?

Another factor that ought to be taken into account when considering planting a vineyard is the question of oversupply. Since the dark days of the last century, when the area under vine decreased by 30 per cent as growers grubbed up their old, mainly Germanic, varieties, the area under vine has risen from a low point of 761-ha (1,880-acres) in 2004 to today's estimated 3,500-ha (8,649-acres) – an increase of almost five times. And, of course, with the increase in vineyard area has come the inevitable increase in production. As can be seen from Table 21, the average annual production of all wines has risen from an average of 2.19 million 75 cl bottles per year in the five years between 2005 and 2009, to a frightening average of almost 12 million bottles per year in the last two years, 2018 and 2019.

However, what is actually more frightening is the amount of wine still in the pipeline of plantings. The 2018 and 2019 production figures above were based upon an average cropping area of 2,288-ha (5,654-acres). This means that, taking the current planted area of approximately 3,500-ha (8,649-acres), there is an additional 1,200-ha (2,965-acres) which will start to crop over the next few years (and that's without any additional plantings). Add on to that the sparkling wines produced and still maturing, and the numbers look a bit frightening. I have calculated that, at March 2020, stock levels of sparkling and still wine in Britain are probably around 30–32 million bottles combined, or between six and eight times the level of sales (sales levels are discussed in the next section). Of course, as has been pointed out earlier, if you are in the sparkling wine business, then by definition you need stocks of wine and, cash-flow issues aside, stocks of four, five and even six years' worth of sales are nothing unusual. In Champagne, the average stock levels of all wines, non-vintage and vintage, currently amount to four years, two months of sales. This includes all the minimum-aged Champagnes with attractive sounding (but mainly made-up) names that are the staple offering of supermarkets around the world which spend a bare 15 months *sur latte*. However, if you look at *Cuvée de Prestige* and *Millésimés* Champagnes (which together account for around 6 per cent of total production by volume), then the stock levels are much higher. Current vintages of vintage Champagnes range from the early 2000s to 2012–13, with nothing much younger than that, in other words at least seven years' worth of stock and for many brands considerably longer. In this context, therefore, stocks of (mainly vintage) English and Welsh sparkling wines of seven to eight times annual sales don't seem that abnormal.

However, the facts are that given the current level of plantings, the production of English and Welsh wines of both types is set to increase. If yield levels of the size of 2018 and 2019 (which averaged 39.11 hl-ha or 5,215 bottles per hectare) are to become the norm, then a total yield based

Meonhill Vineyard – a little bit of Champagne in Hampshire

No. of 75 cl bottles of wine produced per year in Britain	
Years	Million bottles per year
2005-2009	2.19
2010-2014	3.78
2015-2017	4.83
2018-2019	11.79
Source: WS branch of the Food Standards Agency	

Table 21

on the current plantings of 3,500-ha, could amount to over 18 million bottles a year. And that's without plantings to come of 200–300-ha a year over the next few years. The question therefore of 'where will this wine sell?' is one that everybody in the industry is thinking about and anybody who would like to be in the industry would be well advised to give some thought to.

Sales of English and Welsh wines

Nobody really knows for sure how many bottles of English and Welsh wines, both still and sparkling, are sold each year. WineGB's 2018–19 commercial report gave a figure of sales for 2018 of 2.60 million x 75 cl bottles, which included 8 per cent of exports.[4] This implies 2.39 million bottles of domestic sales and 0.21 million bottles (208,000) of exports. Given that the producers surveyed were the larger ones, more given to trade sales and exports than smaller ones, these percentages are almost certainly overstated for the whole industry. It is my guess that what the WineGB survey revealed was that of those producers who export, they export around 8 per cent of their total sales. On an industry-wide basis, therefore, exports account for maybe only 2–3 per cent of the total. HMRC Excise Duty returns for 2018 showed duty-paid sales of 2.76 million bottles. This leaves a slight discrepancy

4 The report also revealed that 53% of domestic sales were trade sales, 31% cellar door sales, and 6% on-line sales (which together with 8% exports, doesn't quite add up to 100 per cent!).

(370,000 bottles) between sales reported by producers and sales reported by HMRC. One would have thought, if anything, that producers might have inflated their sales in their voluntary report to WineGB, rather than under-estimated them. The one fly in this particular jar of ointment is that there have in the past been problems with the HMRC statistics and they have been known to include that other wine of 'UK registered premises', namely British-Made Wine, in their returns. On several occasions I have had to point out what look (to me) like anomalies in the data and at the next three-monthly set of numbers the offending entry has been changed stating 'estimate due to data quality'. The 2019 figures from HMRC show domestic sales of 3.32 million bottles, to which, if you add 3 per cent exports, equates to total sales of 3.42 m bottles.

However, the 2019 WineGB commercial report has reported a total sales figure of 5.5 million bottles which includes exports of 10 per cent. The considerably increased figure of sales has been arrived at after consultation with the major producers who feel that HMRC statistics are not fully reflecting a true sales figure. This is due to wine going directly from producers into 3rd-party bonded warehouses which, when released for sale (at which point Excise Duty becomes payable) is then not correctly recorded as wine from producers in Britain and is lumped in with the majority of wines handled in-bond, which are of course imported. This adds further confusion to the situation and as I said at the start of this section, 'nobody really knows' what the sales are. My best guess is that they might be somewhere between the HMRC's figures plus exports i.e. 3.50 million bottles, and WineGB's (what I feel is an optimistic) total of 5.50 million bottles which means 4.50 million bottles.

Given the production and sales figures above, the question of 'where will it all sell?' is given some perspective. Let's assume that sales in 2020, despite the ravages of Covid-19, are 20 per cent up on the previous year (2019) as they were between 2018 and 2019, and, assuming exports of 10 per cent, that they reach a level of 5.40 million bottles. This is still less than half of the average 11.79 million bottles produced in both 2018 and 2019, to say nothing of the 18+ million that is awaiting us in maybe two to three years' time. Can exports really be expanded to sell this many millions of extra bottles? Whilst a few producers have managed to make a start on exporting, many others speak of the expense of mounting sales trips abroad, the slow nature of the uptake in sales and the need to keep visiting export markets in order to keep importers happy and to keep the retailers and hotel, restaurant, and bar owners who stock it up to date and motivated.

If we look at the domestic market for sales of all sparkling wines, the last two decades have seen a remarkable rise in duty-paid sales. In 2000, Britain's wine drinkers, who then numbered around 25 million adults, plus non-British visitors, who totalled around 15 million adults, managed to put away a modest 71 million bottles or 1.78 bottles per head per year of sparkling wines.

Fast-forward to 2019 and the sales have almost trebled to 203 million bottles, and numbers of drinkers, home-grown and visitors, to 50 million, and consumption has risen to around 4 bottles per head per year. Per year! That's around half a glass a week. Admittedly the sales of sparkling wine, whose dramatic increase has been mainly fuelled by the popularity of Prosecco, has declined slightly in the last three years (2017-19) by a modest 6.4 per cent but even so, Britain is obviously a fertile market for sparkling wines, although maybe only at the right price? The British sales of Champagne, which typically sells at a slightly lower average price to English and Welsh sparkling wine (average shelf price in 2019 for a bottle of Champagne in GB was £23.88), have fallen from a pre-banking crisis figure of 39.10 million bottles in 2007, to almost 27.00 million bottles in 2019. Although this is very slightly up on 2018's figure, it is still a fall of almost 31 per cent from the 2007 peak figure. Exactly why Champagne sales have fallen so much is a question to which nobody has a convincing answer. The Champenoise just seem to think it is the perfidious *Rostbifs* who cannot afford Champagne and prefer cheap and cheerful Italian sparkling wine. Although they don't say it, they also realise that some of their sales have transferred to English and Welsh sparkling wine, although only by about 3 million bottles a year.

The big question, therefore, of 'where will it all sell?' is still unanswered. Maybe Charmat and carbonated wines made from British grapes, cheaper to buy now as there will be a glut of them, will expand to take up some of the slack? But at the current prices of £10–£15 a bottle, can these wines be sustainably profitable? Maybe traditional method sparkling wines will have to meet the market halfway with prices nearer the £15–£20 mark rather than the current £25–£35 level?[5] But at this price, are they profitable, given average yields and production costs? In all probability things will work out as they have with wines from other wine producing regions when there is oversupply. Direct sales in all their different forms will expand, which will give growers bigger margins for some of their sales. When stocks are high and cash-flow tight, then parcels of wine will be offered to those retailers with the capacity to sell large volumes in a short space of time. Producers will get more inventive with their offerings and find new routes to market, and exports will be a sales avenue for some but not the majority. By the time this book is updated again, the answer to this question may have become clearer.

---

5 Waitrose currently list 43 different English and Welsh sparkling wines on their website with an average price of £28.96 a bottle. There are 9 wines under £25, and 14 under £30. The most expensive is £45 a bottle. In 2014 they listed 40 wines with an average price of £31.43, with 9 wines under £25, 14 under £30 and 17 over 30.

Still or sparkling? – a summary

The answer to this question I believe is rooted in several factors. When visiting ('interviewing', I was tempted to write) new entrants to the business of establishing a vineyard, growing grapes and selling the produce, I always ask these basic questions: 'where do you see yourself in twenty years' time? Are you going to be selling all of your grapes? Some of them? None of them? Will you be selling wine? If so, what and where? To wholesalers and retailers? Hotels, pubs and restaurants? Or direct to the public?' You would be surprised how many people have given this aspect of their future enterprise very little thought. However, from the answers to these questions stem a lot of the answers to 'what do I plant?'

There is no doubt that where, as an industry, we once all thought that sparkling wines would come to dominate, there are very positive signs that still wines, with their infinite variety of styles, brought about by using different grape varieties and different winemaking techniques, can play their part in a vineyard's wine offering. Apart from any other consideration, if only for cash-flow reasons, still wines can play an important part in creating a viable business. Still wines can start to return you an income within three years of planting; with sparkling wines it might be five or six. At the volume end of the sparkling wine business, currently occupied by around twenty sizeable producers (but surely to be joined by more to come), my guess is that, across their range, most of them will have quite marked price differences, with prestige cuvées, middle-rank brands and (when necessary) 'on promotion' offerings all under the umbrella of the same label and/or brand. Nyetimber – as an example – currently have their '1086' brand at £130–£150 a bottle; various vintage wines around the £35–£45 mark; and their 'multi-vintage' widely available at £28–£30. I don't suppose the cheaper offering stops people buying their more expensive wines, any more than do Moët & Chandon's offerings across different price bands.

When it comes down to the wire, deciding what varieties to plant, and whether to produce still or sparkling wine, is mainly about your expectations and your budget. If you are proposing to have the all-singing, all-dancing, full-on visitor experience vineyard, then I would suggest planting a spread of varieties to cover both still and sparkling wines, and to create as broad a wine offering as possible in order to satisfy as many palates and as many budgets as you can. If your aspirations – and wallet – run to a bigger operation where, by definition, your sales cannot (or are unlikely to in the medium- to long-term) be mainly over the farm-gate, then establishing a sparkling wine brand, which, unless you are very lucky, will take a minimum of ten years, and probably a deal longer, may well be the way forward. As ever, financial considerations are important as viticultural ones.

Vine variety recommendations

Note: See also the vine variety recommendations in *Chapter 18: Organic and biodynamic viticulture in Great Britain.*

White traditional method sparkling wines

For traditional bottle-fermented white sparkling wines, the Champagne trio of Chardonnay, Pinot noir and Meunier, either singly or in blends, have proved themselves more than capable of producing world-class wines on many sites in Britain. *Blanc de blancs* made from 100 per cent Chardonnay have been particularly successful and are amongst the very best wines that we can produce. However, they do not suit all sites and I would caution against planting these higher acid varieties (especially Chardonnay) on challenging, i.e. marginal, sites. In difficult and late years, acid levels can remain stubbornly high, although with time, even wines with high acid levels can age into graceful swans. On marginal sites, of course, yield levels will be adversely affected which will have an impact upon incomes. Having praised the Champagne varieties, I wouldn't discount using other varieties such as Auxerrois, Pinot blanc, and Seyval blanc, especially if used in blends. Seyval blanc as a single (or dominant) variety certainly has its place in sparkling wines made in Britain, although it is usually going to lag slightly behind the Champagne-variety based wines in terms of quality and retail price (although some producers – Breaky Bottom, Camel Valley, Denbies – are successful with it and their wines fetch good prices). However, the ease with which Seyval blanc can be grown and the level of yields it typically produces may well be more than adequate compensation for these shortcomings. Although there have been a few acceptable sparkling wines made from varieties such as Reichensteiner, and even Müller-Thurgau, usually in blends but occasionally as solo varietal wines, I do not believe anyone planting today should choose these varieties solely for this style of wine. As for other varieties, I would personally avoid any of the aromatic varieties and, worldwide, these seldom make classic sparkling wines (Riesling Sekt for instance). However, there have been a few sparkling wines made from varieties such as Bacchus, Pinot gris and Solaris, but I don't think they have the same gravitas (or sell at the same prices) as wines made from more 'noble' varieties.

Rosé and red traditional method sparkling wines

For rosé sparkling wines almost any combination of Champagne varieties works well, as does single variety wines made from either Pinot noir and Meunier. All can be successful. For many years Meunier was discriminated against (my theory is that the French persuaded us that it couldn't make good wine) but in recent years it has produced some excellent wines, both rosé and red wines, still and sparkling. There is more on Meunier in the individual variety description later in this chapter. For the red element of rosé wines, I prefer wines where the colour has come from red-fermented Pinot noir or Meunier. I can understand why some growers use a small addition of wine made from Blauer Frühburgunder (Pinot noir Précoce), or one of the other red varieties such as Dornfelder, Regent and Rondo, although I don't think they are as good and would not advise it for a new vineyard. Other neutral sparkling wine varieties – Seyval blanc, Pinot blanc, Auxerrois – can be used for blending into rosé wines, but they are rarely as good. Rosé sparkling wines often have their colour tweaked at the time of disgorging and *dosage* and in my experience, red Pinot noir works best, although I have known wine from other high-colour varieties used successfully.

As for sparkling red wines, to date there have been very few that could be called 'red' – a few dark rosés perhaps. However, I personally quite enjoy semi-sweet sparkling reds – some of the Australian Syrah-based ones can be very good and I have also had a Nebbiolo-based red sparkling wine from Asti which was excellent – so I see no reason why Dornfelder, Rondo and Regent could not be used. I would prefer to see them in the *demi-sec* or even *doux* category as British acids and tannins would normally be too high for dry red sparkling wine.

Non-traditional method sparkling wines
– Charmat and carbonated

When I wrote the first edition of this book in 2012–13, non-traditional method sparkling wines barely got a mention. Sure, there had been carbonated sparkling wines produced in the past (the carbonated 1983 Barton Manor Sparkling Rosé won a gold medal in the 1984 English Wine of the Year competition) but they were few and far between and barely got noticed outside the English wine fraternity. Today however, there are both Charmat (tank) method and carbonated sparkling wines made from British grapes on the market. The joint-first Charmat wines on the market (in 2018) were from Flint Vineyard in Suffolk who produced a 'Charmat Rosé' made from 'Solaris, Reichensteiner, Cabernet Cortis and Rondo' which sold well at £22–£24 a bottle and from Fitz, based in Worthing, who produced a tank-method sparkling wine made from 'Chardonnay, Seyval blanc, Reichensteiner, and Madeleine Angevine', selling for around £18–£20. Following the very large 2018 harvest (when grapes were both plentiful and cheap) a brand called Angel & Four (owned by Hayloft Ventures Ltd) produced a tank-method sparkler using '50% Reichensteiner, 25% Madeleine Angevine and 25% Seyval blanc' grapes. This wine appeared under two labels: Angel & Four selling for around £14.99 and under the Masterstroke label selling in Aldi stores at £10. (Aldi is also selling two Denbies traditional method wines at £14.99, just to put their pricing into perspective.) Other Charmat method producers include Chet & Waveney in Norfolk with their Skylark range (£20–£22 a bottle) made mainly from Phoenix and Seyval blanc, and Albourne Estate's *Bacchus Frizzante 2018* (claimed to be the first *frizzante*-style wine from Britain and with only 2 bar pressure) which sells at £14.95–£16.50. These last two producers had their wines 'Charmatted' under contract at Bevtech, David Cowderoy's

winery. I also hear that Mark Dixon's company, MDCV Ltd, who bought Kingscote Vineyards and Sedlescombe Vineyards, extended the lease on Bodiam Vineyard, planted 40-ha of vines on land at Sandhurst, Kent (leased from the same owners as the Bodiam site), planted a new 176.90-ha (437-acre) vineyard at Luddesdown, near Gravesend in 2019 and a further 37.64-ha (93-acres) at Althorne in Essex in 2020 and in total now have 283.5-ha (700-acres) under vine, intend to make and sell a tank-method organic sparkling wine to retail at around £12.99 a bottle. Whether this will come to pass is another matter. Time will tell.[6]

There are also now producers testing the market with sparkling wine in cans. A company called 'The Uncommon' is putting Charmat method wine into 250 ml cans and they have two wines, a Bacchus and a Pinot noir-Meunier rosé, which retail for around £5.50–£4.75 per can. At the equivalent of £16.50–£14.24 a bottle, this probably gives a better return than using glass bottles as there are surely savings on production costs and packaging materials. A few other producers have gone even further down the perceived quality line by producing carbonated wines. Chapel Down have a carbonated Bacchus which retails at £18 (in their shop) which gets mainly 5-star reviews on Waitrose's website, where it is priced at £13.49 a bottle in 2020 and appears to be selling well. I tasted it and wasn't impressed and prefer their still Bacchus. Three Choirs have a carbonated sparkling wine in 200 ml cans, made from Seyval blanc and Phoenix, as well as three still wines in 187 ml cans in three different variants: a white made from Madeleine Angevine, Phoenix and Seyval blanc; a rosé made from Phoenix, Seyval blanc and Rondo; and a red from Regent, Rondo, and Triomphe. If nothing else, this sounds like a good way of using up less popular still wine varieties. An 'urban winery' run by Laneberg Wine, based just south of Gateshead, a fairly unusual place for a winery making wine from English grapes, also have a carbonated 'semi-sparkling' wine made from 100 per cent Seyval blanc called 'This Mortal Angel' (which I guess is a reference to the nearby 'Angel of the North which retails at around £12–£15 a bottle.

How sustainable these non-traditional sparkling wines are is open to question. For producers without vineyards, and therefore dependent on buying their grapes from others, much will depend upon yields, availability and prices. With yields like those in 2018 and 2019, and with the large number of vines planted between 2017 and 2019, maybe these producers are just what the market needs? Soaking up the excess supplies of grapes and knocking out wines in the £10–£15 price bracket. How profitable a £10 sparkling wine is, is another matter. Deduct 20 per cent VAT and £2.86 Excise Duty and you are left with £5.47. Assuming the grapes cost as little as 50p a bottle, production costs are £3 and the retailer wants a margin – say, £1 – and the producer is left with 97p. At £15 retail the figures look better, the producer ends up with nearer to

---

6 As of writing (2020), both Kingscote and Sedlescombe Vineyards have been put up for sale.

£5, but with plenty of traditional method sparklers selling at £15–£18, is this price sustainable for a non-traditional method sparkler? As I said above – time will tell. Where these wines will sell well, I am sure, is at the farm-gate where customers can taste them and at events, functions and in the on-trade where wine by the glass is sold without the customer ever seeing the bottle. Asked 'would you like a glass of sparkling wine' most people are not going to ask 'where's it from?' let alone 'what is the method of production?' assuming that they even know the differences between carbonated, Charmat and traditional methods.

Despite the misgivings of some producers and commentators, who view these non-traditional method wines as somehow debasing 'proper' sparkling wines, it would seem, judging by the rapidity with which they have appeared, that they are here to stay. Their future will surely depend on the availability of grapes at the right price, and how low prices on the high street for traditional method wines will fall. Once Chapel Down starts selling its NV Reserve Brut at £12.99 (a price they are on record as saying they can support) then why buy a Charmat method one at the same price?

Still white wines

Note: White still wine varieties in the top twenty are discussed in order of planted area

Chardonnay for still wines has already been discussed and on suitable sites and in ripe vintages, it can be very good and can sell for high prices. However, it would be a brave grower who based their business plan on growing Chardonnay solely for the production of still wine. Growers with good sites who are currently making sparkling wine, will, in good years, be able to divert some of their grapes to still wines and this will, I am sure, increase as producers value the cash-flow and margins of these wines. For most still white wine producers, Bacchus has to be the top variety, especially if you have a site that really suits it, you know how to grow it and it ripens well. The trade and the public recognise it now as a 'British' variety and given its success in tastings and with a great marketing name, its future is secure. Seyval blanc and Reichensteiner have made a modest comeback in the last decade and their higher yields and more neutral styles make them ideal blending partners. They are also dual still–sparkling varieties and are being used by some of the non-traditional sparkling producers which makes them worth growing. Solaris finds favour in more marginal regions and the area planted has expanded in percentage terms more than any other still wine variety and growers who have it praise it highly. Ortega is a variety that is on the rise and I am sure we will see an increasing number of good still wines made from it. Its earliness can sometimes be a challenge, attracting wasps and rot in equal measure, but its wines have a unique fruit profile which sets them apart. Madeleine x Angevine 7672 has its place in the more marginal parts of Britain but, in general terms, it is not a high-quality variety and whilst I can see vineyards hanging on to it, I cannot see many growers in mainstream regions planting it. Pinot blanc and Pinot gris have already been mentioned and they are most definitely worth growing, although the cropping record of Pinot gris isn't great. However, the quality of the wines made from them can be high (as well as the prices) and their internationally known name makes them easier to introduce to the public. Müller-Thurgau, which was once top dog in Britain's vineyards, has now almost disappeared off the charts, and although it can make good wine, its name, its vigour and its disease susceptibility mean that it's not a variety to recommend. Phoenix has its fans, but it hasn't increased in planted area much since 2012 and I cannot see it really taking off now. Better to grow Bacchus if you want that style of wine. Of the other varieties in the top twenty, Schönburger and Siegerrebe can both produce very good wines in a certain style, but they are difficult varieties to sell through the trade and I would only recommend planting them if the vineyard proposes to sell these wines mainly from the farm-gate. Huxelrebe, the variety at number 20 in the list, is a declining variety and I cannot see any new vineyard planting it, although Davenport Vineyards have done well with it and its 'Hux' is one of its flagship still wines. There are a few producers who have made white wines from Pinot noir, but to date, I have not been impressed with the results. Maybe with time and experience, they will get better and 'Pinot Albinoir' will become a common sight.

Of the minor varieties, Optima and Orion might be interesting on a good site, and their wine quality is good. I have a soft spot for Auxerrois, Gutenborner and Regner, varieties that have done well in the past (I won a gold medal with a Gutenborner in 1983) and from which very good wines can be made. As for other varieties, there are precious few that will make money. One might test the water with Albariño (called Alvarinho in Portugal), but it takes quite a bit of ripening, so a good site, lots of shelter, and patience are needed. There are some growers who believe that Sauvignon blanc can be successful in Britain and maybe, in years to come, they could be right. However, at the moment, except in very exceptional vintages I disagree, and would not recommend it, at least, not as a mainstream variety. Woodchester Valley Vineyard, near Stroud in Gloucestershire, won an IWSC Gold and a Gold at the Drinks Business 'Global Masters' competition for their 2018 Sauvignon blanc, which shows that it can be successful, but which rather goes to confirm my point about it needing an exceptional vintage for it to ripen. Much the same goes for that other 'cool climate' stalwart, Riesling. Denbies and Rathfinny both had quite large plantings at one time, but in both cases the vines were grubbed up. Maybe in a decade or two I will, but at the moment, I cannot advise planting Riesling. Once popular varieties from the past such as Faberrebe, Kerner, Kernling, Scheurebe, and Würzer, whilst they linger on in some of the older vineyards, have all had their day and nobody is planting them anymore. Finally, whilst there are plenty of new varieties available, some of which are discussed later, none have (so far) shown themselves to be better than the varieties already mentioned above.

Still red wines

For still red wines – if REAL quality is the aim – the choice is very limited. The best two are Pinot noir and Blauer Frühburgunder (Pinot noir Précoce), both of which, given a good site, the right crop load, perfect canopy management and, of course, good winemaking, can make very good red wines. I cannot pretend that they match the best from Burgundy, Martinborough, California or other Pinot hot-spots, but the best from Britain have an excellent purity of fruit and a refreshing acidity that is sometimes lacking in wines from warmer climates. Whether they are value for money is another matter. The best sell for anywhere between £20 and £45 a bottle, which is pricey for most customers but as there are (almost) no good cheap Pinot noirs from anywhere, why should ours be any different? Having said the above about Pinot noir and Frühburgunder, there are sadly far too many red wines made from them in Britain that are not great quality and they would be far better off as rosé wines. However, rather like still Chardonnay wines, I would guard against basing my whole business plan on producing still red wine from Pinot noir, but treat it as a bonus in good vintages. As for other red varieties, I believe that Meunier might surprise everyone and turn into Britain's USP for both still and sparkling wines. Of the next four most widely planted reds, Rondo, Regent, Dornfelder, and Triomphe, if you have them in your vineyard you are probably not going to remove them, but they are not being planted widely any more. Rondo, which ripens early, has large crops and low acidity, is being planted in some of the more challenging regions of Britain, but is less common in the southern half of the country. However, if you have a vineyard open to the public and intend to major on farm-gate sales, then a red wine made from a blend of Dornfelder, Regent and Rondo will probably sell and will certainly have better colour than any Pinot-based wine. Acolon, a 1971 Weinsburg cross of Blaufränkisch (aka Limburger) and Dornfelder, has started

to gain a little traction in Britain, although examples of its wine, good or not so good, are hard to find. Next on the list is Cabernet Cortis, a PIWI Cabernet Sauvignon and Solaris cross made at Freiburg in 1982. To date, it's still in the experimental stage and only time will tell whether it's going to be suitable for Britain. the last of the listed varieties, Dunkelfelder, is only used to beef up the colour of other reds, and for that purpose, it does no harm, but I doubt anybody is planting much of it now.

Of the other (very minor) red varieties being grown, Merlot has quite surprisingly gained a very small foothold in Britain, although I cannot understand why, and there are also plantings of Cabernet Franc, Cabernet Sauvignon, Gamay, Pinotage, and Syrah which must be considered as very speculative. The old hybrids Cascade, Léon Millot, and Maréchal Foch, have definitely had their day and are no longer being planted. Of the other reds, I cannot see any that offer any advantages at the moment, although the newish varieties Cabernet Dorsa, a Dornfelder and Cabernet Sauvignon cross bred in Weinsburg in 1971, might surprise us all. There are also a few plantings of Bolero, a Geisenheim PIWI cross, and NIAB-East Malling Research (NIAB-EMR) have been trialling Divico, a complex PIWI cross from the Agroscope Research Centre in Pully, Switzerland (the same place that sent Müller-Thurgau to Ray Brock in 1947). Divico is Gamaret x Bronner which have Gamay, Reichensteiner and Rondo in their parentage. NIAB-EMR are reported as saying that Divico grape could be the 'much sought after game-changer' for red wine production in Great Britain as it produces 'quality red wine and thrives despite the challenges of the British climate.' Time will tell whether this is right, not quite so right, or not right at all.

Still rosé wines

Still rosé wines can be commercially very successful in Britain and I wish that some of the growers currently striving to make palatable red wines would instead turn their Pinot noir and Blauer Frühburgunder (Pinot noir Précoce) grapes into attractive, fruity rosés. The Pinots are to my mind the best varieties for still rosés and have the advantage that in poorer years they can go into sparkling wines and in really ripe years an attempt at making a decent red can be made. They also have the advantage of a recognisable name, a fashionable light-red to grey-tinged *oeil de perdrix* (partridge eye) colour and on the palate live up to the consumer's expectations (which is generally not very high or demanding for rosés). Rosés made from other red varieties – Dornfelder or Regent, for instance – or from blends of white and red varieties can also be successful, but none to my mind have quite the class to match Pinot noir. Again, as with red wines, varietal naming of rosés is not so important and a multi-variety rosé can work well. I would avoid the bulk of a rosé being made from aromatic varieties – Bacchus, Müller-Thurgau, Schönburger, Ortega, Siegerrebe – and use a more neutral variety such as Seyval blanc, Reichensteiner, maybe even Madeleine Angevine

Red varieties		
Variety	Ha 2020 (Estimate)	% of national area
Pinot noir	925	26.43%
Meunier	275	7.86%
Frühburgunder, Blauer	70	2.00%
Rondo (Gm 6494/5)	65	1.86%
Regent	35	1.00%
Dornfelder	18	0.51%
Triomphe	8	0.23%
Acolon	7	0.20%
Cabernet Cortis	2.5	0.07%
Dunkelfelder	1.5	0.04%
Source: Industry estimate	**1,407**	**40.20%**

Table 22

49

7672, or Pinot blanc if your site will ripen it, and colour it with one (or more) of the better reds.

Vine varieties for less favourable sites

For growers determined to plant on less than favourable sites (sites I term 'challenging') which might be at altitudes over 125-150 m above sea level, or with poor shelter from prevailing winds, or just in a less-proven and/or less-favourable part of Britain, their choice of varieties must be made with these disadvantages in mind. There is no doubt that one of the reasons for the survival of Seyval blanc in the top five of British varieties for over fifty years is its ability to produce viable crops in years in which other varieties fail, its relatively good disease resistance, and its ability to hang on and ripen fruit in conditions that would severely test other varieties. Reichensteiner, likewise, whilst not as disease resistant as Seyval blanc, has also survived, mainly because in good years its yields are higher than average and these compensate for the poor years. Its grapes usually get high sugars and the neutral wine is adaptable and can be blended into both still and sparkling wines. In the relatively new (and few) vineyards planted in the more northerly parts of Britain, the selection of varieties is divided into three different sections: the earlier ripening *vinifera* varieties; old hybrids; and new interspecific crosses. The earlier ripening varieties are chosen because they (a) ripen early and therefore stand a chance of ripening on a later site and (b) tend to have lower acidity levels which make them more likely to achieve physiological ripeness. The most widely planted white *viniferas* are: Madeleine Angevine 7672, Ortega, Reichensteiner, Schönburger and Siegerrebe (in alphabetical order). Gaining in popularity is also Acolon, a red *vinifera* variety. Of the older hybrids, chosen because they tend to be resistant to diseases and thus more likely to produce and retain a crop in difficult circumstances, the already mentioned Seyval blanc is, of course, a firm favourite, but so also are the old red hybrids: Cascade, Léon Millot, Maréchal Foch, and Triomphe. It is worth noting that these old hybrids (but not Seyval), have long ago been weeded out of southern vineyards on the grounds of poor wine quality. Of the newer interspecific PIWI varieties, the whites Orion, Phoenix and Solaris, and the reds Regent and Rondo seem popular, again their better (although far from perfect) disease resistance makes them an obvious choice. The problem with many of these varieties is that the wine quality, even in well-sited, southern vineyards, would be considered to be behind that of, say, Bacchus and the Champagne varieties and if you add on to that the additional problems caused by growing vines on less-favourable sites, the wine quality is bound to suffer. Some of the newer vine varieties listed in the next paragraph might also be of interest, although many of them are probably too late for late sites.

Newer vine varieties

When I started growing vines at Tenterden in 1977, the options for 'new' varieties were fairly limited. Although I stuck with Müller-Thurgau and Seyval blanc as my two main commercial varieties (along with most of the other growers at the time), I also planted 1,250 Reichensteiner vines and the same number of Gutenborner, then a variety only being grown by one other British grower (Carr Taylor). This latter variety was recommended to me by Geisenheim's Professor Kiefer who had been helping with the establishment of the nearby vineyards at Lamberhurst, so had some experience of growing conditions in Britain. These four varieties made up the bulk of the 2-ha (5-acres) I planted initially. My almost two years at Geisenheim had not only been spent in learning how to grow grapes and the rudiments of winemaking, but I had also spent a considerable time researching into vine varieties in an attempt to discover if there were any that were then unplanted in Britain, but which might be suitable. I remember very well going to see Professor Becker one afternoon with a list of thirteen varieties I thought of as 'possibles' for Britain, to ask his advice. He more or less grabbed the list and said 'come back on Thursday' which I took to be a signal that he was too busy to discuss the topic, but might have time later in the week. I duly turned up on Thursday to find that his assistant, Ernst Abel (who had been over to Lamberhurst to help make their maiden vintage in 1976), had been instructed to get a bottle of wine from every one of the varieties that I had listed (the research cellar made trial batches of wine from hundreds of varieties each year) and had them lined up in the cellar for me to taste! The result of this was that I planted one hundred vines each of Ortega, Schönburger, and Rabaner, plus twenty-five vines each of Ehrenfelser, Huxelrebe, Kerner, Optima, Siegerrebe and a Riesling x Silvaner cross called Gm 4-46. One of the varieties I had on my list and tried at the tasting was Bacchus and I noted on my tasting notes (which I still have, dated 15 Nov 1976) 'Acid too high. Wine not good if under 75°OE' and was advised not to plant it! In my trial plot I also planted a range of other varieties: Auxerrois, Black Hamburgh, Chasselas Rosa, Chenin blanc, Dunkelfelder, Gewürztraminer, Gold Riesling, Léon Millot, Meunier, Pinot blanc, Pinot gris, and Pinot noir. It was an ambitious list, but I thought that I would try and create a mini-research station of my own.

Today, if faced with the same task – to select some new, untried varieties for planting in Britain – it would be considerably harder to choose a reasonable number, given the huge range of different varieties now available. Most of the work on new varieties has been directed towards disease resistance and breeding new varieties which require no pesticides, even in high disease-pressure regions. Gene mapping of vines has allowed researchers to identify the genes responsible for conferring resistance to the two greatest vine diseases, Powdery and Downy Mildew. The gene conferring resistance for Powdery is called RUN-1, after 'resistance to *Uncinula necator*', and to Downy, RPV-1 after 'resistance to *Plasmopara viticola*'. These genes have been isolated from the wild American vine species *Muscadinia rotundifolia,* which is native to the south-eastern part of the USA and almost totally resistant to these two diseases. By making crosses over several generations, the RUN-1 and RPV-1 genes can be locked into a new variety, conferring resistance. The other major disease, botrytis, is impossible to control in this manner as it is a disease that grows on dying and dead tissue and there is no gene which can confer resistance. The defence against this disease is to change the 'bunch architecture' through selective breeding so that the rachis[7] is bigger and the bunches are looser, with the berries further apart. This will allow for better air circulation and less humidity, both of which will lessen the incidence of *Botyrtis*. In *Les cépages résistants*, the 2013 book from the ICV (*Institut Coopératif du Vin*) Group, there are fifty-six interspecific crosses which are considered to be worthy of a full write up and a total of almost 400 described in table form. The best place to get more information on new, disease-resistant varieties, now known as PIWI[8] varieties, is from www.piwi-international.de. Its website lists all their competition results going back to 2011, so it is easy to see which varieties have performed well.

In order to suggest new varieties for Britain, we have to get over the age-old problem which all new variety recommendations suffer from: none of them has been bred specifically for British climatic conditions and very few have been tested in Britain. We have, therefore, to sift through the claims made for each variety and decide for ourselves how they relate to British problems. Disease resistance is obviously a desirable trait, but I don't feel that it should be the deciding factor. If you had a choice between a variety that cropped well, produced ripe grapes with moderate acidity that made good wine, yet required a modest spray programme, compared to say a variety that cropped well, did not require spraying at all, but whose wine was of a lower quality, which would you choose? For me, the first attribute for a variety to be preferred over the ones we currently grow is the quality of its wine. A reasonable yield is very welcome, disease resistance is admirable, but without wine of sufficient quality to lift it over the problems associated with introducing a new variety to the wine trade and the wine-buying public, the variety is going to struggle commercially.

This problem is not a new one. Varieties already mentioned, such as Faberrebe, Kerner, Kernling, and Würzer, plus others such as Findling and Optima have all been tried in Britain, found lacking and have mainly been grubbed up. Newer varieties that have found favour in vineyards in Britain – Acolon, Orion, Phoenix, Regent, Rondo, and Solaris – which together are planted on 212-ha (524-acres) are, with the exception of Acolon, all modern interspecific PIWI crosses. The reds, Acolon, Regent and Rondo, are successful (or at least, have been quite widely planted) because they are relatively disease resistant and produce wines with plenty of colour. If you are hell-bent on producing red wine in a very cool climate, then these are as good as any varieties you can find (at the moment anyway) and better than the old hybrids Cascade (Seibel 13/053), Léon Millot, Maréchal Foch and Triomphe. The others, all whites, have mainly been planted in vineyards in the more climatically challenged areas of Britain, again because of their resistance to disease and their suitability for cooler climates.

For the future, there are several next-generation interspecific crosses which are already on trial in Britain. Several of them are from the breeding programme developed by a private Swiss vine breeder, Valentin Blattner, who is working with German, French and Spanish vinegrowers to develop ultra-resistant varieties which require no pesticide applications at all. Much of the breeding work Blattner carries out takes place in Thailand, where the climate allows two harvests a year, thus accelerating the development of new varieties. In early September 2013 I visited the '20:20' vineyard – 'twenty hours work a year and twenty tonnes a hectare' – near Béziers in the south of France which belongs to Domaine La Colombette but which Blattner uses as his test-bed. This vineyard receives no pesticide applications, is machine pruned, irrigated with subsoil irrigation, machine harvested and, whilst perhaps not quite 20-tonnes-ha, crops very well and produces very acceptable wine. Varieties from Blattner which are either already on trial in Britain or being considered are listed below, followed by other varieties from German and Swiss vine-breeding institutes already on trial in Britain.

Blattner varieties
- Cabernet blanc – a 1991 white, Cabernet Sauvignon x unknown crossing with Sauvignon blanc characters. Ripens after Pinot blanc, so relatively late in Britain. Has both small and large berries. In 2018 there were 158-ha in Germany, which for a PIWI variety is a fairly good amount.
- Cabernet Jura (VB 5-02) – a 1991 red, Cabernet Sauvignon x unknown crossing with deep red wine. Ripens 'early to mid-season' in Germany, so probably towards the end of the season in Britain.
- Sauvignac (VB Cal. 6-04) – a 1991 white, (Sauvignon blanc x Riesling) x unknown crossing with loose bunches, which ripen 'ten days before Riesling' in Germany. Good resistance against diseases.
- VB 32-7 – a 1998 white, Cabernet Sauvignon x unknown crossing with Sauvignon blanc characters. Ripens mid-season in Germany, so probably late in Britain.
- Satin noir (VB 91-26-29) – another 1991 red, Cabernet Sauvignon x unknown crossing with deep red wine, similar to Cabernet Jura.
- Réselle – a white, Bacchus x Seyval blanc crossing which produces a light, fruity wine. Could be good in Britain

---

7 The 'rachis' is botanical name for the stem (or stalk) of the vine which holds the grapes.

8 PIWI stands for Pilzwiderstandsfähige (Fungus resistant).

for blending with a spicy variety. Probably the best of the Blattnercrossings for our climate.

Varieties from German and Swiss vine-breeding institutes already on trial in Britain include:

- Bolero – a 1982 red, Gm 6427-5 x Chancellor crossing from Geisenheim crossing produced in Professor Becker's era, which is said to ripen early and has lowish acidity. Gm 6427-5 is Rotberger x Reichensteiner.
- Cabernet Dorsa – a 1971 red, Dornfelder x Cabernet Sauvignon crossing (therefore a *vinifera*) from Weinsberg. I fear this might be too late for Britain, but time will tell. In Germany there were 263-ha planted in 2018, a good amount for a new variety.
- Divico – a 1997 red PIWI variety from the Swiss Federal Research Institute at Changins on Lake Geneva. It's a Gamaret x Bronner crossing. Gamaret is Gamay x Reichensteiner and Bronner is Merzling x Gm 6494 (a *Vitis amurensis* cross). Merzling is a complex cross containing Seyval blanc, Riesling and Pinot gris. As reported above, NIAB-EMR rate Divico highly and have been trialling it in Kent for several years. If wines have been made from it, I have not seen or tasted any.
- Muscaris – a 1987 white, Solaris x Muskateller crossing from Freiburg which ripens around the same time as Pinot noir. Has lowish acidity and is said to produce very spicy wine. Solaris (see later in this chapter) is a complex *Amurensis* crossing. In 2018 there were 58-ha in Germany.
- Souvignier gris – a 1983 white from Freiburg (Fr 392-83) originally said to be a Cabernet Sauvignon x Bronner crossing, but now revealed to be Seyval blanc x Zaehringer (Zaehringer is a Traminer x Riesling). It has dark pink grapes and produces 'lightly fruity' wine. Said to have good resistance against both Powdery and Downy Mildew. In 2018, there were 50-ha in Germany and according to the Sibbus website, 'quite big plantings in France and Italy'.
- Villaris – a 1984 white, Sirius x Villard blanc crossing from Geilweilerhof, said to ripen at the same time as Müller-Thurgau and produce similar wine.

As for other varieties that might succeed in Britain, the field is wide open. When I look at the other major *über*-cool-climate European winegrowing regions – Belgium, Denmark, Estonia, Germany's Baltic region, the Netherlands, Latvia, Poland and Sweden – which, between them, have over 1,500-ha (3.700-acres) of vines – varieties such as Bianca, Cabernet Cortis, Garanoir, Gamaret, Helios, Hibernal, Johanniter and Merzling appear quite often. The truth, of course, is that there are far too many new varieties available for any private grower to carry out any meaningful trials. It would be good if an institution (Plumpton College or East Malling Research perhaps?) could plant a small test-bed of possible varieties for the future, but I guess that is just wishful thinking.

Of course, whether Britain actually needs many new varieties is open to question. Perhaps we ought to concentrate on getting better at growing the ones we already have first.

Vine varieties – the legal situation

Note: Post-Brexit, Her Majesty's Government has adopted wholesale the EU regulations on all aspects of winegrowing, winemaking and labelling. Obviously at some stage there will be discussions between the industry and DEFRA about changes and no doubt, in time, things will change.

Until Britain reached a total of 500-ha (1,235-acres) of planted vines, MAFF was reluctant to get involved with the vine varieties growers had in their vineyards, although in theory growers were only meant to plant those varieties on the 'Recommended' and 'Authorised' lists. These lists, which had been drawn up prior to Britain's entry into the Common Market in 1973, were subsequently updated by MAFF's 'Vine Varieties Classification Committee' and brought into line with the varieties actually in the ground. Once a planted area of 500-ha was reached, a fact MAFF reluctantly discovered in the 1987 voluntary vineyard census, growers in Britain were restricted to planting varieties on the lists, but all the while Britain's wine production was under 5,000 hl in any one year, so there was no restriction on the area that could be planted and there would be no planting ban, something that existed in all other Member States with sizeable vineyard areas. But as the area under vine in Britain increased and with it, volumes produced. The limit on wine production before a planting ban would be introduced was upped to 25,000-hl average annual production over 5 years and then again to a 50,000-hl average annual production over 5 years. Since then the area under vine has grown massively and today (2020) the 5-year average stands at almost 60,000-hl.

However, as we are to leave the EU on 31 December 2020 and the industry has yet to have discussions with DEFRA about the way forward, it is probably safe to say that a planting ban is not likely to happen in the foreseeable future and most probably never. Having said that, given the huge (and, of course, uncontrolled and uncontrollable) increase in planting in recent years, some existing growers have suggested that a brake on planting might not be such a bad thing, but this is pure speculation. If changes to the ability to plant vines freely were ever to be mooted, DEFRA would require a substantial majority of votes in favour of it (not just 51:49) and most probably it would be on an area planted basis, not one grower, one vote. Therefore, a few large-scale growers who would be against a ban could easily swing the vote. Certainly, for the time being, a vine planting ban in Britain can be forgotten.

Working under the existing (old EU) rules, the planting situation in 2020 is that, except for six old American and/or hybrid varieties – Clinton, Herbemont, Isabella, Jacquez, Noah, and Othello – which are/were not permitted to be planted anywhere in the EU, there are now no restrictions on the vine varieties one may plant and any variety may be used for winemaking and the variety name may be used on the label. This is, of course, subject to the usual winemaking and labelling regulations which dictate how wines may be labelled.

Categories of wine that may be produced in England and Wales

Following changes to the basic EU wine legislation in 2011, there are now four distinct categories of wine which can be produced in Britain: wines with 'Protected Designation of Origin' (PDO), also called Quality Wines; wines with 'Protected Geographical Indication' (PGI), also called Regional Wines and which are table wines with some additional quality parameters; Varietal Wines which are table wines which have gone through a 'Certification' process; and wines which are none of the above which I have termed 'UK Wines' and which may not bear a vintage or the names of any vine varieties. All PDO and PGI wines have to be tested and tasted before they may be so labelled. Varietal Wines have to be 'Certified' by providing certain information, before they may be so labelled. The benefits of Varietal Wine are that the labels may bear varietal and vintage information, something the final category of wine – wines that are not PDO, PGI or Varietal Wines – may not.

Until these changes took place in 2011, only still wines made in Britain were eligible to be labelled as Quality Wines. To gain Quality Wine status, wines had to be made from *vinifera* varieties, i.e. no hybrids; had to be made to certain quality parameters; and all wines had to be tested and tasted. This system is still current. Somewhat confusingly, sparkling wines made to certain standards could (and still can) be called Quality Sparkling Wines even though they can be made from any variety, including hybrids. This is because the term 'Quality' in 'Quality Sparkling Wine' refers to the method of sparkling, not the base wine from which the *cuvée* is made.

After the 2011 changes, Britain gained the right to make PDO Quality Sparkling Wines which are the sparkling equivalent of Britain's still Quality Wines. The most important difference between PDO Quality Sparkling Wines and non-PDO Quality Sparkling Wines is that the PDO wines can only be made from certain varieties. There are also three more recent PDOs: Darnibole Bacchus, which is a Camel Valley Vineyards *monopole* and the two 'Sussex' PDO wines, sparkling and still, on which a selection of descriptors may be used.[9]

PDO Still Wines
English Quality Wine, Welsh Quality Wine, Darnibole Bacchus Quality Wine, Sussex Quality Wine.
PDO Sparkling Wines
English Quality Sparkling Wine, Welsh Quality Sparkling Wine, Sussex Quality Sparkling Wine.
PGI Still Wines
English Regional Wine, Welsh Regional Wine
PGI Sparkling Wines
English Regional Quality Sparkling Wine, Welsh Regional

Quality Sparkling Wine
Non-PDO/PGI Still Wines
Varietal Wine – allowed to state origin, e.g. Kent, but may not state 'English' or 'Welsh'. May bear vintage and variety name(s).
UK Wine – allowed to state origin, e.g. Kent, but may not state 'English' or 'Welsh'. May not bear vintage or variety name(s).
Non-PDO/PGI Sparkling Wines
Quality Sparkling Wine – allowed to state origin, e.g. Kent, but may not state 'English' or 'Welsh'. May not bear vintage or variety name(s).

Note 1: With regard to which vine varieties may be used for which categories of wine, the current (2020) notices available from DEFRA, WS and WineGB are sadly littered with inaccuracies, duplications and mistakes. Despite having raised these several years ago, they still persist. For instance: on the list of allowable varieties for still PDO (quality) wines you can find Cascade, Gagarin Blue, and Maréchal Foch which are old hybrids; Faberrebe is duplicated as 'Faber'; 'Foch' isn't the name of a variety (or is a duplication of Maréchal Foch); and Blaufrankisch and Kekfrankos are the same variety. There are also around seventeen modern interspecific crosses on the list, not all of which may be cleared for use in PDO wines. Having said that, post-Brexit, does it matter? The list of varieties able to be used for PGI wines has the same duplications and errors, but of course the hybrids, old or new, can be used for PGI wines. There are also a couple of varieties which are being grown in Britain (according to the last variety lists) which are not on any PDO or PGI lists. These are Cabernet Dorsa and Manzoni Bianco. Neither are very important and unlikely to be used on a wine label, so their omission is relatively unimportant.
Note 2: As official figures of varietal plantings have not been produced by WS since 2018 (based upon 2017 data), I have made educated estimates of the 2020 plantings, based upon WineGB surveys and my own database information.

Vine variety descriptions A–Z

The thirty varieties listed in Table 23 are those grown in Britain (2020) on an area of 1.50-ha or more. Together they amount to 3,281.50-ha (8,109-acres) and 93.76 per cent of the planted area.

Notes to variety descriptions:
- **Synonyms:** In general, I have used French variety names wherever possible as these are the names most likely to be found on bottles of wine in Britain and in vineyard literature. Thus, you will find references to Pinot rather than Burgunder or Spätburgunder; Meunier rather than Schwarzriesling or Müllerrebe; Pinot gris rather than Pinot Grigio, Ruländer, or Grauburgunder; although Blauer Frühburgunder, rather than Pinot noir Précoce, as this is a German variety.
- **Madeleine x Angevine 7672:** Following the investiga-

---

9 The label descriptors are: 'Sussex', 'Sussex Sparkling', 'Sussex Still', and 'Sussex Origin'.

Variety	2020 Ha	% age	Variety	2020 Ha	% age
Chardonnay	975.0	27.86%	Siegerrebe	20.0	0.57%
Pinot noir	925.0	26.43%	Schönburger	20.0	0.57%
Bacchus	295.0	8.43%	Dornfelder	18.0	0.51%
Meunier	275.0	7.86%	Huxelrebe	15.0	0.43%
Seyval blanc	145.0	4.14%	Orion	10.0	0.29%
Reichensteiner	75.0	2.14%	Triomphe	8.0	0.23%
Frühburgunder, Blauer	70.0	2.00%	Auxerrois	7.0	0.20%
Solaris	70.0	2.00%	Acolon	7.0	0.20%
Rondo (Gm 6494/5)	65.0	1.86%	Sauvignon blanc	7.0	0.20%
Ortega	55.0	1.57%	Cabernet Cortis	2.5	0.07%
Madeleine x Angevine 7672	50.0	1.43%	Würzer	2.0	0.06%
Regent	35.0	1.00%	Albariño	2.0	0.06%
Pinot gris (Ruländer)	35.0	1.00%	Dunkelfelder	1.5	0.04%
Pinot blanc	35.0	1.00%	Chasselas	1.5	0.04%
Müller-Thurgau	30.0	0.86%	Other varieties	218.5	6.24%
Phoenix	25.0	0.71%	**Total Ha**	**3,500**	**100%**
Source: Industry estimate					

Table 23

tions and DNA profiling of a variety which we have called various things over the years, what I have referred to as Madeleine x Angevine 7672 has now been officially renamed 'Alzey 7672'. At this stage, I am sticking to its old name. In the variety descriptions I use the abbreviation MA.

• **Meunier:** I have used Meunier, the correct name for the variety often called Pinot Meunier or even Wrotham Pinot.

• **Müller-Thurgau:** When referring to Müller-Thurgau in the individual variety descriptions I use the abbreviation MT.

• **Sylvaner:** I have kept to the old spelling of Sylvaner when referring to both 'Madeleine x Sylvaner III 28/51' and 'Riesling Sylvaner' (Müller-Thurgau or Rivaner) in preference to the modern German spelling, 'Silvaner', except where this variety was one of the parents of another variety. The Swiss (and others) continue to use the spelling Sylvaner.

• **German vine-breeding establishments:** References are made in the variety descriptions of the following vine-breeding establishments, all of them in Germany:

   Alzey, Rheinpfalz
   Freiburg, Baden-Württemberg
   Geilweilerhof, Rheinpfalz
   Geisenheim, Rheingau
   Oppenheim, Rheinhessen
   Weinsberg, Baden-Württemberg
   Würzburg, Bayern

• **OsCAR** This French organisation is *'L'Observatoire national du déploiement des cépages résistants'* and has been set up to trial and monitor disease-resistant varieties. It currently has seventeen varieties which it has tested and produced data sheets on, see http://observatoire-cepages-resistants.fr/

• **Sugar and acid levels:** Sugar levels are usually given in per cent potential alcohol. Most British growers and winemakers use degrees Oechsle (°OE). As in almost all other countries – France and French-influenced regions being the annoying exception, where levels in sulphuric acid are used – total acid levels in Britain are expressed in grammes per litre of acidity as tartaric. To convert, multiply tartaric by 0.66 or multiply sulphuric by 1.52.

• **VIVC:** This reference is to the *Vitis* International Variety Catalogue, www.eu-vitis.de

Acolon

Type: *Vinifera*
Colour: Red
Origin: Limberger x Dornfelder

Acolon is a German *vinifera*-cross bred in 1971 at Weinsberg by Bernd Hill and its parents – Blauer Limberger (known also as Blaufränkisch in Austria) and Dornfelder – are both staples of German red wine production. Dornfelder itself is Helfensteiner (Blauer Frühburgunder x Black Hamburg) crossed with Heroldrebe (Portugieser x Limberger), so its antecedents are impeccable. In Germany, plantings of Acolon have grown from its release in 2000 to a total of 461-ha by 2018, down slightly from its 2005 figure. In Germany it is planted because of its very dark red wine which adds colour to Dornfelder wines of which there are plenty – there are 7,498-ha of Dornfelder in Germany (2019) – and it has higher sugars, although a slightly lower yield than Dornfelder.

In Britain, it has been planted for around a decade, New Hall Vineyards being the first to plant, and even now it is only found in around twelve vineyards, including several in the more northerly regions. The area planted has risen slowly from 4.6-ha in 2009 to 7.00-ha in 2020, but that is hardly a vote of confidence and it is no longer being considered as a creditable option for red wine production. As the climate has improved, Pinot noir, with its heritage, class and consumer acceptability, will continue to be the red variety of choice.

Albariño

Type: *Vinifera*
Colour: White
Origin: Original variety

Albariño – or Alvarinho if you are Portuguese – is a staple variety of the north-west of the Iberian Peninsula and found both as a single varietal and in blends throughout this area. It is one of the major varieties found in Vinho Verde. It is a variety that has been 'discovered' in recent decades and is now grown in many different parts of the world. At its best it is Riesling-like, with crisp acidity and good length, an ideal wine to accompany food, especially fish and shellfish.

In Britain it is being grown on only around 2.00-ha in 2020, but it is a variety that could, with a bit of encouragement from the climate, become more widely grown. Sandhurst Vineyards in Kent have been growing 0.40-ha (1-acre) since 2012 and have achieved some success with it. They made a Bacchus-Albariño blend which drank very well and a wine made from 100 per cent Albariño in 2014 surprised many Spanish wine judges when it was tasted at a blind tasting in Madrid. The judges described it as 'full, rich and artisanal' which is hopefully a compliment in Spain. Chapel Down, who are major buyers of Sandhurst grapes, have signed up to take all their Albariño. Its limitations for growers in Britain are a low yield, and

late-ripening and high-acid grapes, but on a good, warm site and with patience, it might be worth trying. This is definitely a variety to watch out for.

Auxerrois

Synonym: Pinot Auxerrois
Type: *Vinifera*
Colour: White
Origin: Original variety

Auxerrois, an ancient variety, related to Chardonnay and Pinot blanc, is widely grown in Alsace where it is usually blended into *Edelzwicker* (although in Alsace it can also be labelled as Chardonnay or Pinot blanc) or used for sparkling wine. It can also be found in Burgundy and Luxembourg in the Old World and Canada, New Zealand and the USA in the New World. Geisenheim has produced some more productive clones, although there are less than 200-ha in Germany. It was included on the original list of 'Recommended' varieties submitted to the EU when Britain joined the Common Market in 1973, even though it was not then widely grown and had not been suggested by the EVA.

Auxerrois is a variety of moderate vigour, fairly resistant to *botyrtis* and ripens its wood well. It is quite a late variety, ripening in Britain after MT but before Seyval blanc. As a neutral Pinot blanc/Chardonnay style variety which would be useful for barrel ageing or as a sparkling wine base, it should be more widely planted in Britain – perhaps on the more challenging sites – as it has lower acid levels than other similar varieties. It is a steady yielding variety and it is a pity that it appears to have fallen out of favour in Britain, but I guess that if you can ripen Chardonnay, why would you want to grow Auxerrois.? There were 9.00-ha in 2009, but only 7.00-ha in 2020, so it is not getting any more popular. However, I believe that it is 'one to watch' and well worth a trial.

Bacchus

Type: *Vinifera*
Colour: White
Origin: (Silvaner x Riesling) x Müller-Thurgau

Bacchus, a crossing made by Peter Morio and Professor Husfeld at Geilweilerhof in 1933, was first registered in 1972 and was known in the Rheinpfalz as the 'Early Scheurebe'. Its parentage is the same as Optima. Bacchus is the second most popular white *neuzüchtung* (new-crossing) in Germany (after Kerner) and in 2019 there were 1,649-ha, down considerably from the 1995 area of 3,449-ha. With global warming, German growers are moving away from their home-grown fruity, sweet wine varieties (Bacchus in Germany is almost always made in a *Spätlese* or even an *Auslese* style) towards more international, food-friendly varieties such as Chardonnay and the Pinots.

Bacchus first appeared in Britain in the Wye College vineyard in 1973 as an experimental variety and was upgraded to 'Recommended' in 1998. Bacchus, as a variety for Britain, is here to stay and the area under cultivation has risen from 76-ha in 1990 to an estimated 295-ha in 2020. Bacchus is the best white variety being grown in Britain today for the production of still wines and it regularly wins many of the major prizes. In growth habit it is similar to MT, although perhaps not quite as vigorous, but just as prone to botrytis. It appears to ripen readily in most vineyards, although in cooler years, acids can be high and care needs to be taken that they are not too high in the bottle, especially with wine made from grapes harvested at lower sugar levels. A little residual sugar often helps.

In the bottle, Bacchus falls into two camps: what one might term the *Sauvignon de Touraine* or pretend *Sancerre* camp, where the wine is light, fruity, with a modest spiciness, but of no great weight; and the full-on Marlborough Sauvignon style with luscious, spicy, even catty, fruit with enough body and weight to carry the flavour. These tend to be the riper (higher natural sugar) examples and are best when they are bottled with perhaps a few grammes of residual sweetness (as are many Marlborough wines). Chapel Down's Bacchus Reserve is often a fine example of this latter (and to me, more preferable) style. In the 2017 Decanter World Wine Awards (DWWA), the 2015 Winbirri Bacchus won a Platinum Award and 'Best Single Varietal Wine under £15' which got huge press coverage and introduced the variety to a whole new audience.

With top Bacchus wines selling for £15–£20 a bottle, this is a variety that can make money. The one *really* good thing about Bacchus is the name – free from the umlauts and unfortunate prejudices associated with Germanic names – just a name everyone recognises and thinks has something to do with wine. Who was that Bacchus fellow anyway, some sort of God?

Cabernet Cortis
Type: PIWI – disease-resistant complex hybrid
Colour: Red
Origin: Cabernet Sauvignon x Solaris

Cabernet Cortis is one of a large number of crossings carried out by Norbert Becker (known in vine-breeding circles as 'the other Becker' in deference to Geisenheim's better-known Professor Helmut Becker) at Freiburg in the 1980s. Solaris is another complex crossing containing (amongst other varieties) a Geisenheim *Vitis amurensis* cross. Cabernet Carol is from the same breeding programme and is a full twin. Cabernet Cortis has been evaluated by the Swiss breeding station at Pully on Lake Geneva and they say that 'it is very resistant to Downy Mildew and botrytis but less resistant [Robinson et al. in *Wine Grapes* says 'highly susceptible'] to Powdery Mildew'. However, OsCAR states that the variety is 'very resistant to downy mildew and powdery mildew. Resistance to botrytis noted in northern region', so there appears to be some disagreement about its qualities. In terms of timing,

it is said to be similar to Pinot noir in terms of bud-burst, but ripens a week earlier. Wine quality is said to be 'rich in colour and phenols', but that, of course, is in climates warmer than Britain's. Denmark is said to have 3.00-ha, Switzerland 2.00-ha and it is also planted in Germany and Italy.

In Britain, it is only grown on around twelve vineyards and occupies a total of 2.50-ha (2020), and the first plantings were in 2011. Very little single varietal red wine has been made from it, although there are some rosés. In terms of its future worth, for most growers in Britain I suspect that it is little better than some of the other PIWI reds already planted, but time will tell.

Chardonnay
Type: *Vinifera*
Colour: White
Origin: Original variety

With its origins lost in the mists of viticultural time, Chardonnay is now found in virtually every grapegrowing region in the world, from the hottest parts of Australia, South Africa and California to the coolest of all growing regions, Britain. With it being the dominant (often the only) variety in white wines from Burgundy and, of course, in most Champagnes, it is not surprising that early vineyard owners in Britain were seduced into thinking that it would do well here. Brock had it in his collection at Oxted, but could never get it to ripen properly. Salisbury-Jones planted it at Hambledon in the late 1950s and had the same problems: excessively high acid levels and low natural sugars. Ian and Andrew Paget at Chilsdown planted Chardonnay and I seem to recall one year (1981?) when the acidity (in grammes per litre as tartaric) was higher than the degrees Oechsle! Ouch. Extreme unripeness was a common finding among those early growers who persevered with it, although most decided to give up and removed the offending variety. Only in really hot years would it produce anything like ripe grapes and tolerable wine.

In the late 1980s, some growers – New Hall, Surrenden and Nyetimber – started to plant Chardonnay for the production of bottle-fermented sparkling wines. In my 2001 book *The Wines of Britain and Ireland* I wrote:

through a combination of good site selection and traditional training systems (aided by a degree of global warming) they have started to produce some interesting results. While acids are still high at harvest (15 g/l is not uncommon) the combination of a full malolactic fermentation and the traditional secondary bottle fermentation, renders them manageable.

Well, the 'interesting results' turned out to be more than interesting and since then some extremely good wines, almost all sparkling, and occasionally (2003, 2009, 2014 and 2018 come to mind) some respectable still wines, have been produced. High acids are a problem and growers, especially those with less than perfect sites, need to make sure they select the right clones and rootstocks, plus prune

Yields 2016-19	2016 T-ha	2017 T-ha	2018 T-ha	2019 T-ha	Av. 2016-19 T-ha	Av. 2016-19 T-acre
Chardonnay	5.58	4.42	8.67	6.61	6.32	2.56
Meunier	5.88	5.24	9.53	6.23	6.72	2.72
Pinot noir	3.00	4.31	7.93	6.18	5.36	2.17
Average of above varieties	**4.82**	**4.66**	**8.71**	**6.34**	**6.13**	**2.48**

Table 24

and canopy manage, to get acids down. Chardonnay plantings have gone from 7.0-ha in 1988, via 33.5-ha in 2002, 119.6-ha in 2007, and 327.0-ha in 2013, to a massive 975-ha (2,409-acres) in 2020 making it Britain's most widely planted variety and accounting for almost 28 per cent of the planted area. Given the way the British climate seems to be heading and the quality of the *blanc de blancs* sparkling wines and more than just the occasional still wines being produced, I can only see Chardonnay going from strength to strength.

As a variety for Britain, Chardonnay appears to be more and more at home. Viticulturally it is no more demanding to grow than most varieties. It buds up quite early and in frost-prone sites this can be a problem. It is susceptible to Powdery and (especially) Downy Mildew and *Botyrtis* and vines almost always need spraying in their first year, i.e. the year of planting. Until it can be demonstrated otherwise, simple *Guyot* single- or two-cane pruning appears to work well. While we might like to have them, we do not get the very high yields common in Champagne (15–20 tonnes-ha is not uncommon), so I don't think cordon pruning is required. It benefits from an open canopy and deleafing both after flowering and pre-*véraison*. It will also get late-season Downy Mildew, which will affect the ripening process, so protection needs to be kept going until the bitter end.

Chardonnay ripens late and is usually not harvested until the third or even fourth week of October. In a very late year – such as 2013 – it was still being picked on 20 November, which must be the latest British harvesting date ever. Of course, 'late' is subjective and having rarely enjoyed sharing Guy Fawkes night (5 November) with my children in the late 1970s and 1980s, as I was always pressing Seyval blanc, today's Chardonnay growers shouldn't complain too much. Potential alcohol levels are usually in the 8–10 per cent region – ideal for sparkling wine – although in cool years and late sites 7–8 per cent is quite common. In really warm years – 2018 being a case in point – natural alcohol levels can reach 12–13 per cent, figures that, forty years ago, Chablis struggled to achieve. Acid levels are typically 12–15 g/l, although in cool years and late sites, 16–17 g/l is quite common and even higher levels are not unheard of. In 2013, some growers harvested Chardonnay at very low sugars and high acids and people wondered what sort of wines these grapes will make. However, with some de-acidification, malolactic fermentation, some blending with lower-acid varieties, extended *sur latte* ageing and an appropriate *dosage*, the wines turned out fine. It just shows what time and patience will achieve. However, I should not want to harvest grapes at this level of acidity every year. In terms of yields, Table 24 shows yields of all vineyards surveyed in Britain, so in well-sited, well-established and well-managed vineyards, yields should be at least 30 per cent higher.

For sparkling wines made in Britain, Chardonnay is an indispensable component and no serious grower can be without it. Ridgeview produce some great Chardonnay wines and, for me, their 100 per cent Chardonnay *Grosvenor Blanc de Blancs* is very often their best wine. Their 2000 and 2001 magnums of this wine, I believe, are absolutely magnificent and I have several still *sur latte* awaiting a suitable occasion to drink them. Ridgeview's 2006 *Grosvenor Blanc de Blancs* will go down, I am sure, as one of their best wines and its winning of the Decanter World Wine Awards 'International Sparkling Wine Trophy' (beating four very prestigious Champagnes in the process) was a real affirmation of the quality of sparkling wines made in Britain. There are today many other British growers making good to very good *blanc de blancs* sparkling wines and this can only enhance its reputation as a must-have variety.

For the production of still wines, Chardonnay has shown that with a suitable site, correct canopy management and, most importantly, the right level of crop, very good wines can be produced in Britain. Chapel Down, Blackbook Winery, Gusbourne, Simpsons and several other producers have produced some very good still wines from Chardonnay, especially from the 2018 harvest which was, of course, a very ripe year. If growers can consistently make still wines that sell at the same sort of prices as middling Chablis and white Burgundy, then we will see even more Chardonnay planted in years and decades to come.

The combination of a classic name, its Champagne heritage and many excellent sparkling (and a few very good still) wines, have made Chardonnay indispensable for any vineyard in Britain with pretensions to seriousness. I am not convinced that it suits all the sites it has been planted on in the last two decades, and some growers may yet struggle to ripen it and/or find that their wines need years (5-plus) in the bottle to come round – not a recipe that I think will guarantee financial success.

Note: See the separate section towards the end of this chapter on: *Clones of Chardonnay, Pinot noir and Meunier*

Wine Growing in Great Britain

Chasselas

Type: *Vinifera*
Colour: White
Synonym: Gutedel (Germany), Fendant (Switzerland), and many other names
Origin: Original variety

Chasselas (and known under many, many different names) is one of the oldest varieties grown around the world and has a multitude of variants including rosé and Muscat versions. As for clones, there are many, and, in France alone, where it is planted on around 958-ha, there are 36 clones for wine production listed on the Pl@ntGrape website. It is also used for table grapes and has been one of the breeding partners for many other varieties. In Germany it is an early ripening variety, mainly planted in the south of the country and in 2019 was planted on 1,115-ha. It produces fairly uninteresting wine wherever it is planted and arouses little enthusiasm amongst wine drinkers. In Switzerland, where it is known as Fendant, it is one of their staple varieties and much of it disappears down the throats of thirsty skiers.

Chasselas was introduced into Britain by Karl-Heinz Johner who ran Lamberhurst and who owned vineyards in Baden, so knew the variety (called Gutedal in Germany) and thought it might be a better variety than MT, as it ripened earlier and usually had heavier crops. In 1990, there were 2.50-ha of the variety in Britain, but slowly over the years it disappeared. In recent years has made a very small comeback, with Bluebell Vineyards planting 1.673-ha in 2014, but, to date, they haven't produced a separate wine from it. Personally, I do not believe it's a variety worth planting from either an economic or wine quality point of view, but I could be wrong.

Dornfelder

Type: *Vinifera*
Colour: Red
Origin: Helfensteiner x Heroldrebe

Dornfelder is one of older 'new' German crossings, bred at Weinsberg in the heart of Germany's red wine producing region, Württemberg. Here, the traditional varieties – Trollinger (Black Hamburg) and Limberger – suffer from a lack of colour and substance and Dornfelder was bred to produce more of both. The crossing was made in 1955 by August Herold and released to growers in 1980. Dornfelder's parents are two other Weinsberg varieties: Helfensteiner (which is Blauer Frühburgunder x Black Hamburg) and Heroldrebe (which is Portugieser x Limberger). Dornfelder is the most widely grown 'new variety' in Germany, and its 7,498-ha (in 2109), although down from what was probably its peak, 8,197-ha in 2012, still occupy 22 per cent of their red vine area and 7.4 per cent of their total vine area. It is capable of producing some good wines, albeit in a spicy Rhône style, rather than a classic Bordeaux style, although some German examples

can be fairly light and insubstantial, probably due in part to high yields: 120–150 hl-ha is not unknown.

Dornfelder first appeared in Britain in the late 1980s and is able to produce tolerable red wines, especially in blends with other varieties. It is no more difficult to grow than other *vinifera* varieties and in good years can produce quite heavy crops. Very few 100 per cent Dornfelder wines have been produced in Britain and it is mainly blended. Plantings expanded from 5.40-ha in 1990 to 18.00-ha in 2009 and have stayed at that level ever since. I suspect that it will suffer a slow lingering death. For true reds, Pinot noir, Blauer Frühburgunder, Regent and Rondo are all preferred and there are a number of new PIWI varieties which are probably better choices. My advice: no need to rip it out, but think twice before planting any more.

Dunkelfelder

Type: *Vinifera*
Colour: Red
Origin: Crossing of unknown parents

Dunkelfelder is a crossing made by Gustav Adolf Frölich in Edenkoben in the Rheinpfalz in the early 1900s and its exact parentage is unknown. It was discovered in the Geisenheim vine collection in the 1930s and clonally selected in the 1970s. It was officially classified for wine production in Germany in 1980 and in 2008 there were 352-ha. It is now no longer listed as a separate variety and it is primarily used as *deckwein* (colouring wine).

Dunkelfelder first appeared in Britain in mid-1980s (I believe I was the first grower to plant it, at Tenterden Vineyards) and in 1990 there were 3.5-ha. Today it is down to 1.50-ha which probably sums up its attractiveness to British growers: small and diminishing. Viticulturally it is undemanding with fairly low vigour, but does not usually run to large crops. It needs regular spraying and *Botyrtis* can be a problem. It ripens very early and is susceptible to bird and wasp attack. However, it is the best *teinturier* variety grown in Britain and can produce massive colour. On its own, the wine is fairly neutral with low acidity and is best blended with other red varieties. As part of a red blend it is fine, but it is never likely to be anything else. Perhaps a variety that red and rosé winemakers should have a few rows of to liven up the colour of their wines.

Frühburgunder, Blauer

Synonym: Pinot noir Précoce
Type: *Vinifera*
Colour: Red
Origin: Original variety

Blauer Frühburgunder is a relatively old variety, planted in the 1800s in the Ahr wine region, Germany's most northerly winegrowing area. The region is named after a tributary of the Rhine which joins the main river just south of Bonn. By the 1900s the variety had migrated to the town of Ingelheim am Rhein, *'Die Rotweinstadt'* (the red wine town) and it became that town's dominant variety for which they became well known.[10] Geisenheim has done some work on it, cleaned it up (the old clones had Leaf Roll and cropped very poorly) and today there are five clones listed that are worth growing (in Germany). In Germany in 2019 there were 237-ha, slightly down from 262-ha in 2012.

When growers started planting Pinot noir in earnest in Britain, the search went out for the earliest ripening clones in the belief that Frühburgunder, as it is generally known, was a late ripening variety (it's not, but that's another matter) and they were offered what many thought was an early German clone of Pinot noir by some British-based vine suppliers, but was, in fact, what the Germans call Blauer Frühburgunder. Vines were imported and planted and the wine was sometimes called by its correct name, (Blauer) Frühburgunder, but was more usually called Spätburgunder or Pinot noir, or occasionally Early Pinot noir. Everything was fine and dandy until some eagle-eyed WS inspector noticed Blauer Frühburgunder on a bottle label and realised that this variety was not on the list of permitted varieties for Britain and could therefore only be sold as UK Table Wine. After some discussion with the industry, the WS agreed that the synonym 'Pinot noir Précoce' would be a permitted name (as it is already called this in other EU member states and a name that is so much better than Blauer Frühburgunder) and growers with this variety were asked to confess their misdeeds and re-register their vineyard parcels containing this 'clone'. This is why it suddenly appeared from nowhere on the variety lists.

Frühburgunder ripens up to two weeks earlier than Pinot noir and on good sites achieves better sugars and better colour. However, it is just as Botytris-prone as straightforward Pinot noir and offers no other viticultural advantages. In the first edition of this book, I was somewhat negative about the variety, suggesting that it was only fit for growers whose sites were not good enough for 'proper' Pinot noir. In my defence I would submit the fact that in 2012–13 there were not a lot of single variety wines made from it, and it tended to get blended with other varieties and/or made into rosés. Since then, no doubt helped by climate change and some good vintages, my opinion of it has changed and I believe it has a future in Britain. Growers also seem to share this view, and the area under vine has gone from 19.94-ha in 2011, to an estimated 70.00-ha in 2020. If you are wanting to make good, still red and rosé wines, and have a good site, then the variety is worth trying. Where I believe it should not be planted is on poor sites and in areas where Pinot noir will not ripen well.

Huxelrebe
Type: *Vinifera*
Colour: White
Origin: Chasselas x Courtillier Musqué

Huxelrebe is a crossing made by Georg Scheu at Alzey in 1927 and is named after Fritz Huxel, a grower from near Worms who first recognised its potential. Courtillier Musqué (also known as Muscat Précoce de Saumur) is a 'selfling', i.e. a self-pollinated seedling of Frühburgunder. In Germany, where the area planted has been falling steadily since its heyday in the 1990s and stood at 424-ha in 2018, it is capable of producing very large yields of grapes (or as the Germans say *trägt wie ein Esel* – 'carries like a donkey'), with high natural sugars.

Huxel – as it is usually called – was first introduced to Britain via the Wye College vineyard in 1972 and on account of its higher yielding ability was quite widely planted. In 1990 there were 43.9-ha, but since then the area has fallen and is today down to around 15.00-ha, although this has barely changed since 2009 and I doubt if anyone is planting it today.

Huxel can be extremely vigorous, with great fat canes and very large leaves and is probably best grown on a spur-pruned system called Geneva Double Curtain (GDC) where vigour can, to a certain extent, be tamed. Bunches are very large and contain an unusually high number of seedless grapes which ripen readily and achieve very high natural sugars, although loved by wasps, unfortunately. It undoubtedly benefits from bunch thinning and pre-harvest deleafing as it suffers badly from *Botyrtis*. When really ripe, the wine can be very fruity with a pronounced Sauvignon blanc character, although acid levels can be high and, to obtain the best wines, the grapes must be allowed to fully ripen. Unripe examples can be rather too herbaceous and catty and can be detected even when disguised in a blend. It has been used for a few late-harvest sweet wines, but only when *Botyrtis* has got the better of it. In 2001 I wrote:

a variety that is probably worth persevering with on account of its good quality wine & higher than average yields. With a new generation of anti-Botyrtis fungicides now available, will Huxelrebe become more viable to grow?

Well, almost twenty years on, I can answer the question: 'no'. It is still a pushy monster in the vineyard, over-cropping when it feels like it and suffering from wasps and rot in equal measure. Its diminishing area sums up its attractiveness to British growers, although organic grower Will Davenport likes it and uses it for his 'Hux' wine.[11]

---

11 The best British-grown Huxelrebe I ever tasted was the late Bill Ash's 1982 Staple St James Huxelrebe, one of the very few English wines I have ever bought a case of. The interesting story about this wine is that Bill sold half the grapes to Lamberhurst (he was always strapped for cash) and they made a wine with it, sweetened with some high-strength Auslese Gewürztraminer süss-reserve that Karl-Heinz Johner had prized out of the local co-op winery where he lived in Baden. This 1982 Staple St James Huxelrebe won the 1983 Gore-Browne Trophy, largely on account of the great süss used, but fell apart after a while and the Staple wine with, as I recall 12 per cent natural alcohol, aged into a really great wine. The süss trick was one I remembered, and when

---

10 Ingelheim, situated on the southern side of the Rhine, about halfway between Mainz and Bingen, is where Charlemagne built a palace (in about 800). According to legend, he noticed that the snows melted first on the south-facing slopes opposite and could see that the area had a unique microclimate. His son, Ludwig the Pious, started growing grapes there and the slopes eventually became the site for what is today Schloss Johannisberg.

Madeleine x Angevine 7672
Synonym: Alzey 7672
Type: *Vinifera*
Colour: White
Origin: Madeleine Angevine x White Riesling

In 1957 Brock at Oxted was sent cuttings of several varieties by Dr Zimmerman from Alzey, and one of them was labelled 'Sämling [seedling] 7672'. Just when this crossing was made was then not known, although the timing suggested that it was while Georg Scheu – responsible for varieties such as Huxelrebe, Faberrebe, Kanzler, Regner, Scheurebe, Septimer, Siegerrebe and Würzer – was the Alzey Institute's director. By 1960 Brock was able to report that it was 'giving large crops which ripen with Riesling Sylvaner [MT]. Considered to be a promising variety.' Not having been given the crossing details of Sämling 7672, Brock wrote to Zimmerman and asked for them. He was informed that it was a 'freely pollinated seedling of Madeleine Angevine'. When Brock started to sell cuttings of the variety (as he did with all promising varieties) he gave it the name Madeleine Angevine 7672 and it was under this name that it was known for many years. As it became quite popular, the name on wine labels became shortened to simply 'Madeleine Angevine'. This was an unfortunate name, as another variety already existed under this name.

The true Madeleine Angevine (also sometimes called Madeleine d'Angevine), is a female only table grape variety, for many decades said to be a crossing of Précoce de Malingre and Madeleine Royale,[12] made by Pierre Vibert at the Moreau-Robert nurseries in Angers in 1857 (some reports say 1859), but described in *Wine Grapes* to be a crossing of a Chasselas-based table grape variety called Circé and Madeleine Royale. It is one of the earliest table grape varieties for open cultivation in France and named after St. Madeleine's Day (22 July), said to be the date on which it can be first harvested. Having only female flowers and being very early, it has been used by plant breeders in a number of crosses over the years. Morio and Husfeld used it for Forta and Noblessa and it is one grandparent of Reichensteiner. Scheu himself used it to produce Siegerrebe and, interestingly, this variety, which at one time was credited with being a Madeleine Angevine x Gewürztraminer crossing, was unmasked by Heinz Scheu, Georg's son, as also being a freely pollinated Madeleine Angevine seedling. It is probable that the variety we now

grow in Britain as Madeleine Angevine 7672 comes from the same crossing programme that produced Siegerrebe. In 1992, I asked Professor Alleweldt (or Professor Dr. Dr. h.c. Diplomlandwirt Gerhardt Erich Alleweldt to give him his full title), then Ddirector of the Geilweilerhof State Institute for Grapevine Breeding, to see if he could discover more about our Mad Angie (as it is often known). He located Georg Scheu's old breeding books for the 1944–45 season in the Alzey archives and found that 'Sämling 7672' was – as Brock was originally told – an 'open pollinated progeny of Madeleine Angevine'. In 2019 I was asked by Geilweilerhof to send leaf samples of 'our' Madeleine Angevine for DNA profiling. Three samples were sent from different vineyards in Britain and, a few weeks later, Erica Maul from Geilweilerhof emailed to say that two of the samples[13] were actually a crossing of Madeleine Angevine x Riesling and that henceforth it should be known as Alzey 7672 and this is how it would be listed on the VIVC. Thus, after 75 years was the mystery of Mad Angie's birth parents was cleared up. (There is also another variety to cause further confusion called Madeleine Angevine Oberlin, a crossing between Madeleine Angevine and Bouquettraube.)

Madeleine x Angevine 7672 (as I will continue to call it) found favour with many of the early British growers and it became planted fairly widely. At one time it was the third most popular variety (after MT and Seyval blanc) and Gillian Pearkes, one of the pioneers of British grapegrowing, who was an influential character in the West Country, thought highly of it. Unfortunately, owing to the confusion over the name and the fact that this variety was not available from any other source other than Brock's original stock or vineyards planted with vines obtained from Brock, some growers were sold vines of the true table grape variety Madeleine Angevine, which was barely suitable for Britain except in the very best years. Unless it was being grown in proximity to other early varieties for pollination, it seldom set a good crop and, in some years, ripened at the end of August. The wines from this variety were flabby and never really acceptable except for blending. The confusion over this variety – foreseen in a rare moment of sanity (as far as rules and regulations governing British viticulture is concerned) by the European Commission in 1973 when MAFF submitted vine varieties for classification – has meant that it gained something of a chequered reputation.

MA is easy to grow, ripens early, gives good crops and the wines have a light Muscat tone and low acidity. In 1988 I took wood from genuine Madeleine x Angevine 7672 vines growing in Robin Don's Elmham Park vineyard in Norfolk, now grubbed up, which Robin had planted with cuttings supplied by Pearkes (who had got hers directly from Brock). Buds from this wood were then sent to France and grafted on to resistant rootstock. The resultant grafted vines were sold to various growers around the country. Sharpham Vineyard in Devon is probably the best known and their wine made using MA (and Pinot gris), the *Sharpham Estate*

I was winemaker at Lamberhurst I used the same Gewürztraminer süss in the 1988 Schönburger which won the Gore-Browne in 1990! They say all's fair in love, war and winemaking. Pilton Manor also produced a very good late harvest Huxel in 1992. This was a complete accident and the crop had been written off, but then owner Jim Dowling was persuaded by his winemaker, John Worontschak, to pick the crop despite it being completely affected by botrytis. It was an excellent wine and I recently read a report of a tasting at which it was still showing very well.

12 Madeleine Royale, a table grape vine variety once thought to be a Chasselas seedling, but now shown to be a Pinot x Black Hamburg cross, has been 'outed' by the gene-jockeys as the father of Müller-Thurgau, with Riesling being the mother.

13 The third sample turned out to be Auxerrois!

Selection, is a regular prize winner. However, the variety reached its peak area at the end of the 1990s with 68.70-ha, but since then it has slowly declined and today (2020) stands at about 50.00-ha.

MA is still being planted by growers in the further reaches of Britain's vinegrowing regions as its low acidity means that it will ripen at lower heat levels than many of our mainstream varieties. For the longer term, however, I suspect that Mad Angie will not survive the Pinot-Chardonnay storm-troopers lining up against her and as her growers get older and give up, sell up or go to that vineyard in the sky, the area will dwindle. Pity, as the variety can make good wine. For growers with more challenging sites, it is probably a variety worth considering, but they are advised to make sure they know where the bud-wood comes from before buying any vines.

Meunier
Synonym: Pinot Meunier, Schwarzriesling, Müllerrebe – and, in Britain only – Wrotham Pinot
Type: *Vinifera*
Colour: Red
Origin: Original variety

Meunier, one of the numerous of the mutations that go to make up the Pinot family, accounts for around one-third of the plantings in Champagne (the actual percentage appears to be a closely guarded secret) and while it is seldom spoken about in the same breath as its nobler companions – Chardonnay and Pinot noir – and is not allowed in *Grand Cru* wines, it is used in some of the finest *cuvées* (Krug for example). In Champagne it is favoured for its higher acidity than Pinot noir, and its ability to withstand spring frosts and higher yields. It is found in the cooler sites and on north-facing slopes. It is characterised by its white tipped shoots and its hairy white leaves, especially on the underside, which give it the 'dusty' look from which the name – which means 'miller' in French – stems. Like many of the Pinots, it can show reversions back to the true Pinot noir and it is quite possible to have vines with both hairy and non-hairy leaves on different shoots in one year, which disappear in another year.

In Britain, Meunier has had a somewhat chequered career. In the early 1950s, Edward Hyams, one of the early pioneers of British grape growing, discovered a vine growing on a cottage wall at Wrotham in Kent which he named Wrotham Pinot and which he gave to Ray Brock at Oxted. Given that this foundling had the classic dusty and red-edged leaves of Meunier, Brock assumed, wrongly or rightly, that it probably was a Meunier, although he reported that it had 'a higher natural sugar content and ripened two weeks earlier' than supplies of Meunier obtained from overseas. Brock sold cuttings and it became quite popular with some of the early vineyards. It is doubtful now whether any cuttings from Hyams or Oxted still survive in vineyards in Britain, and all plantings of Meunier stem from France or Germany (where it is known as Schwarzriesling or Müllerrebe). However, the name Wrotham Pinot is still a permitted synonym for Meunier in Britain.[14] In 1999 there were only 5.5-ha being grown in Britain, most of it fairly old and unloved, but since 2006, when the area stood at 22.5-ha, plantings have shot up and today (2020) it stands at around 275.00-ha. It is Britain's fourth most widely planted variety and will surely, in the not too distant future, claim third place behind Pinot noir and Chardonnay. There is no doubt now that it is seen as an indispensable part of many *classic cuvée* sparkling wines and often accounts for 15–20 per cent of Champagne-variety plantings.

Although it has been grown for over fifty years in British vineyards, Meunier has never really shone as a variety capable of making interesting wines on its own and most of it has been blended with other varieties. However, the rise in the planting of Champagne varieties has meant a resurgence of interest in the variety and my experiences to date, despite my former misgivings about its suitability for our climate, have been favourable. The quality of the Nyetimber *2003 Blanc de Noirs Pinot Meunier,* which won a Gold Medal in both the 2006 and 2008 English Wine of the Year Competition (EWOTYC) was outstanding and for me (and for Tom Stevenson), was the best wine in the *Decanter Magazine* March 2008 mega-tasting of sparkling wines from Britain. When the sparkling wines from the ripe 2018 vintage start to be released, we could be in for some pleasant sparkling Meunier surprises. Table 24 shows that in terms of yield, Meunier can more than hold its own with its Champagne cousins, although the data is skewed owing to the 2018 yields. This is because the Meunier bunches were very, very large – sone growers reported them around the 250-300 gramme range – well over double their normal range.

In most years, Meunier has not been seen as suitable for still red wines, and, given that all Meunier growers will also probably be growing Pinot noir, why would they want to make a red wine from it? Having said that, given its slightly lower cropping level (than Pinot noir) it can get very ripe and colours are no worse than Pinot noir and in recent (riper) years it has been used successfully in single variety still wines. Still red wines from Meunier are extremely rare anywhere in the world, and if Britain can show that Meunier is the equal of its more respected cousin, Pinot noir, then it might become our USP.

Note: See separate section at the end of this chapter on: *Clones of Chardonnay, Pinot noir and Meunier*

---

14 See Appendix VI for the full story of Wrotham Pinot.

Wine Growing in Great Britain

Müller-Thurgau
Synonym: Rivaner, Riesling Sylvaner
Type: *Vinifera*
Colour: White
Origin: Riesling x Madeleine Royale[15]

Professor Dr Hermann Müller, a Swiss national from the canton of Thurgau, near Zurich, produced this crossing while working at Geisenheim in 1882. Returning to Switzerland in 1891 to become dDirector of the Wädenswil Research Institute, Professor Müller asked for 150 of his best crossings to be sent to him, including No.58, which was eventually to become MT. Owing to some confusion with the labelling at the time the crossings were delivered, the true parentage of No.58 was never discovered and it eventually became known as Riesling Sylvaner on account of its wine style, said to resemble a blend of Riesling and Silvaner. Professor Becker, head of Geisenheim in the 1970s and 1980s, attempted to recreate the variety by making multiple Riesling and Silvaner crossings, but failed. He was, however, of the opinion that it more resembled a Riesling x Riesling crossing than any other.

In 1996 an Austrian, Dr Regner, from the viticulture school in Klosterneuburg, had proved (so he thought) that the crossing was between Riesling and a member of the Chasselas family, a table grape vine variety that (so he thought) was called Admirable de Courtiller. It was then discovered that the reference vine of this variety in Klosterneuburg's collection was in fact another variety altogether – it turned out to be Madeleine Royale (a Chasselas seedling variety). In 2001, two researchers in Germany, Erika Dettweiler and Andreas Jung, were able to unravel the DNA in MT and prove that it was indeed a crossing between Riesling and Madeleine Royale: after 119 years MT's parents were finally found.

In 1912, wet-sugaring (the use of sugar in solution to chaptalise wines) was forbidden in Switzerland and MT started to replace Elbling, up until then a widely grown variety, but high in acid. MT then began to find favour in Germany and, in the early 1920s, was taken up by Georg Scheu, then at Alzey in the Rheinland-Pfalz. He subjected it to clonal selection and helped improve its yield. Following the Second World War, when many vineyards were suffering from *Phylloxera* as well as the ravages of the war, MT was widely planted and its large yields, early ripening and soft wines, lower in acidity than either Riesling or Silvaner, were much appreciated. It was once Germany's most widely planted vine variety and in 1999 occupied an area of 20,672-ha which was 20 per cent of the total German vineyard area. By 2019 it was down to 11,736-ha occupying 11.4 per cent of the total German vineyard area.

In Luxembourg – where it is called Rivaner – it is still a popular variety, planted on 295-ha in 2018 which is 23 per cent of their total vineyard area (and cropping at 134

hl-ha or around 8.0 tonnes-acre). Until 1987, it was New Zealand's most widely planted variety and accounted for 42 per cent of their vineyard area although plantings have now almost disappeared. It is popular in high-altitude vineyards in Italy's German-speaking Alto Adige (Südtirol) region, where it can make some very fine wines. It is also to be found in many Eastern European countries, such as Slovakia and the Czech Republic. EU regulations forbid the use of the name Riesling Sylvaner, although this name can still be found in non-EU countries such as Switzerland.

Its introduction into Britain stems from Ray Brock's visit to Switzerland in 1946 when he met Mr Leyvraz at the Swiss Federal Vine Testing Station at Caudoz-sur-Pully. In 1947 vines of various varieties were sent by Leyvraz to Brock, including 'Riesling Sylvaner'. Brock gave cuttings to Edward Hyams in 1949 and they both trialled it for a number of years. Brock first harvested grapes from MT on 14 October 1950. In Hyams' 1953 book *Vineyards in England* Brock wrote in his chapter on vine varieties that 'Riesling Sylvaner is known to give an outstandingly fine wine in cool climates'.

When Jack Ward was looking for vines for the Horam Manor vineyard, planted in 1954, he was recommended to plant MT, and it became a staple of all the early British vineyards. Ward was a very influential figure in the early days of the revival (he was the EVA's first Chairman) and undoubtedly did much to persuade growers to plant MT. Ward's company – the Merrydown Wine Company – sold vines and gave advice, and ran a winemaking co-operative between 1969 and 1979 to which many of the early growers belonged. MT was top of my list when I selected varieties for planting in 1976-7 and it was almost unthinkable to plant a vineyard in Britain then without it. Most German wine and viticulture experts considered England to be a country of mists and not-very-mellow fruitfulness and thought that a high-cropping, early ripening, low acid variety (which is what MT is in Germany) was just what we needed.

MT is a vigorous variety, especially in its early years, and it will grow thick canes and large leaves, leading to excess shading. This often results in poor cropping, especially in years with low light and heat levels in the previous season. It suffers from *Botyrtis*, Powdery and Downy Mildew and requires regular spraying, and must be deleafed after flowering and then again around *véraison*. In heavy-yielding years, stem-rot can be a problem. The wood often ripens poorly and does not overwinter well, often showing active *Botyrtis* in mild winters. It is probably a variety best avoided by organic growers. Since the advent of better anti-botrytis chemicals such as Scala, Teldor and Switch, disease control has become easier (if more costly) and clean crops of MT are the norm, rather than the exception they were in the 1970s and 1980s.

The grapes of MT can have good fruity flavours with light Muscat hints and, at their best, the wines made from them can be very good. However, when less than fully ripe, they tend towards the herbaceous and catty. The

---

15 Madeleine Royale is Pinot x Schiava Grossa (Black Hamburg) made in 1845 at the Moreau-Robert Nursery in Angers. (VIVC)

acidity is usually average to low and with some balancing residual sugar, the wine can be very attractive and fresh when young and will keep well, although probably best drunk within two to three years after bottling. Picked early with sufficient acidity and when the fruit flavours are more neutral, it can be used for sparkling wine, and Chapel Down always used it in their entry-level *Vintage Brut Reserve*, although over the years the percentage in the blend has come down, to be replaced with pressings from Chardonnay and Pinot noir.

In 1984, MT was by far the largest single variety being grown in Britain, with 149.0-ha, accounting for 35 per cent of the vineyard area. Over the intervening years, however, it has slowly fallen to 61.5-ha in 2009 and is now (2020) down to only 30.0-ha. It would appear that a few growers are hanging on and no doubt, in these warmer times, getting good enough crops to justify its existence. However, no one is planting it any more and, over time, it will continue to dwindle. With hindsight, it was a variety that Britain could probably have done without, although one must not forget that the most popular wine at the time it was being planted was the light, fruity *Liebfraumilch* style that MT is so suitable for. Its extreme vigour, especially in the early years, coupled with its on–off cropping pattern and disease problems, make it a difficult variety for British conditions.

Orion

Type: PIWI – disease-resistant complex hybrid
Colour: White
Origin: Optima x Seyve-Villard 12-375 (Villard blanc)

Orion is one of the many crossings made by Professor Dr Alleweldt at Geilweilerhof. It dates from 1964 and was first registered in 1994. It is a crossing of another Geilweilerhof crossing, Optima, which is ((Silvaner x Riesling) x MT), and Seyve-Villard 12-375 which is known as Villard blanc. Orion is one of the new breed of complex interspecific crosses (now known as PIWI varieties)[16] which have been bred for both wine quality and disease resistance. It is not as resistant to fungal attack as some of the older hybrids (Seyval blanc for instance) and usually requires spraying against Powdery Mildew, but is good against botrytis

and ripens its wood well. It also shows better resistance to winter frost damage than MT (not often a problem in Britain) and its basal buds are said to be very fruitful.

Orion was introduced to Britain in the late 1990s and by 1999 there were 8.4-ha being grown. The area planted expanded very slowly and got to 10.8-ha in 2009 from which point it has not changed much, being still around 10.0-ha in 2020. Wines made from 100 per cent Orion seldom surface, so it is difficult to judge its true worth, but the wine is said to be fruity and aromatic. I don't see the area planted with it expanding much more, although some of the more northerly vineyards seem to like it.

Ortega

Type: *Vinifera*
Colour: White
Origin: Müller-Thurgau x Siegerrebe

Ortega is a crossing made in 1948 by Dr Hans Breider at Würzburg between MT and Siegerrebe. Siegerrebe is a freely pollinated Madeleine Angevine seedling with Gewürztraminer now confirmed as father. Ortega was first registered in 1971 and is named (somewhat curiously) after the Spanish philosopher José Ortega y Gasset, (but good for pub quizzes when 'what's the name of a famous Spanish philosopher' comes up.)

In Germany there are 440-ha (2018) mostly in Rheinhessen and the Pfalz. However, it is a declining variety there (down from 951-ha in 2001), so falling in popularity. In Germany, it ripens early, achieves high natural alcohol levels and low acidity levels, and with its rich, spicy tones, makes a good blending partner for higher acid varieties such as Riesling.

In Britain, Ortega was introduced by Jack Ward, who planted it at Horam Manor in 1971, and it has slowly grown in popularity. It ripens early – usually just after Siegerrebe and Optima – has high sugars, low acids and plenty of flavour. An early bud-burst makes it susceptible to spring frost damage and it is sensitive to difficult flowering conditions and will suffer from *coulure* in poor years. It can be quite vigorous and canopy management needs to be good to get the best fruit. It will also get botrytis towards the end of ripening which will turn to 'noble rot' if sugar levels are high enough. Given good canopy management and timely attention to spraying, this variety can provide high-quality grapes, useful for both normal still wines and dessert wines. Growers report that its bunches tend to get tangled up with each other and with the canes and wires and picking therefore can take twice as long as for other varieties.

When fully ripe, wines made from Ortega are rich and zesty with good balance, although warm years may result in wines with rather low acidity and care needs to be taken to pick at the correct time. Biddenden Vineyards have won numerous gold and silver medals with Ortega wines and won the Gore-Browne Trophy with one in 1987. Surprisingly, it takes to new oak well and Chapel Down

---

16 In the late 1980s, German vine developers, who had bred complex interspecific crosses which we now call PIWI varieties, were being denied the right to produce Quality Wines with them. They asked the German Bundes-sortenamt (the regulatory body that grants plant breeder's rights and classifies new cultivars) whether they could tell, from their growth habits, their leaf and fruit shape and the taste of the wine, whether they were viniferas or non-viniferas. The Bundessortenamt decided that, as the vines resembled viniferas in every respect – apart from their superior natural disease resistance – then they could be classified as pure viniferas and thus be made into Quality Wine. This judgment was accepted by the EU, and German winegrowers were then allowed to make Quality Wines from them. After five years of lobbying by the UKVA and a lot of heel-dragging, MAFF/DEFRA finally agreed that what the Germans had done was legal and the four PIWI varieties then being grown in Britain – Orion, Phoenix, Regent and Rondo – could be made into English and Welsh Quality Wine.

used to make a very good barrel-aged *Condrieu*-like wine from Ortega. Growers of Ortega seem to like the variety and the area has risen from 29.5-ha in 1990 to 55.0-ha in 2020 which must be seen as a vote in its favour. I suspect that it will remain as one of our important, albeit minor, varieties and, as vineyards creep ever northwards, might find favour there.

Phoenix
Type: PIWI – disease-resistant complex hybrid
Colour: White
Origin: Bacchus x Seyve-Villard 12-375 (Villard blanc)

Phoenix (and not Phönix) is another of the many complex hybrid crossings made by Professor Dr Alleweldt at Geilweilerhof and is Bacchus x Seyve-Villard 12-375 (Villard blanc). It was first registered in 1984 and listed for general growing in 1992. In Germany there were 24-ha in 1999 and it has grown slowly to 48-ha in 2018. Although not as resistant to fungal attack as the older hybrids, it ripens its wood well.

Phoenix first appeared in Britain in the late 1980s and has risen very slowly from 1.9-ha in 1999 to around 25.0-ha in 2020, which, all things considered, is relatively respectable. Although only planted on a few sites in Britain, the wine quality can be good (Three Choirs' is usually the best), with higher sugars and lower acids than MT and it is Bacchus-like, although not as powerful. It is one of the several complex hybrids that can be made into Quality Wine in Britain. Whether Phoenix really has a place in British vineyards when Bacchus is now easier to grow, given the warmer summers and better fungicides, is open to question. Its rate of increase over the last decade has been steady, suggesting that it has some positive attributes and I suspect that it is better suited to more challenging sites and. as such, will remain a minor variety.

Pinot blanc
Synonym: Weißer Burgunder, Pinot Bianco
Type: *Vinifera*
Colour: White
Origin: Original variety

Pinot blanc is one of the most widely distributed of varieties across Europe and is one of the vast family of Pinots. It is often confused (one suspects mostly on purpose) with Chardonnay – the style of wine they produce can be similar – and in general terms it is less demanding than Chardonnay, will ripen more easily and has a higher yield. In Germany it is increasing in popularity and there were 5,747-ha in 2019, making it the fourth most widely planted white variety. It is to be found in France, Alsace especially, and in many Italian regions in great quantity and in many other cooler regions.

In Britain it was first grown in 1979, but on only one vineyard and didn't really start to be more widely planted until 2003 when global warming enlightened us to its possibilities. Since then it has grown steadily from 6.2-ha in 2006 to around 35.0-ha in 2020. Like Chardonnay, it requires a good site and careful management to ripen it fully and get the acids down, although it is no more difficult to grow than Chardonnay. Pinot blanc bunches tend to be very large and more compact than Chardonnay, and in order to avoid botrytis it needs timely deleafing and spraying.

Initially it was used for the production of sparkling base wine, but in the last 10-15 years a few 100 per cent varietal still wines have been made. Chapel Down's, made from grapes grown in Hampshire, Essex and Kent, is one of their best still wines and more growers should follow their example. Acids can be at the higher end of the spectrum, but by allowing the grapes to ripen fully, and then putting the wine through lees-ageing, malolactic fermentation, and maybe a couple of years in bottle, some very good still wines can be made. As a useful variety that will find a good home in both sparkling wine blends and high-end still wines it deserves to be planted more. It has a good pedigree, an acceptable name, and is gaining a track record. I expect the area under this variety will continue to expand.

Pinot gris
Synonym: Ruländer, Grauer Burgunder (or Grauburgunder)
Type: *Vinifera*
Colour: White
Origin: Original variety

Pinot gris is another variety from the large Pinot family and it appears to have almost as many clones as there are winegrowing regions that use it. In Alsace, its most respected home (where until a decade or two ago it was known as Tokay d'Alsace), it is capable of producing both good to very good dry and off-dry wines, as well as exceptionally fine late-harvest dessert wines. In Germany there were 7,069-ha in 2019, making it the third most widely planted white vine variety, and the area under vine has almost trebled since 2000 and looks set to overtake MT at some stage. In Germany it is known as Ruländer for wines with residual sugar and Grauer Burgunder or Pinot gris for soft, fruity dry wines. In Italy (as Pinot grigio), where there were almost 25,000-ha of the variety in 2017, it makes a very neutral, quite crisp wine, ideal with food. It seems to be able to change its style to suit the region. There are numerous clones of Pinot gris. VCR, the biggest vine nursery in Italy, has around fifteen different ones, and there are three listed for France.

In Britain, where it has been grown since the late 1970s, it is making slow progress from 6.0-ha in 2007, to today's total of 35.0-ha. It is not the easiest variety to ripen and requires a good site and careful canopy management to make the best of it. Yields are not the highest and need to be kept under control to produce the best grapes. Some reports talk of a prolonged flowering, resulting in picking in two or more passes. There are probably some clones which would suit our conditions more than others, but (again) who is doing any trials? As plantings have increased over the last decade or so, the number of good to very good wines made from it has also increased, helped a bit by warm vintages such as 2014 and 2018. It has also been used in sparkling wines, but, except in a blend, I cannot see it becoming an important variety for this type of wine.

Assuming that the climate continues to improve, and as more single varietal wines become available and achieve success, I can see Pinot gris making steady progress and becoming a minor, but useful, variety for producers in Britain.

Pinot noir
Synonym: Blauer Spätburgunder
Type: *Vinifera*
Colour: Red
Origin: Original variety

Pinot noir is one of the most ancient of varieties and probably has more clones and variants than any other variety. While its home is often thought of as Burgundy, where undoubtedly many of the finest examples can be found, it seems to thrive in both very warm and very cool climates and good examples can be found in Australia, California, Chile, Germany, New Zealand, South Africa, Spain and even Norway. In Germany, the planted area has gone from 7,607-ha in 1995, to 11,717-ha in 2019, making it the most widely planted red vine variety, occupying 34 per cent of the red wine area and 11.4 per cent of the total vine area. It is also, of course, not only used for still red wines, but is also one of the classic sparkling wine grapes, found not only in Champagne, but wherever good sparkling wines are made.

In Britain it was one of the earliest varieties to be grown. Brock trialled it at Oxted, but was not pleased with it and by 1961, when he issued *Report No.3 Progress with vines and wines,* he stated that Pinot noir 'appears to be very much later [than on the continent] in this climate... and has been discontinued'. His lack of success was probably due to the fact that, at the vineyard's elevation of 125–137 metres above sea level, the site was just too cool. In addition, Brock was not, at least not in the early years, looking to make sparkling wines, although he did make some later on. Jack Ward was likewise somewhat dismissive of the variety and also thought it ripened too late.

Despite the reservations of these pioneers, it seems to have been planted quite widely, if not in any great quantity. Many of the early growers seemed to limit their investigation of continental vineyards to a quick visit to Champagne and returned enthused with the idea that Pinot noir, and for that matter Chardonnay, were naturals for our climate. The truth of the matter is that in those early days they were not, and good wines made from them were *very* few in number. Whatever Bernard Theobald claimed in the 1970s and 1980s, Pinot noir was *not* the best variety to be growing then. What I wrote in 2001 (I hope) bears repeating:

The rise in interest of bottle-fermented sparkling wines and the results to date, from those vineyards that have gone into the job seriously, is impressive. There have also been a number of very creditable red wines made from the variety in recent years – only it has to be said in the warmer years – but it is an exciting trend. There are a number of factors which might make Pinot noir a more acceptable variety for Britain. There is no doubt that vine breeders have done much to change Pinot over the last 50 years. Clones are available now that ripen earlier, are more consistent, produce better quality wines and are more disease resistant. The upsurge in interest in making bottle-fermented sparkling wines seems set to continue and Pinot noir must be considered as a major variety in this respect. Today's fungicides, especially for the control of botrytis are markedly better than they were only a decade ago and this undoubtedly allows growers to leave their grapes to hang for longer and to ripen more fully.

Whether or not global warming is affecting Britain now – or will do in years to come – is open to debate. However, most winegrowers sense that weather patterns are changing and we are now experiencing earlier bud-bursts (although this is coupled with a higher incidence of spring frost damage) and therefore longer growing seasons. This does mean that the more marginal varieties – of which group the Pinots and Chardonnay are members – may fare better in the next decade than the last. One other factor to be considered is that of marketing. There is no doubt that the public will buy wines bearing names that they recognise and Pinot noir is one that has its devotees.

The upsurge in plantings since 1984, when 11-ha were recorded as being planted in vineyards in Britain, via 46.6-ha in 2002, to today's total of 925-ha (2020) and occupying 26.40 per cent of the planted area, is impressive and speaks volumes about the ease with which it can be grown, its relatively good crops and its suitability for turning into fine sparkling wines. It is also due to three other factors: what I have called the 'Nyetimber effect', the weather and the name. The first factor showed that good, even great, sparkling wines could be made using the classic varieties, of which Pinot noir was an important component (although the first Nyetimber release was actually a *blanc de blancs*). Since then, all of the major players in Britain's sparkling wine business have shown that this was no fluke. The second factor has meant that the acids are manageable and – if you want some colour – colours are acceptable. It has also meant that it crops adequately – another not-to-be-forgotten aspect of the variety. Thirdly, the very name 'Pinot noir' has a magic which cannot be denied, partly because it can make some of the best wines in the world, but also because it is such a chameleon, making both great red still wines and great (white) sparkling wines. Everyone talks about it being a 'difficult' variety and a 'heartbreaker' which further adds to its mystique. It is also a name which almost every wine drinker will know.

On the red wine front, I am more convinced now than I used to be about its suitability to make good red wines in Britain. Pinot noir wines have won the Bernard Theobald

65

Wine Growing in Great Britain

Trophy for the Best Red in the EWOTYC in sixteen of the eighteen years between 2002 and 2019 (five times by Chapel Down, three times by Gusbourne, twice by Titchfield and twice by Sharpham, and once each by Bolney, Biddenden, Plumpton College, and Sandhurst), although these wines were the exceptions rather than the rule. Too many of our Pinot noir wines are just thin, acidic and poor value for money. The best are generally light in colour (although no lighter than many top-class Burgundies) and have simple red-fruit bouquets – perhaps with a touch of oak – adequate tannins and a fresh, fruity crispness about them that says 'attractive' rather than 'serious'. They are not usually very substantial wines and must be put in context with wines from other cooler growing regions: the Loire (Sancerre especially), Burgundy and even Alsace. Against wines from warmer climes they stand no chance. Many British-grown Pinot noirs would be much better presented as light, fruity rosés with a touch of sweetness and a very light *petillance*. However, the top Pinot noir reds from producers such as Chapel Down and Gusbourne show that it is possible to make serious wines from this variety in most vintages and wines that sell for serious prices: £20–£45 a bottle.

Viticulturally, Pinot noir presents no great problems to the seasoned grower. It buds up quite early, making it susceptible to damage on frost-prone sites, but it will shoot from secondary buds. It gets all the usual ailments and requires good management to keep it botrytis-free, an important factor if the grapes are needed for whole-bunch pressing or for fermenting on the skins as a red wine. With a good spray regime and deleafing at the appropriate times, 100 per cent clean fruit is achievable in every year. On the cropping front, I have been very impressed by the level of the yields and whilst cooler years with difficult flowering conditions might set the averages back, 10 tonnes-ha (4 tonnes-acre) would appear to be achievable in 'normal' years and even higher in better years, which is fine for sparkling wine. For still red wine production, the quality of the site and the year will play a major rôle in fruit quality, but it is probably wiser to aim for nearer half this figure if good wines are to be made.

Clonally, Pinot noir is very diverse and I have no doubt that in Britain we have only scratched the surface of this aspect of growing. It is a great pity that very little work (i.e. none) has been done in Britain on clones suitable for our unique growing conditions (another job for Plumpton College perhaps?) We can only, therefore, take anecdotal evidence gleaned from existing growers in Britain, plus take notice of recommendations offered to overseas growers.

Pinot noir is undoubtedly here to stay and, given its success in making some great sparkling wines and the beginnings of what may well be some great still red wines, I can only see its importance to growers in Britain increasing.

Note: See the separate section at the end of this chapter on: *Clones of Chardonnay, Pinot noir and Meunier*

Regent

Type: PIWI – disease-resistant complex hybrid
Colour: Red
Origin: Diana x Chambourcin

Regent (pronounced in German *Ray-ghent* with a hard 'g' as in Ghent, not as in Regent Street) is another of Professor Dr Alleweldt's Geilweilerhof crossings (see Orion and Phoenix). Its parents are Diana, a Silvaner x Müller-Thurgau crossing (and not the native Catawba variety grown in the USA of the same name) and Chambourcin, a Joannès Seyve crossing of Seyve-Villard 12-417 and Chancellor. The Regent crossing was made in 1967 and in Germany it was released in 1995. It has become popular with growers and, at 1,754-ha in 2019 (down from 2,047-ha in 2012) is their seventh most widely grown red vine variety. It has very good colour, is more disease resistant than *viniferas* and the wine quality is usually quite good (always a help). In 1996, despite not being 100 per cent *vinifera,* it was authorised for the production of Quality Wines in Germany, the first modern interspecific cross to be allowed to do so. According to *Wine Grapes* it is named after Le Régent, the 410 carat diamond found by a slave in the Kollur mine at Golkonda in India, and eventually placed in the crown of the French King Louis XV when he was crowned in 1722.

In Britain, Regent has only been around since the early 1990s and by 2006 there were 12.6-ha. Today, it has grown to around 35.0-ha and is the third most widely planted purely red variety (i.e. ignoring Pinot noir and Meunier). It requires a good site, good management and deleafing at the right times, but no more so than other red varieties. It is certainly less disease-prone than Dornfelder and Rondo, so on that score alone it could be a better bet for a deep red wine variety, but is susceptible to Downy Mildew. It usually yields quite well and 7.5-10 tonnes-ha (3–4 tonnes-acre) are achievable on a good site. For deep red wines, it has a higher quality than wines made from Dornfelder or Rondo, although it probably best used in a blend of all three. It takes oak well, has great colour, soft tannins and its acids are manageable. Its continuing expansion would suggest that it is a valuable variety for those wanting to produce deep coloured, blended red wines.

Reichensteiner

Type: *Vinifera*
Colour: White
Origin: Müller-Thurgau x (Madeleine Angevine x Calabreser-Froelich)

Reichensteiner is a Geisenheim cross produced by Professor Henrich Birk in 1939 by crossing Müller-Thurgau with a cross between two other varieties: Madeleine Angevine, the female French table grape used by Scheu (among others) for a number of his crossings, and Calabreser-Froelich, an early white Italian table grape. Together, these have combined to produce an early ripening variety, capable of producing large crops of relatively neutral grapes, high in natural sugars and low(ish) in acidity.

In Germany, Reichensteiner was seen as a substitute to the practice of 'wet-sugaring' (where up to 20 per cent by volume of water could be added to the must during the enrichment process in order to lower the acidity) – a practice that only came to an end in the late 1970s. In Germany, the area of Reichensteiner has fallen to only 43.0-ha in 2019 and it is certainly not being planted there any more. New Zealand once had 74-ha of it (where it cropped at 23 tonnes-ha) but it is now no longer listed as a separate variety in their annual vintage reports. It is found in many of the cooler vinegrowing regions – Belgium, Britain, Canada, Denmark, the Netherlands and Sweden – where its relatively high yields (higher than more classic varieties) mean it is an economically viable variety. Its low acidity means it can be used as a blending partner for varieties with more acidity.

Reichensteiner was introduced into Britain by Jack Ward when he extended his Brickyard vineyard in Horam in 1971. Early results showed that it was a consistent cropper, with good yields and high sugars. Although it can be vigorous in its youth, it settles down after a few years and is less vigorous than MT at the same age. It has a good open habit and a leaf-wall that is less crowded and better ventilated than MT, both of which help disease control.

As a varietal wine, it does not have the character of fruitier varieties, but has lower acid levels, more body and extract, and higher natural sugars and is probably best used in a blend. If picked early when acidities are high it can be used for sparkling wines (Camel Valley, Carr Taylor, Chapel Down, Davenport, Meopham), but, if left to fully ripen, it can also be used for the production of *süss-reserve*. Although it has fallen in area, from 113.9-ha in 1990 down to around 75.0-ha in 2020, I believe it still has a valuable part to play in Britain's wine industry on account of its reliable cropping record. It is certainly a variety that the more climatically challenged regions should not ignore. Vines are difficult to obtain as no one is grafting them except to order.

Rondo

Synonym: Gm 6494/5
Type: PIWI disease-resistant complex hybrid
Colour: Red
Origin: Zarya Severa x St Laurent

Of all the varieties with complicated histories attached, Rondo must have one of the most complex. The original *Vitis amurensis* vines, from which Rondo is partly derived, were wild vines that came from Manchuria, in the north of China, where the River Amur marks the border between China and Russia. Here, on account of the early onset and severity of the winters, wild vines need be able to withstand deep winter temperatures and therefore colour up and ripen early. In 1910, a Russian vine breeder, Ivan Vladimirovich Michurin (1855–1935) – pictured here – selected an

open-pollinated seedling of Précoce de Malingre, which was probably pollinated by a *Vitis amurensis* in Michurin's vine collection, which he called Seyanets Malengra.

Michurin was a famous plant breeder whose work was recognised by Lenin and whose private research station at Tambov, about 435-km south-east of Moscow, became the Michurin Central Genetic Laboratory in 1934. Seyanets Malengra is a female vine and is used in Russia and Ukraine as a table grape variety. In 1936, two Russian plant breeders, Yakov Potapenko and E. Zakharova, working at the All-Russia Research Institute of Viticulture and Winemaking (later called the Potapenko All-Russia Research Institute of Viticulture and Winemaking), situated at Novocherkassk in Rostov Province, crossed Seyanets Malengra with another *Amurensis* vine to produce a variety called Zarya Severa. This crossing then found its way to Lednice in Moravia, Czechoslovakia where Professor Dr Vilém Kraus crossed it, in 1964, with St Laurent, an old Austrian wine variety. This crossing then travelled to Geisenheim where Professor Helmut Becker improved it and gave it a breeding number Gm 6494/5. With me so far? It was then trialled at Geisenheim and proved capable of producing good crops of early ripening, deeply coloured grapes. However, it was never widely planted in Germany and in 2019 there were only 11.0-ha.

In 1983 Professor Becker gave me 50 vines to plant at Tenterden, as well as 50 each to Ken Barlow at Adgestone and Karl-Heinz Johner at Lamberhurst, to see what the results would be. From the first harvests it showed itself to be well adapted to our conditions, and plantings have been increasing since then. From 2-ha in 1990, it has continued to grow in popularity and stands at around 65.0-ha in 2020. Ignoring Pinot noir and Meunier (from which little truly red wine is made) Rondo is the second most widely planted red variety in Britain. In Germany there used to be a small hectarage (11-ha in 2003), but this appears to have gone now. In other cool climates such as Belgium, Denmark, the Netherlands, and Sweden it is one of the most important varieties. Ireland's largest vineyard, the 1.0-ha Thomas Walk Vineyard near Cork, first planted in the 1980s, is 100 per cent planted with Rondo, which they call 'Amurensis Walk'.[17] For the first decade of its life in Britain it was called Gm 6494/5, but eventually we got it a name – Rondo – and persuaded DEFRA to classify it as a *vinifera* (as the Germans had done five years earlier).

Initially, Rondo had all the makings of a good variety for Britain: early to ripen, with a good crop of deep red grapes with low(ish) acidity and wine quality that was certainly better than anything else seen in the mid-1980s – Léon Millot, Maréchal Foch and Triomphe d'Alsace (as it was

---

17 Thomas Walk, a German national, knew Professor Becker (who loved Ireland) and was also given some of the first Rondo vines by him.

then called) – being the alternatives. Chapel Down made two versions: a straightforward wine and a Reserve, both from grapes from Anthony Pilcher's Chapel Farm vineyard on the Isle of Wight. The Reserve, with enough oak and some time, could be very good, although the flavour was always very slightly unusual. I always likened it to a Syrah-Tempranillo blend – the American oak helped – and I tried to persuade myself for years that it had a future. However, reluctantly, and in the face of the onslaught from Frühburgunder, Pinot noir and Regent, I don't think the wine quality is high enough for it to succeed in the long term.

It is a very vigorous variety, with big leaves (almost like the leaves of kiwi fruit) and canes that can stretch for several metres if allowed to run. Consequently, canopies tend to be crowded and shaded and although an interspecific cross, it is still fairly susceptible to botrytis. The wine colour is excellent, deeper than Dornfelder and Regent and the quality is passable, but not of the first order and certainly not what the consumer expects of a wine of that appearance. Although the area planted has expanded, most of this additional planting is in vineyards in the more challenging parts of Britain and I do not believe that any mainstream vineyards view it as a serious variety for the future. Although I value the direct link back to Helmut Becker and those first 50 vines, sentimentality never helped make good wine and plantings of Rondo, although they will probably expand, will do so only slowly.

Sauvignon blanc
Type: *Vinifera*
Colour: White
Origin: Original variety

Sauvignon blanc is one of the world's most popular and widespread white vine varieties and can be found in climates both cool and warm. Great wines can be produced in a variety of styles: Marlborough's high-acid, zesty pure-fruit wines; the classic lean and minimalist wines from Sancerre and Pouilly-Fumé; and the full-bodied, oaky offerings labelled as Fumé Blanc from California are all fine examples of this versatile variety. In Germany it has risen in popularity from nothing in 1999 to 1,498-ha in 2019 and is also popular in Austria, northern Italy and the Adriatic countries. It is also just starting to be grown in Luxembourg where in 2018 there was 0.80-ha. There are numerous Sauvignon blanc clones – 21 listed for use in France alone – and plenty more listed for use in Italy. Ninety per cent of New Zealand's Sauvignon blanc is the Californian UCD1 clone (said to have originated at Château d'Yquem in the 1880s) and Riversun, the major NZ vine nursery, lists six further clones as being of interest.

In Britain, Sauvignon blanc has risen in area from 0.5-ha in 1990, via 4.6-ha in 2013, to around 7.00-ha in 2020. Hardly a meteoric rise, but at least a positive one. My experience with it has been slightly negative and, except on the very best sites and in the very warmest of vintages,

it struggles to set fruit and ripen that fruit. Denbies, who claimed, in 2010 when they planted 1.21-ha of it, that it was a 'British first' (it wasn't), were very bullish about their 2013 harvest, which was picked on 31 October at 9.2 per cent, 11.4 g/l acidity and had a pH of 2.94. These are exceptionally good figures (especially given the year). The wine, however, was very grassy and perhaps 2013 was not the year for it. They ended up blending it with Bacchus. In the end they grubbed the vines up and as their winemaker, John Worontschak said 'wrong clone, wrong rootstock, wrong country'. In more recent times, Woodchester Valley Vineyard, near Stroud in Gloucestershire, won an IWSC Gold and a Gold at the Drinks Business 'Global Masters' competition for their 2018 Sauvignon blanc, which shows that it can be successful, but which rather goes to confirm my point about it needing an exceptional vintage for it to ripen. Ideally, if Sauvignon blanc is to be successful in Britain, someone needs to do some clonal trials, as out of the many dozens of different clones grown around the world, there are bound to be some more suitable for our conditions than others.

As a commercial variety for Britain I am very doubtful about Sauvignon blanc. But of course it has a recognised name and style, always an advantage when it comes to selling wine. Maybe blended with, say, Bacchus in less good vintages or even with another, more neutral variety, it will find a niche. I used to say that Chardonnay was impossible in Britain for still wine, so who knows? Maybe in fifty years' time, Britain will be growing all the Sauvignon blanc and New Zealand will be sending us table grapes!

Schönburger
Type: *Vinifera*
Colour: White
Origin: Pinot noir x IP1

Schönburger is a crossing made at Geisenheim by Professor Dr Heinrich Birk in 1939 and first registered for use in 1980. Like many of the crosses from this era, it combined a classic winemaking variety, Pinot noir, with an early ripening table grape variety, in this case a crossing from an Italian I. Pirovano called IP1 which is Chasselas rosa x Muscat Hamburg. The hope was that a unique German table grape could be produced. There is very little Schönburger planted in Germany, only 16.0-ha in 2018, and countries such as Canada, New Zealand and Britain have more, and some can even be found in Australia and South Africa!

It has been grown in Britain since the late 1970s and on account of its high-quality grapes, its ability to be turned into award-winning wines, its low acidity and its regular crops, it became quite widely planted. Colin Gillespie at Wootton Vineyard made it his speciality and won the Gore-Browne Trophy with it in 1982, and Lamberhurst won the same trophy with Schönburger wines in 1982 and 1990. By 1990 there were 75.3-ha, making it then the fifth most planted variety. It is an undemanding variety, more disease-resistant than MT and not so on–off in its cropping habits. Yields are never large, but what is picked is almost always of excellent quality, with high sugars and low acids. As they ripen, the grapes change colour from a light lime-green to, at first, light pink and then, when fully ripe, to an almost tawny brown colour. This gives a very good visual indication to the pickers as to which bunches are ripe and allows for some selective picking. This can only aid wine quality. The wines are almost always light and very fruity, with some good Muscat tones (some resemble a less powerful version of Gewürztraminer) that are best balanced with a slight amount of residual sugar. These wines age well and have been known to keep for up to seven years. Schönburger is one of the few varieties we grow that can be eaten with pleasure and although the berries are small and have seeds, they are packed with flavour.

Despite all these attributes, the area being grown has fallen to 20.0-ha in 2020. Why? Well the name is tricky – an umlaut and '—burger' on the end never helped any vine variety become popular – and in style it's just a bit too – well, it has to be said – a bit too weak and wishy-washy. I say this even though a well-made Schönburger, with a touch of residual sweetness, nicely chilled and served as an apéritif is as good an English wine experience as you could want – not serious, but light, and slightly frivolous. Why has this counted against it? Who knows, fashion in wine is a fickle thing and while I like Gewürztraminer, Viognier, Torrontes, and other aromatic, flowery varieties, I know that I am in the minority. However, I do sense that there is an increasing interest by producers in still wines, mainly because the cash-flow is so much better than long-aged sparkling wines. Maybe Schönburger is due for a modest revival? It's definitely on my 'ones to watch' list.

Seyval blanc
Synonym: Seyve-Villard 5276
Type: *Hybrid*
Colour: White
Origin: Seibel 5656 x Seibel 4986

Seyval blanc, a crossing made by Bertille Seyve (the younger) at the nursery of his father-in-law, Victor Villard, in St Vallier on the Rhône in 1921, was originally known as Seyve-Villard 5276. In some references, the crossing is given as Seibel 4996 x Seibel 4986, but Pierre Galet (who was probably the most famous ampelographer in the world) assured me in a letter that Bertille Seyve's son stated in the *Viticulture Nouvelle* of 1961 (p. 182) that the crossing was Seibel 5656 x Seibel 4986. S. 4986 is known as Rayon d'Or. In 1958 there were 1,309-ha in France, but plantings have declined since then and were down to only 90.0-ha in 2018. It can be found in many different parts of North America, most notably in New York State and Ontario, and it is also grown in Switzerland.

It was first grown in Britain by Edward Hyams who, in 1947, obtained vines directly from the Seyve-Villard nurseries (as they had become known by then) and planted

them in his small vineyard at Molash, near Canterbury, Kent. Brock also imported the variety from the Swiss Federal Vine Testing Station at Caudoz-sur-Pully a year later, in 1948. These two pioneers tested it and soon found that it was a very suitable variety for our climate and. together with MT, it became the standard variety for almost all the early vineyards. When Sir Guy Salisbury-Jones planted his vineyard at Hambledon in 1952 he chose Seyval blanc on the advice of a helpful Burgundian grower whom he sat next to at a *Confrérie des Chevaliers du Tastevin* dinner.

Seyval blanc has many attributes. It sets good crops even in cool years, is not vigorous and has a good open habit with small leaves and only a few side-shoots. It is unusual in that the flowers appear and are very prominent before many of the leaves really develop. Its disease resistance is good, although with climate change and the rise in natural sugar levels, it has become a bit more prone to botrytis. Last, but by no means least, it yields well – in some years very well – and 20 t-ha (8.0 t-acre) is not unknown. Although the grapes are never very high in natural sugar (I remember plenty of years in the 1970s and 1980s when only 5–6 per cent natural alcohol was reached), wine quality can be good *if* the winery knows what it is doing. On the downside, apart from the low natural sugars, the wine can be very neutral, high in acidity and, if not pressed with care, can get a rather grassy herbaceous tone. However, it takes to oak-ageing well, is a good foil for lees-ageing and *battonage* and is great for sparkling wine. What's not to like about it one might ask?

Seyval blanc suffers from a bad press and – rather like Meunier in Champagne – it's a useful variety, but don't let's frighten the horses by talking about it too much. Most wine writers and many wine merchants, faced with the word 'hybrid', seem to recoil into their shells a bit, strike an arrogant pose, and declare without a *shred* of evidence, let alone experience in winegrowing or winemaking, that 'of course you cannot make good wine from a *hybrid*'. The truth is, of course, that you can and people do. Wines made from Seyval win their share of medals and awards: I won the Gore-Browne Trophy with Seyval twice (in 1981 and 1991) and Colin Gillespie at Wootton won it in 1986 with a Seyval blanc. It can still be found in some good sparkling wines (Breaky Bottom, Camel Valley, Denbies, Monnow Valley and Stanlake Park. for instance) as well as a lot of good still wine. Many growers like it and although the area fell from a high of 122.7-ha in 1990 to 88.9-ha in 2000, it has now staged a modest comeback to around 145.0-ha in 2020. It is Britain's fifth most widely planted variety.

When I imported my first Seyval, I obtained stock from one of Professor Becker's friends, Ernest Pfrimmer who had a nursery at Hurtigheim, near Strasbourg, and the results were always good. When I started importing a lot of Seyval vines from France, I spent a considerable time searching for a good source-block of vines and together with François Morisson (from the Morisson-Couderc Nursery) found an ideal one in Blois on the Loire. From this vineyard he took wood and grafted many thousands of vines for me. Having seen a fair few Seyval vineyards, both in Britain and overseas, I am convinced that there is quite a bit of clonal variation and some of Britain's vineyards are planted with inferior clones. This might account for some of the grassy, herbaceous notes that are sometimes encountered although that could just be harsh pressing. Recently I have seen some Seyval that is plainly another variety altogether – possibly another Seyve-Villard hybrid – and clearly unsuited for British conditions. In 2013, I travelled to Switzerland to see an excellent mother-block of Seyval and I now only import vines produced with scion wood taken from this vineyard.

The slight revival in Seyval's fortunes, and the plantings that I know to be in the pipeline, give me some encouragement that its future is assured, albeit as a minor but useful variety. It is especially suitable for organic and biodynamic producers, plus those in the further reaches of our islands. The sight of a heavy crop of Seyval gently turning golden in late October was always a welcome one and it is such an *easy* variety to grow and turn into wine! Too easy – perhaps that's the problem?

Siegerrebe
Type: *Vinifera*
Colour: White
Origin: Freely pollinated seedling of Madeleine Angevine

Siegerrebe is another Georg Scheu crossing from Alzey, made in 1929 and released to growers in 1958. In the *Taschenbuch der Rebsorten* (2010 edition) the breeding details are given as follows. It was originally said to be Madeleine Angevine x Gewürztraminer, but it was later revealed by Scheu's son, Heinz, to be a freely pollinated seedling of Madeleine Angevine (which is the same as Madeleine x Angevine 7672). More recently it has been confirmed in VIVC as MA x Gewürztraminer. However, in *Wine Grapes* it is given as Madeleine Angevine x Savagnin Rose. Savagnin Rose (without an é) is a 'non-aromatic version of Gewürztraminer', which, in Germany, is often called Roter Traminer (Red Traminer). In Germany there were 97-ha of Siegerrebe in 2012, but it has gone down to 75.0-ha in 2018. Siegerrebe can reach very high natural sugar levels and is used for making sweet dessert wines, although the acid levels are often very low. The grapes have a strong Muscat character and when ripe, can be very concentrated and almost overpowering. It is more often used for blending with other, less distinctive varieties.

It has been grown in Britain since 1957 when Brock was sent cuttings from Alzey. By 1960 he was able to report 'exceptionally early and the grapes have a strong bouquet' and he highly recommended it as suitable for both the table and winemaking. Despite its attributes of early ripening, low acidity and strongly flavoured grapes, it has really failed to be planted in anything like serious amounts and although one or two vineyards do make interesting varietal wines from it, it mainly gets lost in blends. Maybe because

it does ripen very early and falls so far outside what most growers consider their normal harvest time, it is seen as rather disruptive. Wasps are a particular problem and good wasp nest control needs to be practiced if real damage is to be avoided. Birds, likewise, tend to be a nuisance.

The area planted in Britain has risen slightly: 9.30-ha in 1990, 12.8-ha in 2009, 14.3-ha in 2013 and approximately 20.0-ha in 2020, which makes one think that it has a future. Three Choirs consistently make an interesting semi-sweet wine with the variety (it's usually one of their better wines) and their 2006 Estate Reserve Siegerrebe won the Gore-Browne Trophy in 2008 their 2007 Three Choirs Siegerrebe won a gold medal in 2009 and their 2010, 2011 and 2012 all won silver medals. Siegerrebe can add a nice piquancy to an otherwise dull blend and is probably worth considering as a minor variety. The rise in the area would suggest there is some interest in the variety and it will probably hold its own and possibly expand slightly.

Solaris
Type: PIWI disease-resistant complex hybrid
Colour: White
Origin: Merzling x Gm 6493

Solaris is a Freiburg interspecific cross, once known as Fr. 240-75, bred by Dr Norbert Becker in 1975 and released in 2001. Its parents are Merzling and Gm 6493. Merzling is a crossing between our old friend Seyval blanc and a (Riesling x Pinot gris) crossing. Gm 6493 is a Geisenheim *Vitis amurensis*-influenced crossing and is Zarya Severa x Muscat Ottonel, Zarya Severa being one of the parents of Rondo! Complicated or what? The *Taschenbuch der Rebsorten* says that it is very vigorous, with very large leaves, and that it ripens early with very high sugars, usually over 100°OE (13.8 per cent potential alcohol) and often nearer 130°OE (18.4 per cent potential alcohol). It also states that the acidity is less than Ruländer (Pinot gris) – which would mean that its acidity is towards the bottom of the acid spectrum, i.e. 5–6 g/l as tartaric acid. Being a complex hybrid, it has a good resistance against the major diseases, especially botrytis to which high-sugared varieties are usually prone and its loose bunches help in this respect. It sounds very similar in many ways to Reichensteiner, but with better disease resistance. In 2012 there were 101-ha planted in Germany (in Baden, the Pfalz and the Mosel), and by 2018 this had grown to 160.0-ha. It is also planted in the Trentino-Alto Adige region of Italy, and in Switzerland. It is also found in the cooler climates of Europe such as Belgium, Britain, the Netherlands and Sweden and is by far the most widely planted white variety in Denmark.

In Britain it is planted on around 70.0-ha (2020) which is up from 4.70-ha in 2009. This is a greater rate of expansion of any variety, including Chardonnay and Pinot noir. There are plantings spread across the country, a relatively small number in the South East and East Anglia, some on organic and biodynamic vineyards, with the majority being in the more northerly and westerly regions where its disease resistance is valued. In terms of wine quality, it has done well in competitions with at least one gold, several silvers and many bronzes in the WineGB annual competition. As vineyards expand ever further north – productive Scottish vineyards cannot be far away – I see Solaris being planted more widely still. Another minor, but important variety.

Triomphe
Type: *Hybrid*
Colour: Red
Origin: Open-pollinated seedling of 101-14 *Millardet et De Grasset* x Knipperlé,

Triomphe is an old French–American hybrid bred by Eugène Kuhlmann in Alsace in around 1911 and released in 1921. Its breeding number is Kuhlmann 319-1 (although some references state 319-3) and it was originally known as Triomphe d'Alsace and for many years was labelled as such. When, in 1992, Britain tried to register it under that title, the EU Commission objected (because it contained the name of an *appellation contrôlée* region – Alsace) and the name was shortened to Triomphe. Its parentage is slightly uncertain, but would appear to be an open-pollinated seedling of the rootstock variety 101-14 Millardet et De Grasset crossed with Knipperlé, an ancient Gouais blanc x Pinot crossing from Alsace. (This is the *Wine Grapes* version of its parentage; Wikipedia gives a different version.)

Triomphe was first introduced into Britain by Brock at Oxted, quite widely planted by the early growers and together with Cascade, Léon Millot and Maréchal Foch, became one of the most widely planted red varieties. It is a vigorous variety and needs a bit of canopy management to keep it under control. However, disease resistance is fairly good and it has the small berries and dark red skin, red flesh and red juice of the other Kuhlmann hybrids. However, its wine quality is not that good, certainly below that of Dornfelder and Regent, with a slightly non-vinous tone about it. It is probably best blended with other varieties. There is said to be a lot of clonal variation.

The planted area in Britain expanded from 7.5-ha in 1990 to 14.7-ha in 2013, but since then has declined and in 2020 stands at 7.00-ha. It will no doubt linger on in a few older vineyards, but its days are numbered as a variety for quality wine production in Britain and we now have several much better varieties.

Würzer
Type: *Vinifera*
Colour: White
Origin: Gewürztraminer x Müller-Thurgau

Würzer is a crossing made by Georg Scheu at Alzey in 1932 and released to growers in 1978. It is another of Scheu's spicy crosses (Scheurebe, Faberrebe and Huxelrebe being the other ones used in Britain) and in a similar mould. There were 54-ha in Germany in 2018, an area that has fallen from 108-ha in 1999.

For Britain it ripens really rather too late for most sites, has high acids and does not show any real advantages over several other similarly flavoured varieties (Bacchus for instance) and the decline in the area from 14.3-ha in 1999 to around 2.0-ha in 2020 is probably a good indication of its usefulness. Not a variety for the future.

Clones of Chardonnay, Pinot noir and Meunier

Until Nyetimber started winning medals and prizes (in 1997), the question of which clones of the Champagne varieties to plant had never really been discussed much amongst vineyard owners in Britain. In around 1998, when I was winemaking at Chapel Down, I planted 1.77-ha (4.37-acres) of Pinot noir and selected clones 115, 666, 667 and the *teinturier* clone Tête du Nègre. The intention was to produce red wine, not sparkling. This was the first time that I had really given much thought to clones. The next occasion I had to choose some clones was in 2001–2 when I was asked to advise on the planting of the initial vineyards at Hush Heath Estate which were solely for the production of a bottle-fermented rosé sparkling wine, which became *Balfour Brut Rosé.* This time I made sure I had done some homework before I put pen to paper. I therefore did as much investigative work as I could with: (a) the vineyards in Britain then making the best sparkling wine, (b) the CIVC, and (c) some of the *pépiniéristes* selling vines to Champagne and other sparkling wine producers.

From sparkling wine producers in Britain, principally Chapel Down, Nyetimber, and Ridgeview, I discovered some of the clones they were using, although about others there was a certain amount of secrecy. The CIVC sold me their *Viticulture raisonée guide pratique* which, in a separate chapter headed 'Clones', lists all the clones allowed to be used in Champagne: eleven clones of Chardonnay, eighteen clones of Pinot noir, and eleven of Meunier, all listed according to their ability (or otherwise) to crop, produce sugar and succumb to botrytis. From vine nurseries I was able to find out a bit more about other clones, especially ones with lower acid levels and better disease resistance, two attributes I felt were important in Britain. As with many other things viticultural, one always has to remember that when taking data from abroad, Britain is different, in some respects very different, from other vinegrowing regions.

Compared to Champagne for instance, yields in Britain are very much smaller. The average yield in Champagne since 2000 has been near to 15.00 t-ha (6.00 tonnes-acre) whereas in Britain, in the vineyards surveyed between 2016 and 2019, it has been around 6.00 t-ha (2.40 tonnes-acre). Of course, the planting density in Champagne of around 8,000 vines-ha (3,238 vines-acre) compared to a typical British vineyard of around 3,500–4,000 vines-ha (1,416–1,619 vines-acre) makes a huge difference and the unprofessional nature of many of Britain's smaller vineyards accounts in part for our low yields. However, the truth is that high yields are what the Champenoise want and high yields is what they get. Their selection of clones therefore takes this into account.

In Britain, the natural sugar levels we achieve are, in all but the best of years, towards the lower end of those achieved in other cool climate regions and distinctly off the scale compared to anywhere warmer. Chardonnay and Pinot noir is generally picked at between 8.5 and 10 per cent natural alcohol, although in recent warm years – 2003, 2014, and 2018 – sugar levels have been at least 10 per cent and, for still wines, even higher. However, these are still low levels and make many an overseas grower wonder how we manage to make wine so good! Likewise, our acidity levels. Even in Germany where Riesling's high acids can test even the best winemakers, Britain's acid levels which, with Chardonnay, can often be around 14–16 g/l as tartaric, with pH levels routinely below 3.0, take a bit of getting used to. In Champagne, natural alcohol levels are around 10.00 to 10.50 per cent,[18] with acidity levels (as tartaric) 11.00 to 11.50 grammes per litre, figures that are not too dissimilar to those achieved on recent years in vineyards in Britain, although, of course, one has to remember that yields in Champagne are almost three times ours.

One also has to remember when looking at clones of two of the three varieties in question – Chardonnay and Pinot noir – that these are also major varieties for the production of still wines of different qualities: high-volume, entry-level quality; mid-range, generally higher volume quality; and low-volume, top-notch quality. Chardonnay and Pinot noir are also varieties found in some of the warmest parts of the winegrowing world – California's Central Valley and some of Australia and South Africa's hottest regions – as well as some of its coolest, Champagne, Chablis and of course, coolest of all, Britain. Therefore, opinions about which clones suit which regions are always coloured by the type, style, and quality of wine being produced, plus the yield levels wished for.

Owing to our historic links with Germany in relation to viticulture, some vine suppliers seem keen to recommend and supply German clones grafted on to rootstocks common to German vine nurseries. This practice is something that concerns me. I take the view that if you are trying to produce a top-quality sparkling wine, then you have to have a very good reason not to copy what happens in the region that produces, without any argument, the best bottle-fermented sparkling wines in the world: Champagne. Champagne has the advantages of scale – there are around 35,000-ha (86,500-acres) of vines in the region – and of time – they have been growing vines in the region for many centuries. They also have the unique advantage that every individual vinegrower, all 15,000 of them, growing vines in 270,000 individual plots,

belong to one organisation – the CIVC – whose job it is to oversee every aspect of production, including research into what clones suit the wine being produced. Therefore, my practice since 2002–3 has been to stick very firmly to selecting French clones – mainly, but not only, CIVC-recommended clones – for producing sparkling wines in Britain.

Although I am sure there are very many good German clones of both Chardonnay and Pinot noir, there are several factors which make me wary of using them in Britain. To start with, Chardonnay is still a relatively new variety in Germany, and although they had 2,100-ha (5,189-acres) in 2018, making it the sixth most widely planted variety (up from fifteenth place in 2013), most of it is used for making still wines and very little is used for sparkling wines. Likewise, Pinot noir. Although a much more popular variety than Chardonnay, the 11,762-ha (29,064-acres) grown in 2018 make it the most widely grown red variety and third most widely grown variety overall, very little of it goes into bottle-fermented sparkling wines.[19] Although there are some very serious growers of Pinot noir, mainly in the southern growing areas, Baden especially, producing some excellent Burgundy-style red wines, much of the Pinot noir is used for making relatively light-character, fruity wine, which finds favour with the German consumer. An additional factor in Germany is the cost of production, and yields, both of which are high. For a vine variety and/or clone to be taken up in Germany, it needs to be easy to grow, show good disease resistance and yield well. Low-yielding, disease-prone clones are soon weeded out.

Clonal development of Pinot noir in Germany started in the 1930s in three principal institutes: Freiburg, Geisenheim and Weinsburg. To start with, high yields and high natural sugars were the key goals, with disease resistance a distant aspiration. In the second generation of German clones, those with loose bunches were preferred as they had a greater resistance to botrytis and these have found favour with growers. Consequently, most German vine nurseries concentrate on producing these clones, and Freiburg and Geisenheim clones such as Fr 18-01 and Gm 20-13 are the most widely planted. Taking all these factors together, I am not persuaded that German clones are to be preferred for the production of sparkling wines in Britain.

As for recommending which clones to plant, as I have already said above, this is a confusing and unfathomable task. The clones I recommend in this section are ones used by many of the most important still and sparkling wine producers in Britain and all can be planted with confidence on suitable sites. This is not to say that they are the only clones that can be successful, and I am sure, in time, others will be tried and will prove suitable. My advice would be to plant at least two clones of each of varieties

---

18 It is, however, worth noting that until the mid-'90s, grapes in Champagne only had to reach a potential alcohol level of 7 per cent before they could be harvested and that today it is only 9 or 9.5 per cent, depending on the vintage.

19 I visited Volker Raumland, who produces what are probably the best Champagne-style sparkling wines in Germany, and asked him which clones he uses. He looked at me a bit quizzically and said he 'wasn't sure, but they were all French'.

you have chosen, and on a vineyard of more than 2-ha (5 acres), you might plant more clones. On a small vineyard, around 0.4-ha (1-acre) of each clone makes a good-sized block and, if you are making your own wine, allows you to make separate *cuvées* of each clone. Of course, on a much larger vineyard, your blocks and areas of each clone will be much larger.

As a general rule, clones will, in most instances, vary only a little. Some will be earlier than others, some will have bigger and/or better crops than others, but these differences might well be much less important than the influences of the site, the rootstock, the weather during the year and of the management of the vineyard in all its aspects. Finding THE clone for your vineyard, one that will produce you perfect wine, year in and year out, is probably not possible, so play safe, and use a sensible range for the size of vineyard. Over the years, on your site, you will get to notice the differences and when you come to plant more, you can take a more educated decision. For still wines from Chardonnay and Pinot noir, much of the quality of the grapes will depend on the quality of the site, the yield level, canopy management and the degree of risk you are prepared to accept to leave the grapes to ripen

Chardonnay – still wine

For still wines, I regard clones 75, 76, 95 and 548 as suitable for Britain. Clones 121 and 131 have also been recommended for still wine, but acids are said to be higher. However, given that, in France alone, thirty-one clones of Chardonnay are registered for use and in other countries there are tens of dozens more (some of them duplicates, admittedly), I am sure there are other clones suitable for our vineyards yet to be trialled and proved.

Chardonnay – sparkling wine

The following CIVC clones have performed well in vineyards in Britain for sparkling wine: 75, 76, 78, 95, 96, 118, 121, 124, and 131. I also like these non-CIVC clones: 119, 277 and 548. Clone 809, which has a slight muscat tone, might also add some interest to a blend and is also used by Gusbourne to make a well regarded still wine.

Meunier – sparkling and still wine

The following CIVC clones have performed well in vineyards in Britain for sparkling wine: 817, 865, 900, 977 and 978. Very little still wine has been made from Meunier, and where it has been, it has been made from sparkling wine clones. As with all still red wines, crop level and canopy management are the keys to success.

Pinot noir clones in general

Pinot noir clones are loosely divided into four main types:
- Clones with compact bunches, also often with small berries. These are favoured by still red wine producers. Yields tend to be lower than other clones. They are also called Pinot Fin clones. 375, 667 and 777 are the main ones planted in GB vineyards.

- Clones with loose bunches, often with larger berries, and higher yields. German growers have planted these widely as they are more botrytis resistant. Probably not the best for sparkling wines. Also sometimes known as Mariafeld or Wädenswil clones which stem from Switzerland. Pépiniéristes Guillaume have their own proprietary clone of loose-bunched Pinot noir called their GL (*grappe lâche*) clone.
- Upright clones (Pinot droit) have more erect foliage, favoured by some growers, and often have tight bunches. They tend to be less botrytis-prone. Good for sparkling wines, although can be later to ripen than other clones.
- Clones with small berries. German research stations (principally Geisenheim and Freiburg) have developed these clones which tend to have large bunches and high yields. They are good for still red wine in the German style, but not much favoured in other regions.
- There are also several 'mixed' clones, as well as various *selection massale*[20] which many Burgundian growers now prefer.

The question of botrytis resistance has to be understood and taken into account when considering descriptions and recommendations in publications and on websites, all of which are based upon non-British conditions. As I point out in *Chapter 16: Pest and disease control*, the incidence of botrytis in well-managed British vineyards is actually far less than it is in the regions of Germany and France where they grow Pinot noir. In Germany, for instance, where yields are high and grapes are often very tightly packed together inside the canopy, botrytis is a major concern. As will be seen in *Chapter 16*, some German growers practice *traubenteilen* – the cutting of the bunch in half – which causes the shoulders of the bunch to droop, loosening the bunch and opening it up. Many clones have therefore been selected for their resistance to botrytis, mainly based upon the clone having large, loose bunches, so that individual grapes are not packed tightly together and air and light, plus chemical sprays, can penetrate the bunch easily.

Another aspect which also needs to be borne in mind when looking at Pinot noir clones is the question of colour. Deep-coloured red Pinot noir wine is not something that traditional Burgundian growers set much store by and in fact many old hands in the wine trade, on seeing a dark red Burgundy, will immediately suspect that it has been 'Hermitaged' – the practice of beefing up wines by the addition of Syrah-based Hermitage wine. Many German and New World growers believe that light coloured, slightly brick-red Pinot noir wines are to be avoided and thus select clones giving good colour. They also like clones producing wines with plenty of good fruity characters and one often sees the word 'cherry' mentioned in relation to non-French clones. Whether British growers require botrytis resistant

20 Selection massale vines are produced by taking scion wood from (usually old) vineyards which are not planted with clones and contain vines whose ancestry is typically from a local, well-known and successful vineyard. In this way, clonal variation and local typicity are maintained in new vineyards.

Vines with chlorosis planted in soil with a high active Calcium Carbonate level

Pinot noir clones producing fruity, dark red wine is open to question. For the production of high-acid, white sparkling wine, the answer is probably 'no'. For non-French clones, the German nursery Reben Sibbus has some interesting data: www.sibbus.com

Pinot noir – still wine

For still red wines, the most successful Pinot noir clones in vineyards in Britain appear to be the compact-bunch French clones 113, 114, 115, 667 and 777. There are also good reports of two Geisenheim clones, Gm 20-13 and Gm 20-19, and an Italian clone, SMA 201. I personally prefer 667 and 777. There is also a *teinturier* clone of Pinot noir, called (not unsurprisingly) Pinot teinturier – also known as *Pinot tête-de-nègre* – which a few British vineyards have tried and which gives both good wine and very good colour. Grafted vines of these clones are hard to come by, but it is worth trying if still red wines are the aim. There are several sub-clones of *teinturier* clones including Bury, Guicherd, Mourot and Marsellein clones.

If you start looking into Pinot noir clones, you will find a huge amount of information, much of it confusing and contradictory. Given that there are forty-five Pinot noir clones registered for use in France (including eighteen in Champagne), eight-hundred in total in collections in Alsace, Bourgogne and Champagne, dozens (if not hundreds) in other countries and that many nurseries have their own 'selections', this is a vast, unfathomable subject. Find a great site, get the density right, plant a couple of the clones recommended above, and when they start cropping, manage the yield and canopy properly and, given a fair year, you will produce something interesting – maybe a light red, perhaps a decent rosé. Only time will tell you if you have the right combination for great still red Pinot noir. Of course, with the 85:15 rule applying to labelling i.e. if the wine is at least 85 per cent of one variety, it can carry the name of that variety, there is ample scope to add a wine made from one of the better colour red varieties in order to beef up the colour of a Pinot noir varietal.

Pinot noir – sparkling wine

For sparkling wine, I am not as concerned about the question of clones as I am with still wine. There is so much that happens in sparkling wine production between picking and drinking that can substantially alter the nature, flavour, style and quality of the wine (blending, *sur latte* ageing and *dosage* being the main ones) that the clone is of secondary importance. It would be nice to plant those clones with the lowest acidity and highest yield and in time these will hopefully make themselves known. The following CIVC clones have performed well in British vineyards for sparkling wine: 292, 375, 386, 521, 779, 792, 828, 870, 871, 872 and 927. I also like non-CIVC clones 114, 115, and 459.

Note: The best source of information on French varieties, clones and rootstocks is the Pl@ntGrape website: http://plantgrape.plantnet-project.org. The USA National Grape Registry, based at the University of California, also has a good website for varieties, although clones have all been given FPS (Foundation Plant Services) numbers: http://ngr.ucdavis.edu.

Rootstocks for Great Britain

Note: The planting of ungrafted 'own-root' vines is covered in *Chapter 16: Pest and disease control* in the section on *Phylloxera* and at the end of *Chapter 17: Trunk diseases*.

The question of which of the many dozens of rootstocks used around the world are best for vineyards in Britain is an interesting one. Like the choice of varieties, this task is made harder by the fact that no rootstock has been developed for British conditions and that all rootstock recommendations are based upon data derived from climates which are both warmer and drier than Britain's, from regions where yields are typically higher (often much higher) and more regular, and in soils which may differ considerably from ours. The behaviour of a rootstock in a lean, dry soil, where there is little summer rainfall, where temperatures are often over 30°C and where yields are both higher and more consistent than in Britain, will not be the same as in a deep, fertile soil, well-watered with summer rainfall and where yields might be modest one year, negligible the next and well above average the year after. Likewise, a rootstock in a region where the climate is hot, vines are irrigated and yields are consistently high. Therefore, whenever looking at recommendations for rootstocks to use in Britain, these differences in climate and growing conditions need to be taken in to account.

Rootstock tables (such as the one at the end of this chapter) will usually give you several recommendations concerning not only their active calcium carbonate (CaCO$_3$) tolerance and root growth habit, but also their 'vigour' and 'influence upon maturity'. As I have said above, these recommendations are not based upon British data (another job for Plumpton perhaps?) and therefore a 'health warning' needs to be issued before relying too much upon the tables you can find on the internet and issued by vine nurseries. In my experience, there are other factors which can also have an influence upon the behaviour of the vine, not only the rootstock. The weather during the season undoubtedly has a huge impact upon the vigour of vines: a cool, wet season will promote growth; a hot, dry season will almost certainly curtail vigour. The other very important factor is yield. A vine bearing 5 tonnes-ha (2 tonnes-acre) is going to be more vigorous than one carrying 10–15 tonnes-ha (4–6 tonnes-acre), especially if this is coupled with a warm, dry year (i.e. the sort of year likely to lead to high yields). In 2013 British vineyards were subject to very unusual circumstances. We had a very late spring, an ideal flowering period which gave high yields, a warm spell in the middle of the summer, followed by a cooler, wetter, less sunny October and the latest harvest ever experienced. How could you expect a rootstock to conform to a description based upon completely different circumstances?

In the early days of modern British viticulture, with the exception of a few growers who obtained their vines from France, the bulk of vines planted up until the early 1990s were German varieties, of German origin and consequently, on German rootstocks. The four standard German rootstocks are 125AA, 5BB, 5C and SO4, with Teleki 8B and Börner bringing up the rear. All of these are classified as 'high' to 'medium-high' in terms of vigour and, coupled with already vigorous varieties such as Bacchus, Huxelrebe, Kerner, Müller-Thurgau, Ortega, Reichensteiner and Schönburger, tended to produce, in our cool climate and our fresh, loam-rich, well-watered soils, some exceptionally large, leaf-endowed canopies which did little to help the problems associated with shading and over-vigorous canopies. Of the 'old' varieties, only Seyval blanc, being a naturally low-vigour variety, did not suffer from over-vigorous canopies, one of the reasons why, of almost all of the varieties popular in the early days, it has maintained its position high up in the list of varieties being grown in Britain.

The arrival in Britain of the Champagne varieties has somewhat altered the rootstock situation and, in my view, made a significant impact upon the vigour of our canopies. In France, the origin of the majority (but by no means all) of the Chardonnay, Pinot noir and Meunier vines now being grown in Britain, the most popular rootstocks for these varieties are SO4, Fercal and 41B in high active $CaCO_3$ soils (with 161-49 as another option) and 101-14, 3309C and 420A in soils with 20 per cent or less active $CaCO_3$. With the exception of SO4 and Fercal (both classified as having 'medium' vigour) all of the others are classified as 'low' or 'low medium' in terms of vigour. On paper, this may seem a very marginal difference, but it is my experience that these marginal differences, in the vineyard, make an appreciable difference to vigour of shoot growth and therefore to the density of the canopy. It worries me sometimes that

Rootstock	2018	2019
SO4	40.0%	50.0%
Don't know	12.0%	12.6%
Fercal	9.0%	10.6%
3309C	1.0%	8.4%
41B	6.0%	7.8%
5BB	1.0%	3.1%
Own roots	5.0%	2.8%
Binova	1.0%	2.0%
125AA	2.0%	1.1%
5C	1.0%	0.6%
161-49	1.0%	0.3%
Gravesac	1.0%	0.3%
Others	20.0%	0.6%
Total	**100.0%**	**100.0%**
Source: 2018-19 ICCWS-WineGB Yield Surveys		

Table 25

growers prefer to get their Chardonnay and Pinot noir from Germany and on German rootstocks without realising the difference the rootstock can make.

The first thing you need to know before you can select a rootstock is whether you are likely to need lime-tolerant rootstocks. In soils of pH 7 or lower, and without any obvious signs of chalk or limestone, the level of active $CaCO_3$ in your soil is likely to be low. In soils of over pH 7, the best and safest advice is to test for the level of active $CaCO_3$. This is NOT the same as the level of total $CaCO_3$ – they are related, but not in a linear fashion and it is not, in my view, safe to rely on taking 50 per cent of the total level of $CaCO_3$ as being the level of the active $CaCO_3$ (something I have seen suggested is acceptable) as this is not accurate enough. The reason that the level of active $CaCO_3$ is important is that if the rootstock selected does not have the right level of tolerance, the vines will get chlorosis, a condition where they are unable to access iron in the soil and are therefore unable to produce chlorophyll, necessary for photosynthesis to take place. Vines suffering from chlorosis will turn light green, then yellow, and then almost white, before starting to die. Vines suffering from chlorosis can be treated by spraying or soil drenching with chelated iron, iron that has been treated so that it is soluble (Ferleaf and Fersoil are the two best known products), and whilst this may be a temporary solution, it is much better to select the right rootstock in the first place. In France, rootstocks are given an 'IPC' rating (l'Indice de Pouvoir

Rootstocks for vineyards in Great Britain					
Rootstock	Crossing	Active CaCO3 tolerance	Root growth	Vigour	Influence on Maturity
Riparia Gloire (RGM)	Riparia	6%	Shallow	Low	Advances
101-14	Riparia x Rupestris	10%	Shallow	Low	Advances
3309C	Riparia x Rupestris	11%	Semi-deep	Low-Medium	Advances
Gravesac	161-49 x 3309C	15%	Semi-deep	Medium	Advances
Börner	Riparia x Cinerea	15%	Semi-deep	Medium-High	No effect
125AA	Berlandieri x Riparia	17%	Deep	Medium-High	No effect
5C	Berlandieri x Riparia	17%	Deep	Medium-High	Slightly Advances
420A	Berlandieri x Riparia	20%	Semi-deep	Low-Medium	Slightly Advances
SO4	Berlandieri x Riparia	20%	Semi-deep	Medium+	No effect
Binova	SO4 Mutation	20%	Semi-deep	Medium++	No effect
161-49C	Riparia x Berlandieri	25%	Semi-deep	Low-Medium	Advances
41B	Chasselas x Berlandieri	40%	Shallow	Low-Medium	Delays
Fercal	Berlandieri x Riparia	45%	Deep	Medium	Advances
Note: Recomendations on root growth, vigour and influence on maturity are based upon non-UK data					

Table 26

Clorosant) which runs from 0 to 120, with rootstocks at the lower end (0–20) being safe in soils with no or low levels of active $CaCO_3$ and those with higher ratings (30–120) being better in soils with high levels of active $CaCO_3$. I have never used this index myself and in practice find that the active $CaCO_3$ measurement is more than adequate.

When it comes to active $CaCO_3$ levels, I would prefer to err on the side of caution as, however well done, sampling a field can only tell you about a small section of the whole field and there may well be patches of ground where levels are higher than that shown by sampling. Therefore, for levels of active $CaCO_3$ of 15 per cent of more, I would only plant on 41B or Fercal. 41B to me is the better rootstock, even though it is listed as having a delaying effect upon maturity. I find it less vigorous than Fercal. Unless you are really sure about the accuracy of your soil sampling, I would avoid SO4 at active $CaCO_3$ levels of 15 per cent or more, even though the tolerance of SO4 is generally stated to be 20 per cent. Rootstock 161-49 is also an option, but whilst in some situations I have found this to be excellent, producing a low-vigour canopy and good crops, in other situations it has struggled, especially when paired with Chardonnay. With Pinot noir and Meunier it has proved to be better. Rootstock 161-49 suffers from what I call 'tylosis' (and which the French call *thyllose* or sometimes *folletage*) which can result in a higher percentage of early vine deaths than would normally be expected with other rootstocks. I would therefore advise caution with 161-49, especially in situations where the vine might be put under stress in its first two years. Such situations might be poorly drained sites, adverse soil conditions, drought, weed competition, or early cropping. If 161-49 is used, I would recommend keeping the young vines as weed-free as possible and

cultivating the inter-row areas for the first two full growing seasons before grassing down.

In situations where levels of active $CaCO_3$ are lower than 15 per cent, my first choice would be 101-14, 3309C and 420A, all of which are at the lower end of the spectrum for vigour and have a slight advancing effect upon maturity. If these were not available, I would select 5C, 125AA and SO4, listed in increasing order of vigour, with SO4 being the most vigorous. However, in soils with low levels of active $CaCO_3$, there is nothing stopping you using 41B or Fercal and I have planted plenty of vines in soils with pH 6.5 using these rootstocks. For very acid soils i.e. pH 5.5 or below, the only rootstock that appears to be recommended in the literature is Gravesac. In some sources it is suggested that 101-14, 3309C and SO4 might also suitable, although in other sources this is contradicted, and before one were to embark on a large-scale planting on soils with this level of pH, some further research on rootstocks would be advised. Luckily there are few soils in British fruit-growing regions with this level of pH and, in any event, soils with pH levels below 6.0–6.5 must be limed in an attempt to bring them up towards neutral (pH 7).

Two other factors must be considered when selecting rootstocks: soil and vine variety. With deep, free-rooting soils, the best rootstocks are going to be ones with lower vigour. This does not mean that more vigorous rootstocks cannot be used, but in order to have an open and airy canopy, you will probably need to do more canopy management (trimming and deleafing) than with lower-vigour rootstocks. Likewise, with vigorous varieties such as Bacchus, Reichensteiner, Müller-Thurgau and other German varieties, I would always aim for lower-vigour rootstocks, although 101-14, 3309C and 420A are seldom available from German nurseries unless custom-grafted. Pinot noir can also be quite vigorous and is best on a lower-vigour rootstock, although Chardonnay appears to do well on both. Only with Seyval blanc, which is naturally low in vigour, might I look for a rootstock with a slightly higher vigour potential, and I have known it do well on SO4 and 5BB. On deep, vigorous soils, 420A also suits it.

Apart from those mentioned above, there are many other rootstocks which may be perfectly acceptable in British vineyards. I have used both Gravesac and Riparia Gloire de Montpelier (RGM) in vineyards and they performed well, although RGM does not appear to be that low in vigour. However, they are not widely used and are not always easy to obtain. Yalumba Nursery has both an excellent 'Rootstock Selector' as well as a lot of information about individual rootstocks on its website (www.yalumbanursery.com), although of course aimed at Australian growers. Mercier Nursery also has good information in English on their Californian website (www.mercier-california.com) under 'Varieties' and then 'Rootstocks'.

A final warning concerns vine nurseries. They are in the business of selling as many vines as possible from the vines they graft and seeing those vines survive and thrive in their client's vineyards. Some rootstocks graft more easily than others, giving a higher percentage of 'take', meaning that once lifted and on the sorting table, far fewer are rejected. Rootstocks with higher vigour also tend to root more easily, giving better roots at lifting from their seven to eight months in the nursery and also tend to put down roots more easily once planted out in the vineyard. Fercal and SO4 are well known as falling into this category and are well liked by nurseries. However, that does not mean that these vines always produce the best vineyards and the best crops in the long term. Under British conditions, with most situations and vine varieties, I prefer a rootstock with a slightly lower-vigour potential.

Chapter 5

Pruning and trellising systems for vineyards in Great Britain

A short history of vine training in Great Britain

There are not only dozens of completely different pruning and trellising systems in use in commercial vineyards throughout the world, but even within a general style of pruning, there are also an infinite number of individual variations based upon personal choice, the different materials employed to build the trelliswork and the type of equipment used in the vineyard. Having said that, in certain regions – old world ones such as Bordeaux or Burgundy – there is an almost universal uniformity about how vines are pruned and trained, although row widths, inter-vine distances and therefore planting densities, will still vary. However, drive along the St. Helena Highway, which runs up the centre of the Napa Valley, and you can see almost every system and every vine density known to man being used. Which, therefore, is the correct system for British vineyards?

In the very early days of the revival, most vineyards looked very French. Ray Brock at his Oxted Viticultural Research Station planted most of his vines at 3-feet (0.91 m) between the rows and the vines 4-feet (1.22 m) apart, giving a vine density of 9,007 vines-ha (3,645 vines-acre). The vines were always *Guyot* trained and very low to the ground. Brock remained convinced throughout the life of his Oxted vineyard that close planting encouraged root competition and that vines close to the ground ripened their fruit better as they absorbed heat from the soil.[1]

In 1952, Major-General Sir Guy Salisbury-Jones, who had visited Oxted and knew Brock well, planted his vineyard at Hambledon in a very Burgundian style, with 4 feet (1.22 m) between the rows and the vines 3 feet (0.91 m) apart, giving the same densities as Brock was using at Oxted. With such narrow rows, specialist equipment was needed and as the vineyard got bigger and spraying by hand was no longer an option, Salisbury-Jones bought a Jacquet *tracteur-enjambeur* (straddle tractor) to make life easier. As the vineyard expanded, so did the row width and inter-vine spacing. In 1982, an experiment was tried with 8 feet (2.44 m) wide rows and 4 feet (1.22 m) between the vines, with the vines trained to a single high trellis – they called it a 'modified Geneva Double Curtain system' – and eventually, after the Major-General's death in 1985 and in the hands of a new owner, additional vineyards were planted using a full-blown GDC system at 3.5 m row width and 2.0 m between the vines with 1,429 vines-ha (578 vines-acre). The vineyards stayed that way, being gradually reduced as owners came and went, until by 1999 only around 2-ha (5-acres) remained. By 2020, under new ownership, Hambledon had been replanted and greatly expanded with 90.29-ha (223-acres) of vineyards, mostly planted at 2.20 m x 1.2 m and 1.1 m.

Jack Ward, who spoke fluent German, started planting vineyards at Horam Manor in 1954 and took a much more Germanic approach to British viticulture. He visited Geisenheim on at least one occasion, where he saw Professor Helmut Becker who introduced him to Geisenheim varieties and growing techniques. Ward, as has already been mentioned, was the Chairman of the EVA, and also ran what was known as the Merrydown Co-operative Scheme which made wine under contract in exchange for a proportion of the grapes which were taken in lieu of payment and used to make a blended English wine called 'Anderida'. He always made time for new growers, myself included, tirelessly promoting the cause of English and Welsh wine. Merrydown, the company he co-owned, also

---

1 In June 1961, Brock had heard from Nelson Shaulis, inventor of the GDC training system (and see next section for his biographical details), who said that he would be coming to England in September and wished to visit Oxted, which he duly did. In a letter sent in November that year, Shaulis thanked Brock for his hospitality, saying that the noon meal: 'was very delightful, especially the ginger beer shandy'. Shaulis also said that he was just concluding the harvest at Geneva and yields had been between 3 and 9 tons per acre. What Brock must have made of Shaulis' views on trellising is not recorded, but it would be interesting to know.

imported vines and he was instrumental in popularising what were then newer varieties such as Bacchus, Reichensteiner and Huxelrebe in Britain.[2]

Ward's initial plantings were with 5' (1.52 m) rows and 4' (1.22 m) between the vines, a density of 5,393 vines-ha (2,182 vines-acre) and his book, *The Complete Book of Vine Growing in the British Isles* (published in 1984), says that for those with specialist 'vineyard tractors' these distances and densities were the right ones. He also said that farmers who were 'not prepared to invest in a vineyard tractor' would need to plant at 7'6" (2.29 m) row width. On inter-vine distances, he advocated planting according to the vigour of the rootstock, with weaker rootstocks at 4' (1.22 m) and on 'richer soils' and with 'strong rootstocks' an inter-vine distance of 6'6" (1.98 m) could be used.

High and wide – the rise and fall of GDC

Jack Ward also favoured Dr Lenz Moser's eponymous system of high training (*Hochkulture*) which was developed in Austria in the 1920s and found favour both there and in Germany in the 1950s, '60s and '70s when the price of wine came under pressure and growers sought ways of lowering input costs.[3] The system was essentially a wide-rowed, single curtain system, very similar to the 'ballerina' variant of the Smart-Dyson system (which was developed in the early 1980s) and based upon a row width of 3.0 m to 3.5 m with a fruiting wire at around 1.2 m to 1.3 m from the ground – hence the name 'high culture'. These spacings enabled growers to convert existing narrow-rowed *Guyot* systems to Lenz Moser by removing every other row and re-training their vines. Nelson Shaulis, Professor of Pomology and Viticulture at Cornell University, Geneva, New York, (and Richard Smart's mentor and PhD supervisor) developed

GDC vineyard at Pheasants Ridge near Henley-on-Thames, Oxfordshire, now grubbed

his GDC system in the 1960s[4] borrowing freely from Lenz Moser and the experiences of producers who had used it. GDC was developed for growers of the vigorous *Vitis labrusca* variety 'Concord' which in New York State was (and still is) mainly grown for making low-value grape juice and grape jelly; high yields, low establishment costs and low running costs (including machine harvesting) were required. It was never envisaged that it would be suitable for growing high-quality *vinifera* wine grapes, although in some countries it is now used for this purpose.

Which vineyard was the first in Britain to be established using GDC is open to question, but the best known were New Hall, near Maldon in Essex, planted by Bill Greenwood in 1968 and Westbury, near Reading, Berkshire, planted by Bernard Theobald in 1969. These two growers, together with David Carr Taylor, who planted his vineyard at Westfield, near Hastings, East Sussex, in 1973 and A. H. Holmes at Wraxall, near Shepton Mallet in Somerset, who planted in 1974, were not the only growers using GDC, but certainly amongst the most vocal in their support for the system.

Standard GDC has rows at 12-feet (3.66-m) wide and vines spaced at 8-feet (2.44-metres) giving a vine density of 1,120 vines-ha (453 vines-acre). Its attractions – on the surface – are obvious: the ability to use standard farm tractors, mowers and sprayers down its wide rows; the low vine density, meaning that a GDC vineyard was much cheaper to plant, establish and trellis than a high-density *Guyot* vineyard; and its relatively low labour input, once it was established and cropping. These were all attractive attributes, especially to growers with little experience of growing vines on any system. On the downside, GDC vineyards, especially those planted on marginal, windy sites, were often very difficult to establish and the wide rows and sparse vines do not create the microclimate seen in a more densely planted *Guyot* vineyard. In addition,

---

2 Ward also employed two people who each in their own way helped change the face of English and Welsh wine: Christopher Lindlar and Greg Williams. Lindlar (usually called Kit) was one of the most experienced winemakers working in Britain in the late 1970s to mid-1990s. After Ward retired and the board of Merrydown decided to end their involvement with English and Welsh wine in 1980, Lindlar ran a contract winemaking business based at Biddenden Vineyards, and then built his own winery, High Weald Winery, at Grafty Green, near Maidstone, Kent where he co-made the first three vintages of Nyetimber (1992–1994) as well as a large number of other wines. After he closed the winery, he became winemaker at Denbies for two years before retiring from winemaking to take holy orders. I guess that his experience of praying for good weather and good winemaking conditions gave him something of an advantage at theological college! Williams looked after the selling of vines whilst at Merrydown and, after their involvement with English wine ended, he moved to Tenterden and continued to sell vines for a number of years. In 1988, tragically, he fell downstairs and was killed. A fuller account of Jack Ward and Horam Manor can be found in Appendix VIII.

3 Dr Lenz Moser III (1905-78) came from a long-established family of Austrian winegrowers and developed his system of high training in an effort to reduce both establishment and running costs. He wrote and self-published a book called Weinbau Einmal Anders which translates as 'Viticulture will one day be different'. Professor Becker used to say the book's title meant 'try the new system, but only try it once' as once you had tried it, you never wanted to use it again!

4 Smart and Robinson in Sunlight into Wine say that a paper describing the GDC system was published in 1966 and that it 'was a milestone in the development of canopy management theory, elegantly establishing that shade within the canopy was the principal limit to production of quality grapes'.

most GDC vineyards, even well-managed and well-sited ones, take at least five years to come into full cropping and, although crops can be good, the vineyards are often not quite so labour-saving as many claimed.

The pro-GDC growers, notably Theobald and Carr Taylor, would hear nothing against the system and encouraged growers to plant what, at the time, were considered to be huge vineyards, almost all of which eventually disappeared. Karl-Heinz Johner, who ran Lamberhurst from 1976–88, was also attracted to GDC when it came to advising growers, many of whom were under contract to supply grapes to Lamberhurst. Whether this enthusiasm on the part of these people was because they thought it was the best way of growing grapes in Britain or was a way of producing grapes as cheaply as possible so that grape prices were low, I will leave to speculation. However, whatever the reasons, the GDC system was popular with growers who wanted to get into owning a vineyard with (a) the minimum of capital cost and (b) the minimum of effort in growing grapes. The results often showed.

GDC growers often wanted an 'easy-care' system of growing vines – 'ranching' I called it – and were not always really committed to the attention to detail that each individual vine calls for in any climate, let alone a challenging climate like Britain's. Consequently, many GDC vineyards established poorly – if at all – with multiple misses and re-plants and some taking forever to fill out the wires and achieve a full canopy of fruiting wood. With each vine occupying almost nine square metres, one missing or poorly performing GDC vine has four times the impact that a missing or poorly performing *Guyot* vine has. Whether the failure[5] rate of GDC vineyards was any higher than that of high-density *Guyot* vineyards it is impossible to tell – so many vineyards from the 1970s and 1980s of both types have disappeared – and the reasons have been varied.

Many of the failed GDC vineyards were quite large and prominent and therefore stood out more than the others, perhaps, but either way, there are almost no GDC vineyards left from the 1970s (Carr Taylor being a notable exception). The roll-call of large GDC vineyards that came and went is long and these are only a few of the largest, all in the South East: Ashburnhams (7-ha), Barkham Manor (14-ha), Chiddingstone (14 ha), Great Shoesmiths (7-ha), Honeybee (11-ha), Lamberhurst (20-ha), Leeford (14-ha). I cannot claim complete immunity from the desire to plant a GDC vineyard and in 1980, whilst at Tenterden Vineyards, having planted around 4-ha (10-acres) of 2.0 m wide *Guyot* vineyards, I did succumb and planted 2-ha (5-acres) of GDC in order to see just how it performed. Sadly, it took so long to establish that I sold up and moved on before I had a chance to make comparisons.

The main problem with GDC, at least as it is practised

in Britain, is that the canopy is very often a tangled mass of shoots and side-shoots and the grapes are often hidden below several layers of foliage and require considerable hand-work to expose them. Although the buds for the coming year are produced at the top of the GDC canopy, and therefore exposed to air and light, because most sprayers are not made to spray from the top down and from the bottom up, it is more difficult to keep disease under control in GDC vineyards. Apart from damage to grapes, the effect of disease – especially botrytis on the fruiting buds for the next year's crop – is also an issue. Although GDC is sold as a 'low input' system, it is my experience that the vines are often just as vigorous as their cane-pruned compatriots (so much for the 'big vine' theory which suggests that the bigger the vine the lower will be the vigour) and in order to manage GDC vines well, a considerable amount of shoot positioning, shoot removal and 'combing-out' of shoots between the curtains is required to expose the fruit and buds/canes. Given that some growers only used GDC because they had been advised that it was a 'plant and forget' system, this work often did not get done.

As you can tell from the above, I am not really a fan of GDC-trained vines! Having said that, I knew of two (albeit fairly small) vineyards planted on GDC, which used to use my consultancy service, where crops were produced and results were acceptable and sometimes very good, so it is possible to make GDC work in Britain. However, apart from my own small experiment whilst at Tenterden, I have never advised a grower to plant a vineyard using GDC and cannot really see any circumstances in which I would. I am also not aware of any GDC vineyards having been planted in Britain in recent years.

Vine density

Vertically Shoot Positioned (VSP) training

By the time I came to plant my first vineyards in 1977, having spent almost two years working in German vineyards and studying at Geisenheim, I was firmly of the opinion that a 2.00 m row width was right for our conditions, with vines spaced at between 1.1 m and 1.3 m in the row – classic cane pruned Vertically Shoot Positioned (VSP) training – and I bought a David Brown 885 Narrow tractor which, at 1.25 m wide, fitted nicely down the rows. At the time, the ferocious GDC vs *Guyot* debate was going on between growers and every new edition of the Grape Press seemed to contain a letter or two, usually one from a 'pro' and one from an 'anti' extolling the virtues of their pet system and spacing. As a newcomer, I held back at first, but it wasn't too long before I waded in on the side of the anti-GDC brigade. It seemed to me that although GDC had some obvious advantages, the results didn't really show it to be a system that could produce the best quality grapes (and therefore wine) and I felt that whatever the savings in establishment and running costs, if the end result was worse wine, I was against it. I have always

---

5 I use the term 'failure' to mean any vineyard that was grubbed out before the end of its useful life. It could have 'failed' because the vines died, the wine it produced was poor quality, or the producer had problems selling the wine. Whatever the reason, the vineyard is no longer.

taken the view that the establishment and running costs of a vineyard must always take second place to the single most important and long-lasting factor: the production of the highest quality wine that any particular site is capable of. Of course, establishment costs and running costs have to be taken into account, as well as yields, which are also an important factor in the cost of production, but GDC certainly didn't seem to me to offer enough advantages to outweigh its failings. With grapes worth at between £1,500–£3,000 per tonne, and considerably more if you are selling wine, it doesn't take much of an increase in yield (which I believe is possible with cane pruned VSP-trained vines) to pay for any additional costs. Add in the unquantifiable bonus of better wines which, in the long term, has to be every grower's aim, and the question of which pruning and trellising system to use in British vineyards is, I believe, easily answered. In order to justify why I consider that a high-density *Guyot* system most suits vineyards in Britain, I offer the following evidence:

- Almost all northern European winegrowing regions – most of Germany, all of the Loire, Champagne, Burgundy and Bordeaux – as well as many other cool winegrowing regions around the world, use high-density *Guyot* cane pruned systems. Yes, there are a few exceptions ('Chablis' cordon pruning on Chardonnay in Champagne, for example) and with winegrowers in some places being free to do exactly what they want, there are bound to be other systems used. But these are very much in the minority. In many cases, narrow rows and cane pruned vines have been the norm for centuries and whilst this is no reason

Classic 2 m wide VSP training at Yotes Court Estate, Kent

not to innovate, it ought to make one stop and think.

- Narrow rows and high densities help create a better vineyard microclimate than you get in low-density vineyards. The rows are closer together and, as the fruiting wire is nearer the ground, there is less air movement below the vines. The idea that air movement beneath a vine's canopy is necessary 'to keep disease away' (something I have heard so many times from vineyard owners) is just wrong and in VSP *Guyot* trained vines with a fruiting wire at, say, 700–800 mm, the air movement can be kept to a minimum. The aim is to keep as much heat in the vineyard as possible. If there are disease problems, open up the canopy and spray.

- It is said that having the grapes nearer the ground is beneficial as they get some heat from the soil, which is released as the sun goes down. This is actually very hard to prove and is probably just an old wives' tales we accept as being true. Heat absorption by the soil depends almost entirely on what the soil type is and/or what the soil covering is. Bare dark soil will absorb heat and, in some regions, the Mosel for instance, many soils are bare and covered with dark slate. Light soils, chalk and limestone, for instance, will reflect light and heat back into the canopy, but of course only when the sun is out. Typically, in Britain we have grass cover on most of our soils which, as the sun goes down, gets wet with dew; the amount of heat absorbed by the soil and which might be released later is minimal. However, I do think that the immediate area beneath the vines should be kept as weed-free as possible. This reduces the amount of moisture below the vines, which will help with disease control, and will help keep the soil temperature in this area as high as possible. Of course, the narrower the rows, the higher will be the percentage of the vineyard which is (in theory) weed-free.

- An individual vine is able to nurture and ripen a smaller quantity of grapes more easily that a larger quantity. A single vine occupying 9 m² (GDC spacings) needs to ripen 9 kg of grapes at 10 tonnes-ha. A vine occupying 2.4 m² (standard VSP spacings) carries 2.4 kg of grapes for the same yield per hectare. In addition, a vine in a low-density vineyard has a large amount of permanent wood to support and the grapes are often several metres from the roots.

- The fruit from cane pruned vines in cool climates appears to be intrinsically better than the fruit from trellising systems which are spur-pruned. Whether this is because the fruit comes from buds borne on canes or whether it is because the canopy is easier to manage is open to debate. If the rows in a VSP *Guyot* trained vine are orientated north–south, then each side will get sunshine during the day. With GDC trained vines this is not the case, as the fruit is on one side of the curtain or the other. Low-density GDC trained vines, in order to carry the same weight of grapes for the area they occupy, have to be spur-pruned. Vines planted at 2.44 m apart cannot grow canes long enough to be cane pruned.

- The higher the density of the vines, the greater will be

Vine density	Median yield 2018	Median yield 2019	Average 2018-19 t-ha	Average 2018-19 t-a
Less than 2,000 vines per ha	3.00	2.51	2.76	1.11
2,000-3,000 vines per ha	6.20	5.02	5.61	2.27
3,000-4,000 vines per ha	9.19	6.38	7.79	3.15
More than 4,000 vines per ha	10.00	5.94	7.97	3.22
Source: ICCWS-WineGB Yield Surveys 2018-19				

Table 27

the element of root competition for the available water and nutrients. In cool climates, especially in 'ultra-cool' climates like Britain's, we usually have more than enough water to support vine growth and our soils are typically well supplied with nutrients. Remember that the rainfall per hectare is the same whether there are 1,000 vines-ha or 10,000 and therefore the higher the density, the less water each individual vine has. Remember also, that leaf wall height dictates the number of leaves per vine, so a high density, plus the maximum height of leaf wall (dictated by the row width) will give you many more leaves from which moisture can transpire, thus lessening water availability. Many of the problems in our vineyards are associated with over-vigorous vines which lead to crowded and shaded canopies. Root competition for the available water and nutrients is not something that happens the instant a vine is planted, but after the first few years this kicks in and helps towards overall vigour control. A GDC vine, despite being physically several times larger than a high-density planted vine, has much more soil into which to expand until it meets competition from its neighbours and, in my experience, may never get to the point where its roots are in competition with other vines. In *The Grapevine* (p.51), the authors state: 'Higher density planting generally results in a reduced root system per vine but a higher density per unit soil volume – this causes more rapid water extraction in the spring. Increased planting density tends to produce deeper roots where soil properties allow this to occur'.

- The leaf-wall and the canopy of a high-density, VSP trained vine is relatively thin, hopefully no more than four to five leaves from one side to the other, so that the grapes can be plainly seen and there are plenty of gaps. A canopy like this allows air and light to reach the grapes being ripened, as well as the buds being produced for the following year's fruiting. This is probably THE most important factor in canopy quality.
- A relatively thin canopy also enables chemical sprays to penetrate it, keeping both grapes and buds healthy. Getting chemicals, especially anti-botrytis chemicals, into the flowers and into the grapes before bunch-close,[6] is a

major factor in disease control.
- A VSP system also allows you to have a good height of leaf-wall which will help in the battle against high vigour. The higher the leaf-wall, the greater the number of leaves and, therefore, the rate of transpiration is greater thus putting more demands on the available water. (The height of the leaf-wall is discussed in some detail in *Chapter 9: Trellising systems – construction* in the section on *Posts*.)
- It is much easier to carry out the summer canopy management tasks on a relatively thin VSP trained leaf-wall, as operations such as deleafing and 'hedging' (trimming top and sides) can be done efficiently with a machine. Deleafing can also be carried out by hand more easily on a VSP than a GDC system. This is not to say it cannot be done with GDC trained vines – it's just more time-consuming and therefore less likely to get done well or even be done at all.
- Fruit thinning, whether post-flowering, post-*véraison* or pre-harvest, can really help improve quality, and is much easier on a vertical plane where the fruit faces the operator, than on GDC trained vines where the fruit is all over the place. Again, it's not that it cannot be done, it's just that it's more difficult and more time-consuming.
- A high-density VSP *Guyot* pruned system is simple to install and manage and is easy for those looking after the vines to get to grips with and there are no difficult concepts to understand.
- As part of the 2018 and 2019 ICCWS-WineGB Yield Surveys, questions about yields in relation to both vine density and row width (which of course are related) were asked and the results analysed. As can be seen from Table 27, in the two years surveyed, as vine densities increase, so do yields.

'Blondin' training system

In 1998 I was sponsored by the Canterbury Farmer's Club (Canterbury in Kent, that is) to travel to New Zealand for three weeks to study their viticultural practices. One of the nuggets I returned with was the idea that a 'high-trained,

---

6 Bunch-close is the point where the immature grapes on a bunch expand so that they start touching each other and make access by fungicides to the interior of the bunch virtually impossible.

Pruning and trellising systems for vineyards in Great Britain

Sylvoz' training system, such as I had seen at Matt Dicey's vineyard off the Felton Road in Central Otago, might suit British growers for its ease of management and lower costs of production. I had already tried this system with a few rows of Seyval blanc in my Scott's Hall vineyard (planted in 1988) and had found it to be a low input system giving good results with regard to both quantity and quality. This system uses rows at around 2.40 m wide and an inter-vine distance of 1.80 m if spur-pruned (giving a density of 2,315 vines-ha or 937 vines-acre) or 1.20 m if cane pruned (giving a density of 3,472 vines-ha or 1,405 vines-acre). I named this the 'Blondin' system (after the first person to walk across the Niagara Falls on a single-wire tightrope). The single wire needs to be at around 1.50–1.80 m off the ground, depending on the variety grown and the size of your tractor, although if you have your single wire at 1.80 m (assuming you are not very tall) it also enables you to walk across the vineyard at right angles to the rows, ducking under the wires – very useful for inspecting the vineyard! By planting the vines in pairs, one either side of an intermediate post or planting stake (one vine trained in one direction and one in the other direction), the number of posts and planting stakes can be kept to a minimum. The only problem with this type of irregular planting is that it precludes machine-planting.

This system has all the advantages of GDC – lower establishment costs, ease of working, and downward facing canes – without the tangle of shoots and side-shoots and the work involved with GDC canopy management. Pruning, whether spur or cane, is much easier than with VSP as most of the wood to be pruned will be hanging free of the wire and will simply drop to the ground. As the vines are high off the ground, the weed control need not be quite so good and indeed, I have seen vineyards trained like this where 'mow-and-throw' has been used (see *Chapter 14: Weed control* for an explanation of this technique). Taken together, this is a lower cost system – lower establishment costs and lower annual management costs – which could be useful for growers selling lower-value wines (still wines) or to producers selling just grapes. Bolney Vineyards have several hectares trained to this system and regularly win prizes for their wines, so there don't seem to be any ripening or quality issues. I will not pretend that it suits all varieties and it needs correct management, as with any training and trellising system. For some varieties and situations, however, it might be worthwhile trying.

Row width

My experience in growing vines in Britain over the last forty years is that rows should be, within reason, as narrow as possible. We do not need the ultra-narrow rows that they have in some European vineyards – 1.0 m wide is not unknown in areas where *tracteur-enjambeur* (straddle tractors) are used – but in Britain, rows of between 1.75 m and 2.25 m ought to be the aim. In order to have enough space between the tractor wheels and the canopy, allow

Wide-rowed vineyards like this one are only viable in hot climates where irrigated, spur-pruned vines deliver high yields

around 350 mm between the edge of the tractor wheel and the centreline of the row, i.e. add 700 mm to the width of your tractor for a safe working row width. For instance, for rows (accurately) planted at 2.00 m wide, a tractor of no more than 1.30 m overall width ought to be used. When looking for a tractor, always see if the wheels can be moved in (the rims can often be reversed so that they stick in rather than out) and often narrower tyres will shave a few centimetres off the width. Once you have selected a tractor, make sure all other equipment (mowers, sprayers and trailers) are no wider (or very little wider) than the tractor. Having rows with more distance between the outside of the tractor and the vines themselves is just a waste of good vineyard space and, for reasons set out below, will not make for as good a vineyard.

Remember also when choosing a row width that in cane pruned vines, where the fruiting wood is trained along the fruiting wire, the narrower the row, the greater will be the number of fruiting buds and therefore the higher will be the potential yield. In a vineyard with a row width of 1.75 m, there are 5,714 running metres of fruiting wire (and therefore of productive canopy) per hectare. For the sake of the argument, if one says that 90 per cent of the wire can be filled with cane, and says that there is one bud per 100 mm (ten per metre), then the number of buds in the vineyard per hectare is 5,714 x 90 per cent x 10 = 51,426 buds. However, take a vineyard with 2.25 m rows (only 500 mm wider) and the running metres of fruiting wood is 4,444 per hectare. The bud number equation is therefore 4,444 x 90 per cent x 10 = 39,996, over 22 per cent less than in the vineyard with a 1.75 m row width. In other words, if the vines in both vineyards are planted at the same inter-vine distances and both pruned identically to two flat canes, each vine in the wider rowed vineyards has got to work that much harder and carry 22 per cent more crop for the same overall yield. At 10 tonnes-ha (4-tonnes-acre) in the vineyard with 1.75 m rows, each vine will carry 2.10 kg of grapes; in the vineyard with 2.25 m rows, each vine will have to carry 2.70 kg of grapes to bear the same yield per hectare. This is not likely to happen.

Row width	No of parcels	Median yield 2018	Median yield 2019	Average of median yields 2018-19
Below 2.0m	39	5.40	5.00	5.20
2.0m - 2.1m	72	10.00	5.59	7.79
2.2m - 2.3m	98	8.40	6.02	7.21
2.4m - 2.5m	47	7.89	5.70	6.80
Over 2.6m	20	5.75	3.39	4.57
Source: ICCWS-WineGB Yield Surveys 2018-19				

Table 28

Where are these extra grapes to come from? Are the buds going to be more fruitful because the rows are wider? No. Are the bunches of grapes going to be more plentiful and larger? No. You cannot lay down more canes as you do not have the space along the row. You could adopt an arched-cane system – arched *Guyot* or *Pendlebogen* – but for reasons which I will point out later, this is not to be advised except under certain circumstances. You cannot make the vines work any harder than they already are. In addition, the narrower rows will create a better microclimate, there will be more root competition (and it will arrive sooner) and you will get better vigour control. For these reasons, the vines in the narrower rowed vineyard will be happier and more productive than the vines in the wider rowed vineyard. On the downside, the vineyard has more vines and more trellising, so, yes, establishment and running costs will be higher, but it is my contention that the benefits of narrower rows far outweigh the negatives and, over the 30–40 year life of the vineyard, the additional investment to begin with and the additional running costs should not be an issue. As can be seen from Table 28, the highest yields in the two years surveyed are found between 2.00 m and 2.10 m and that for each 20 cm of additional row width, yields drop by about 500 kgs per hectare.

A row width of 1.75 m is quite narrow for Britain, but I had a small vineyard planted at this width at Scott's Hall, near Ashford, Kent which did well (the sparkling rosé won the President's Trophy in 1994). In recent years, Henry Laithwaite planted both the Marlow Estate vineyards at Pump Lane, Marlow, Buckinghamshire in 2010 and the Mezel Hill vineyard in Windsor Great Park in 2011 with 1.80 m-wide rows and 800 mm and 1.0 m inter-vine distances and he has been achieving both good yields and good quality. His is one of the first Champagne-variety vineyards to flower and go through *véraison*, and he starts harvesting around one week earlier than other local vineyards. For me, 2.00 m is the optimum width for British vineyards, giving a high enough density of vines, coupled with enough length of fruiting wire to give the crop aimed for; there are plenty of tractors available at or under 1.30 m overall width. Now that the question of row width has been debated, the question of inter-vine distances needs to be resolved.

Inter-vine distance

The distance between vines – the inter-vine distance – is, of course a major factor in overall vine density, so row width and inter-vine distances need to be considered together. With cane pruning, the main question to be decided before you set the distance between vines is whether you have single-cane (single *Guyot*) or two-cane (double *Guyot*) pruning. The canes which are going to become your fruiting wood for the coming year typically stem from a two-bud spur (or 'thumb') that was left at pruning the year before. Alternatively, they come from a suitably positioned point around the 'head' (also sometimes called the 'crown') of the vine. The length of any cane is dictated by (a) the height of the leaf-wall, as all canes will be trimmed to this length and more importantly (b) how much of the cane is good, ripe wood containing viable fruit buds. Typically, the last 25 per cent of the cane, which is of course the youngest part of the cane, will not have ripened well (if at all) and will be too thin and weedy to be part of next year's fruiting wood.

In recent years, inter-vine spacings have been coming down in many British vineyards for a variety of reasons. When we were only planting German crosses plus Seyval blanc, inter-vine distances on cane pruned vines were typically 1.20 m to 1.30 m, with some growers opting for even greater inter-vine distances, and almost all VSP vines were two-cane *Guyot* pruned. Since the mass planting of Chardonnay and Pinot noir, and other harder-to-ripen varieties, inter-vine spacings have come down, with 1.00 m in many vineyards and even as close as 850 mm in others, with much more single-cane *Guyot* pruning to be seen. The reasons for this have already been aired in the section on *Why high-density VSP training?* so no need to go through them again. Therefore, for single-cane pruning, I would advise an inter-vine distance of no greater than 1.00 m and for two-cane pruning, a maximum of 1.40 m, with my preferred distance being 1.20 m.

Arched canes

In the early 1970s and '80s, many growers in Britain, myself included, used arched canes. My first vineyards, planted when I returned from Germany in 1977, were spaced at 2.00 m x 1.10 m to 1.30 m and trained to two-cane *Pendlebogen* (pendulum arch) training which was an exaggerated arch where the cane is arched over one fruiting wire and tied down to a lower one, with the two fruiting wires around 300 mm apart. In Germany this is a very common pruning system and, in vineyards where a high yield is required, is almost universal. The reason is that, depending on the space between your two fruiting wires, the length of cane increases by between 10 per cent and 25 per cent, thus increasing the number of buds on it, which of course leads to higher yields. In Germany, high yields are important to many growers, especially those selling lower classification wines and/or belonging to co-operatives where their payments are weight-based. Yields of 100–150 hl-ha (14–21 tonnes-ha or 5.7 –8.5 tonnes-acre) are typical and yields 50 per cent higher than this are certainly not unknown. In order to get these types of yields, the bud count is critical; narrow rows and arched canes help achieve this. However, in Britain, where these yields are the stuff of dreams (except in exceptional years) and where our light levels and temperatures are so much lower, arched canes only aggravate the situation. It is not more buds that are required for higher yields, but fewer, more fruitful, buds – a critical difference. Remember, it's not _more_ buds you need, but _better_ buds.

The length of a single arched cane on a vine planted at an inter-vine distance of 1.30 m and pruned to a two-cane _Pendlebogen_ system may well be around 750 mm to 850 mm long which means that the last half of it will have been growing right in the thick of the canopy, well above the region which is typically deleafed (or at least partially deleafed). In addition, they will be growing later in the season, when side-shoots may well have started contributing to the canopy mass, when disease pressures will be higher, and when crop load may well be testing the vine's resources. In British vineyards, where light levels and temperatures are often low, which in itself is reason enough to worry about bud fruitfulness, we just do not need any extra stress placed upon fruiting buds. In addition, as I have pointed out above, arched canes are only necessary where high yields can easily be obtained. In fact, leaving more buds in order to get higher yields in British vineyards

Two-arm Pendlebogen (pendulum arch) trained vine

might actually be doing the very opposite. Leaving more buds just places a higher strain upon the winter reserves of the vine and in the spring and early summer, when the vine depends almost entirely upon winter-stored reserves to both grow leaves and canes and – most importantly – to lay down the inflorescences for next year's harvest, the vine will struggle. In _The Grapevine_ (p.66) the authors state that:

yield variation from season to season in a vineyard or region is largely the result of bunch number per vine (60 per cent), followed by number of berries per bunch (30 per cent), followed by berry weight (10 per cent). The variation in bunch number is a consequence of conditions that occurred in the previous growing season.

In other words, it's not the <u>number</u> of buds that are left at pruning that determines yields, but the <u>number of bunches per shoot</u> that emanate from those buds. In poor-yielding British vineyards, it is common to see many shoots with no, or only one, good bunch. This is a result of low light and temperature levels at the critical time in the preceding year. In high-yielding British vineyards, the difference is that there are fewer blind canes bearing no bunches and plenty of canes bearing two and even three good bunches. As I have said above, what we actually need is fewer buds but more importantly, buds able to produce both more and larger bunches of grapes.

Having said that I do not favour arched canes, in a well-sited, sheltered vineyard, perhaps one which had been planted slightly too wide (say over 2.30 m row width) which had already proved itself capable of producing good yields (7.5–10.0 tonnes-ha, 3.0–4.0 tonnes-acre), it might be worth experimenting with a lightly arched cane – say over a second fruiting wire around 200–250 mm above the first fruiting wire – which would increase the bud count by about 10–15 per cent. With a little bit of extra deleafing and careful attention to shoot positioning, a higher yield might be achieved without any problems.

Other pruning systems for Great Britain

As for other pruning and trellising systems for British vineyards, there are several that have been tried, but there are none that I consider superior to the single- or two-cane flat _Guyot_. Several growers use the Scott Henry system, which has a split canopy with half the leaf-wall growing vertically and (in theory) half growing downwards. Whilst I am sure it can be made to work, I ask myself why would we need this system in Britain. The system was developed to deal with over-vigorous vines (often on their own roots and usually irrigated) in Oregon, and later in New Zealand, where growing conditions are very different to Britain's. It was developed with the hope that an increase in bud numbers per vine would increase yields and these yield increases would have a devigorating effect upon the vines. The reason I do not favour its use in Britain is that, to start with, we are not able, in most years, to get the types of yields that they can achieve in Oregon or New Zealand, and that the level of yield we do get in Britain have a negligible impact upon the vigour of our vines in most years. The choice of a devigorating rootstock has much more effect. Downward facing shoots, even supposing you can achieve the type of downward facing canopy that the textbooks show, also causes problems with late-season weed control and, in my experience, just leads to a damp jungle of vine foliage and weed growth where there ought to be clean, dry soil. This makes disease control much harder and poor disease control impacts not only upon the current year's crop, but also on the viability of next year's buds. I firmly believe that loading one individual vine with so many buds must place a strain upon that vine which in our cool growing conditions must be a negative. If you are hell-bent upon Scott Henry, my advice would be to visit as many British growers as you can that use it (and maybe a few that have given up using it) and then just try a few rows to start with. Having said the above about Scott Henry training, Simon Day uses it with great success at various vineyards he is involved with, just proving that it can be made to work!

Of the many other pruning systems used throughout the world, I do not believe there are any that are superior

Walkway in the middle of long rows formed from two end posts joined together at the top with wire

to the ones I have recommended above. Some have been tried – Lyre, Single Pole and Sylvoz, for instance – but have not lasted the course. If cane pruning is what a cool climate needs – which I believe it does – then the options are in any event somewhat limited. The simplicity of *Guyot* systems, from both the installation and management points-of-view make it the de facto system for most British growers.

Chapter 6

Site planning and preparation

Site planning and preparation

So, the site has been chosen, the varieties, clones and rootstocks selected, the row width and inter-vine distances decided upon and now it's time for the next stage: getting the site planned and prepared.

Note: See *Appendix II - Vineyard pre-planting check list*. Also available at: *www.englishwine.com/wgigb*

Site layout

The first task is to decide how the vineyard should be laid out. For many sites, the decision is an easy one. The shape or the slope of the site dictates that there is an obvious way the rows should run. In the case of sloping sites, the easiest way to work the vineyard will be with the rows following the slope and this will also help with air drainage in case of frost. If there is a gentle slope across the site, then this is not usually a problem. However, a steeper slope across the site could possibly cause tractors and machinery to start 'crabbing' (i.e. sliding) and once you get trapped, it is often very difficult to escape without vines and posts being damaged. Reversing in the row, especially when towing a sprayer or trailer, is not usually an option. A way to cope with a slight slope is to grade the soil so that gentle terraces are created. This needs to be done after the site has been planted (assuming machine-planting – if hand-planting it could be done before) and before the soil firms up and grass starts growing. A few passes along each row with a cultivator fixed at an angle to the slope so that the soil is levelled up will help create a more comfortable working platform. Creating terraces, much as you see in other countries, is not something that has ever been done in Britain, mainly because our sites tend not to be that steep, and with relatively wide rows, wider (and more stable) tractors are used which can cope with the slopes that we do have. On quite steep slopes, it will often be wise to flatten out the headlands at both the top and bottom of the site in order for tractors to turn safely. Turning on a bank is when a tractor is most likely to overturn.

The question of row orientation is one that often perplexes those new to the task of laying out a vineyard. The literature will tell you, of course, that in order to get the maximum sun exposure and heat on to the vines, the best orientation is north to south or possibly north-east to south-west. Maybe on paper this is the case, but my preference (if it is possible) is to orientate the rows so that they run at 90° to the prevailing wind. Why? – because in Britain one must attempt to provide as much shelter in the vineyard as possible and in laying out the rows at an angle to the prevailing wind, the first few rows will act as 'sacrifice' rows, offering shelter to the remainder. Typically, in British vineyards, the prevailing wind comes from the south-west, so the rows would want to run north-west to south-east for maximum shelter. Rows facing slightly to the east will also get the benefit of the morning sunshine and dry out sooner than westerly facing rows, although westerly facing rows will get the benefit of the warmer, afternoon sun.

Of course, on many sites there will be one fixed boundary which will dictate the direction of the rows, otherwise you will end up with too many short rows and awkward corners. Following the line of this boundary will ensure that the rows are as long as possible (although see below about row length) and the number of short rows – and therefore of end-posts and anchors – will be kept to a minimum. What you will need to do, especially if using a GPS-guided planting machine, is to be able to strike a very accurate right angle to the line of your first row. This is discussed in *Chapter 8: Planting your vineyard* in the section on *Machine-planting*.

Measuring the site

After the decision about row orientation has been taken, the next task, before the correct number of vines can be ordered, is to measure the site as accurately as possible, a job that is sometimes easier said than done. If you do not intend to fill the whole field, then the question of how accurate you need to be is less relevant and, as long as you don't massively over-order, then the exact number of vines does not matter that much. Either way, you will need to peg out the area to be planted, taking into account space for fencing, windbreaks and internal roadways (more about these below). On sites that have good, straight sides, no awkward angles or curious corners, a decent tape measure (of at least 30 m or even 60 m length) will suffice. I always use a Leica Disto D5 laser measure. This device needs to be mounted on a camera tripod and can (with practice) measure up to 200 m, quite far enough for most British vineyards. Unless you are measuring quite short distances and there is some solid object to bounce the laser off, you need special 'bounce plates' that reflect the laser back to the device. I have two of these mounted on metal posts which can be easily pushed into the ground. Take measurements of the sides of the site, plus diagonals across the site. These can then be plotted on to Google Earth Professional (not Google Maps) and, using its polygon facility, you can draw a line around the area to be planted and it gives you the number of square metres in the plot. Divide this by the area occupied by each vine and you will have a fair estimation of the total number of vines to be planted. Remember, however, that in a vineyard 100 m wide, there are 51 rows of 2.00 m wide, not 50 so budget for these extra vines. For larger and more complex areas, employ a surveyor equipped with GPS surveying equipment so that much more exact area measurements can be arrived at, although at greater cost. GPS surveyors typically charge £300–£500 a day. I am sure that one day mobile phones will be accurate enough to plot field areas, and whilst they can currently give you a reasonable approximation (www.fieldmargin. com is good) they are not accurate enough to give you the area to be planted.

However, before the planting area can be measured, there are several things to be taken into account: the exact position of rabbit and deer fencing must be settled as this will take up space; if there are to be windbreaks, both external and often internal, then they have to be allowed for and also remember that they have to be trimmed, so allow enough space for the trimmer; and finally, the area devoted to headlands (the space between the ends of the rows and the fence or windbreak), and internal roadways and/or turning spaces have to be factored in. Fencing and windbreaks are discussed later in this chapter, but headlands, internal roadways and turning spaces can be dealt with here. I always peg out the whole vineyard using bamboo canes with different colours of fluorescent tape tied to the tops which makes lining up straight lines much easier. For larger sites and/or where there are undulations which obscure the direct line of site, I use 2.0 m long red and white surveyor's ranging poles.

Most tractors suitable for commercial viticulture, i.e. not garden or hobby tractors, will require at least 6.0 m (and quite often a bit more) of clear headland to get them out of one row and into another. The exact turning circle will be part of the manufacturer's specifications. Some tractors have better turning circles than others – New Holland tractors with 'SuperSteer', for instance – and almost all tractors have independent braking on the rear wheels which allows you to spin round on one wheel. However, this is not easily done if towing a sprayer or trailer and will be done at the expense of your brake pads and usually with some scuffing of the grass. Whether a tractor can turn directly into the next row will depend upon row width and, with most tractors and in most vineyards, this is not possible without stopping and reversing, something generally to be avoided. It must also be remembered that a tractor seldom works without an implement of some sort mounted or being towed behind it – a mower or sprayer or a piece of machinery or a trailer – all of which will make the whole unit longer and thus require the headland to be wider. If you are proposing to have a trailed sprayer, a dung spreader, or to tow a trailer or any other long implement, then you will typically need at least 8.0 m of headland for comfortable working. Tractors with a three-point linkage (three-point hitch) and PTO (power take-off) at the front, which will allow a front-mounted mower to be used whilst spraying at the same time, will naturally require even more space. You should also bear in mind that you may one day want to use contractors to spread manure, compost, and fertiliser (especially lime, which is very bulky) and they generally favour high capacity spreaders that can hold maximum volume, but which are narrow enough to fit down vineyard and fruit orchard rows. These spreaders tend to be much longer than the average. My rule of thumb is therefore to leave at least 9.00 m (and often 10.0 m) between the boundary – fence line or edge of where you expect your windbreak trees to grow to – and the first planted vine. This allows for around 1.40 m to be taken up by the end-post and anchor and 600 mm for the final vine to spread its cane into. This will leave you with a clear 7.50–8.00 m of headland. Obviously if your inter-vine distance is greater than 1.20 m then this distance ought to be slightly larger. You may also need to leave additional space at the bottom of a slope to allow frost to drain away. This is covered in *Chapter 13: Frost protection*.

The question of row length will very much depend on the shape and size of the site. Rows much longer than 150 m can be daunting to work in. Get halfway along, and you are 75 m from the end and that's a long walk. It's easier today when almost everyone has a mobile phone, so none of the shouting 'where are you?' and loud whistling that used to be the case. When picking, 150 m is also a long way to have to take out the full bins or boxes of grapes and return with empties. This can take up an awful lot of time and leave the pickers waiting for boxes. If your measurements show you that you have an appreciable number of rows of 150 m or more long, I would advise considering two options: creating a mid-point walkway through which you can walk and/or drive a quad bike or motorcycle, or splitting the block into two with an internal roadway. Creating an internal walkway is simply a question of splitting the row with two extra-high posts, typically end-posts as they need to be strong, and tying them together at the top with wire. You need to make sure that the wire is high enough for a person to walk under without decapitating themselves, but not so high that a leaf trimmer cannot raise itself up to clear them.

If you decide to split the block in two, then of course you will double up the number of end-posts and anchors required, plus lose a fair amount of space. You have to decide whether your internal roadway will be just enough to get a tractor through – around 3-4 m between the anchors ought to be enough for this – or whether it is to be enough to turn a tractor round in – in which case you will need 6.5–8.0 m between the anchor wires, which means 10.0 m between vines.

Ordering vines

Note: The planting of ungrafted 'own-root' vines is covered in *Chapter 16: Pest and disease control* in the section on *Phylloxera* and at the end of *Chapter 17: Trunk diseases*. In *Chapter 17* there is also a full explanation of the process of producing grafted vines.

Almost all vines used in commercial vineyards consist of a scion – the variety to be grown – and a rootstock[1] – the part of the vine in the ground – and these are joined together by the graft. Vines are bench-grafted during dormancy onto rootstocks in the early spring of one year, spend a few weeks in a greenhouse to let the graft form a callus (i.e. become hardened off) and to get the roots to start growing, and are then planted out in nursery fields for the remainder of that growing season. Once they are dormant in November and December, they are lifted, taken back to the nursery, sorted, cleaned, trimmed, disinfected and placed into a cold store until needed for planting.

Most European growers plant vines in the spring of the year, although winter planting is not unknown. In regions with plenty of winter rainfall and the possibility of both winter and spring frosts – such as Britain – there is absolutely no advantage in planting in the winter. For one thing, most nurseries will not have stock available and therefore getting hold of vines can be a problem. Also, I have known vines planted in early March and April which have started to grow and then been hit hard with a late spring frost in April or early May. In some regions, rooted

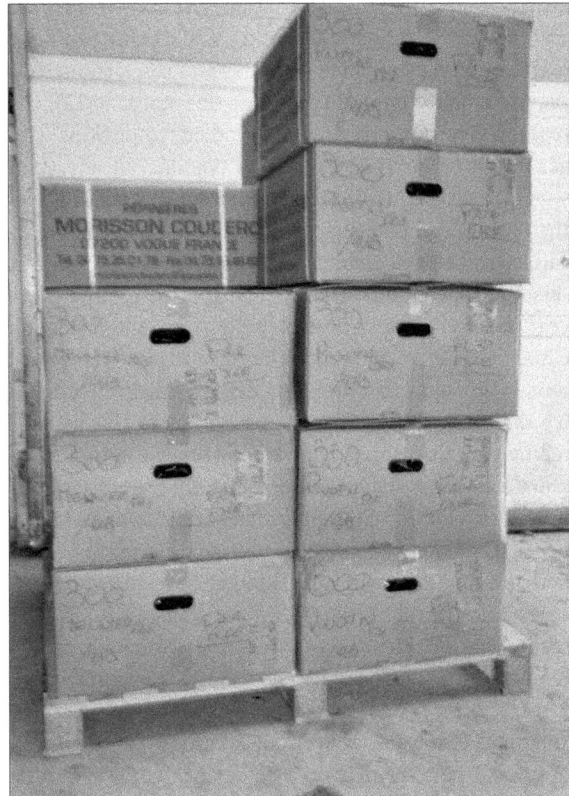

Vines in cold storage awaiting planting. Each box holds 300 grafted vines

cuttings of rootstocks are planted out into position in the vineyard for a year or two, after which the scion variety is chip-grafted (also known as field grafted), on to the rootstock. This system of planting is usually only found in warm climates and where experienced labour is available to carry out the chip-grafting.

To my way of thinking, the best time to plant vines in Britain is in the first two to three weeks of May. Correctly planted into well-prepared sites and looked after properly, there will be absolutely NO difference in the amount of growth by the end of October between vines planted in November, December, March, April or May (and I have known vines planted in June do just as well). In fact, quite the reverse, and I would put money on vines planted in the first half of May being the best of the lot. By then, sites will have started to warm up and root growth will start soon after planting. By the time green shoots first appear (late May to early June) the dangers of spring frost will be over and they can spend the summer putting on weight.

Although vines can be ordered at any time, the 'high season' for ordering vines in Europe is late September to the end of October. This is the time when growers in the major vineyard regions will have a good idea of the size of their harvests and can decide if they are able to afford to plant any vines. Also, the nurseries will have a good idea of what percentage of 'take' their vines will have had and how many of any one variety, clone and rootstock they will have for sale. Not all the vines grafted and placed out

---

1 Whilst in some countries vines are grown on their own roots, in most parts of the world Phylloxera makes this impossible. In the British Isles there have been many outbreaks of Phylloxera (at least one or two are reported to DEFRA each year) and you would only plant ungrafted vines in exceptional circumstances and where you appreciated the risks.

into the nursery fields will be saleable and losses from poor graft callusing, poor size matching, and disease can be as high as 50 per cent. Some varieties and some rootstocks are particularly prone to losses at this stage, one of the reasons why vine prices vary. My advice to growers is to get orders in for vines as early as possible. Ideally vines should be ordered 14–18 months in advance of planting and the nursery can then supply vines of the exact rootstock, variety and clone combination requested. In practice this seldom happens and therefore August or September of the year prior to planting is usually soon enough, although earlier is better. If you leave ordering until later, then some nurseries will start a vine-hunt amongst other nurseries to try and source the vines you want. If they do this without telling you, then this is regrettable as they cannot always know the circumstances under which the vines have been produced. Unfortunately, it is a fact of life and one of the penalties of later ordering, of ordering from nurseries you perhaps don't deal with very often or nurseries you don't really know at all. Although I have a vested interest in saying this, of course you should always order through a reputable vine supplier who knows his or her nurseries.

Most vines these days are prepared for machine-planting and their roots will have been cut to the right length. If you propose to hand-plant using a water lance, you will want to have the roots cut shorter, although this is something that can be done in the field on the day of planting. All vines will normally have their scions and grafts re-waxed before shipping and will be shipped in stout boxes, or sometimes in plastic bags. Either way, they ought to be moist – sometimes they are packed with damp peat to keep them moist – and well-wrapped in plastic to stop them drying out. They will be stored in cold storage at the nursery until shipped to Britain and hopefully will not get too warm en route. Depending on the weather conditions, I usually use refrigerated transport to bring vines in. I also always try and avoid the period around Easter when, given both the French national and British bank holidays over this period, vines in transit can spend four to five days on their journey. Once in Britain, vines should go straight back into a cold store, running at about 4–5°C, until just before planting. I always keep vines packed in the boxes or bags in which they have left the nurseries, and they stay in those boxes or bags whilst they are delivered out to the vineyard site and until the moment of planting.[2]

One question which new growers always ask is 'how many vines are there in a box?' and 'how big are the boxes?' The answer is that a box of around 650 mm x 500 mm x 300 mm (about the size of a large suitcase) will hold between 300 and 600 vines, depending on how much root they have and how many the nursery needs to pack in. If the vines are in bags, a bag the size of a typical 25 kg fertiliser sack will hold around 250–300 vines.

Sometimes when you open the boxes of vines, you will find that they have grown white sprouts, sometimes 25 mm long or even longer. This is usually a result of the vines having been kept at too high a temperature and/or being exposed to light whilst in storage or in transit. Whilst this is probably not a good idea, I have planted plenty of vines with white sprouts and – so far – have not experienced problems. The sprouts usually blacken and wither, to be replaced in the normal course of events with fresh green shoots.

One other way in which vines are produced is known in Germany as *kartonage-reben* or *topf-reben* and in France as *sept á neuf* (seven by nine – the size of the pots they are planted in). I have also heard them called *vignes Jiffy pots.* These are vines produced in the spring of the year of planting and instead of being placed into nursery fields after grafting and callusing, they are placed, immediately after grafting, into 7 mm x 9 mm fibre pots (Jiffy pots) filled with peat and then rooted in trays and sent out to growers as soon as the grafts have callused and the roots have started to emerge from the sides of the pots. The advantages of these are that they can be ordered and supplied in the year of planting. The disadvantages are that they are expensive to buy (around twice the cost of a normal vine), difficult and expensive to transport as they take up much more space than bundles of dormant vines, more difficult and expensive to plant, and losses will be higher. Because they are typically supplied later in the season, the chances are that the weather will be warmer and drier and therefore these vines may need to be watered after planting. Although I have used them in desperation and known growers who have used them, the results, both short-term and long-term, have never been that satisfactory and my advice is to avoid them.

All plants produced in the EU have to bear 'plant passports' and on vines, these are small (70 mm x 110 mm) plastic labels, usually blue (other colours exist for different categories of vine) and usually attached to each bundle of twenty-five vines or each larger maxi-bundle of one hundred vines. On the passport will be details of the variety and its clone, the rootstock and its clone, the details of the supplier, lot number and other relevant information. It is your guarantee that the vines are what they say they are and that they have been inspected and are free of viruses and *Phylloxera*. What they do not do is guarantee that they are free of disease or that they will necessarily grow. Quite what to do with plant passports after you have planted the vines is open to question. Unless you track every bundle and know which passport relates to which block of vines, I cannot see that they are of any help if you find that some vines have, for example, a virus. For years I had dozens of plant passports hanging on a hook in my workshop gathering dust, but eventually threw them away. I was once told by a WS inspector that 'one has to keep them for six months' after planting, but I am fairly sure this is not true. (You are meant to register your vineyard with WS within six months of planting, but that is a different issue). Currently

---

2 As I only use nurseries whose packing and storage methods I am happy with, I feel quite comfortable leaving all boxes and bags sealed until the moment of planting.

(2020) the EU Plant Passport is all that is needed to import vines into Great Britain and any additional checks or extra paperwork should be minimal. This, of course, may change once an EU–GB trade deal is hammered out (or we exit on WTO rules).

High-graft vines

By far the majority of vines produced in the world are grafted on to 300–400 mm long rootstocks which means that the primary shoot has to grow from just above ground level in order to produce the stem (the trunk) and then the fruiting wood. Nurseries do, however, produce 'high-graft' vines, called '*hochstammreben*' (high stem vines) in German and '*greffe haute*' (high-graft) in French. These are grafted vines with a long rootstock – around 550–600 mm in France and 700–900 mm in Germany – so that, at planting, the vine's stem is already grown and all the vine has to do is to grow the fruiting wood. The advantages of these vines are that you save a year's growth and (in theory) can get the vines to fruit a year early; you can use weedkillers as soon as the vines are planted without the need for spray guards; you do not need rabbit guards as the young shoots are beyond the reach of rabbits (but not deer); and – most importantly from a cost-saving point of view – you do not have the annual spring work of removing shoots from the stems. The disadvantages are that they are slightly more than twice as expensive to buy; they are more expensive to transport; they can be machine-planted, but the work is slightly slower; and you have to have the fruiting wire installed more or less as soon as the vines are planted otherwise the vines will blow about in the wind and the grafts can get damaged. And also remember that, even though they may have a pre-built stem, the roots of a high-graft vine are no more developed than those of a standard vine.

Whether the advantages of high-graft vines outweigh the disadvantages is still an unanswered question and, given that the technique is not that old or widespread, many growers view high-graft vines with some suspicion. The most obvious advantage – that of an earlier crop – is one I feel unhappy about. Given that the vine's root system expands with age, not with the length of the rootstock, even though with high-graft vines you might have grown a fruiting cane at year two, can the root system take the stress of fruiting? Given a dry season and no irrigation, a high-graft vine with a small root system might suffer considerable stress. From what I have seen, high-graft vines do not establish as well or as evenly as normal 'low-graft' vines, possibly because high-graft vines start off by waving around in the breeze and gain nothing from being close to the ground during their first, all-important growing season.

There is also the question of Trunk Diseases (TDs). The classic technique for invigorating a vine infected by Esca, one of the oldest and most widespread of the TDs,

High-graft vines on the left. Standard vines on the right

is to take a water shoot[3] from as low down on the trunk as possible (just above the graft being ideal), cut the old trunk off just above the water shoot and re-grow the trunk. With high-graft vines this is not an option. Some growers also think that having vines with twice as much rootstock is probably best avoided just from the TD aspect. Given that you can (if you really want to), in favourable circumstances, get normal vines to crop at the end of their second summer, I cannot see that high-graft vines offer a real advantage and, with one exception (see below), for the time being I would avoid them. TDs are dealt with in detail in *Chapter 17: Trunk diseases*.

Having said all of the above, high-graft vines are very popular in Germany, and in some regions are almost universally used for new vineyards. Quite a few vineyards in Britain have used them – Langham and Rathfinny have to my certain knowledge – although the latter I understand have now reverted to standard length vines, but that could be because they now have over 90-ha cropping, so can afford to wait an extra year. It would be interesting to survey vineyards in Britain to get grower's perspective on high-graft vines.

The one exception to using high-graft vines is where you are 'gapping-up', replanting the odd missing vine in and amongst already established vines, in which case they are useful. If you plant standard length vines when

---

3 A water shoot is a shoot that seemingly comes from nowhere, i.e. directly from the trunk and not from a piece of one year old wood such as the spur (thumb) left to provide replacement canes. Water shoots seldom produce viable inflorescences.

gapping-up, unless you are careful to keep the foliage from neighbouring vines away from them, they often get overshadowed and will take several seasons to grow to full size. High-graft vines start at fruiting wire level and stand a better chance of making it to full fruiting size in a season or two.

Site preparation

The aim of site preparation is to get the ground into the best condition possible to take the vines. The task falls into two categories: physical and chemical. On the physical side are the following: earthmoving, water to the site, drainage, green manuring, additions of humus, the actual cultivations which will prepare the soil for planting, fencing and windbreaks. The chemical side involves the addition – if required – of chalk to correct the pH and of fertilisers and this is dealt with in *Chapter 7: Vineyard nutrition*.

When to prepare the site

The sooner the better is the quick answer and if you have the luxury of one (or more) growing seasons before you plant, so much the better. Having said that, I have seen fields turned from grass, arable, orchards and even hop gardens into vineyards in the space of a few days.

Starting early gives you the chance to have a go at reducing the weed spectrum by spraying the site with herbicide, thoroughly cultivating it, perhaps growing a

Tanks for quick filling of vineyard sprayers

cleaning crop for a season or two, and making sure that the soil has settled well after any earthworks or drainage has been carried out. It also gives you the chance to get windbreak trees planted a season or two before planting, meaning that they will start to provide shelter as soon as the vines are in the ground.

In the normal course of events, most sites will be available in the late summer after the previous crop has been harvested, in which case all major preparatory works – additions of manure, humus and fertilisers, earthmoving, and drainage – can be carried out in late August, September and October whilst the land is still dry, followed by subsoiling and deep ploughing, after which the land can be left to overwinter before tackling it again in the spring prior to planting. The site may need re-ploughing after the winter, or just disc-harrowing followed by power-harrowing. If you are using a competent agricultural contractor to do the work for you, then he or she will probably have knowledge of the type of soil you have and advise accordingly. It is important to state that, of course, no two sites are the same and therefore how they are handled will vary considerably. One would therefore approach chalk downland, heavy clay and light sandy soils in different ways. One thing to remember, and this applies to every site, whatever the soil is like, make sure that all ploughing and harrowing is carried out in the direction of the planting line as far as is possible. If you do it at ninety degrees to the planting line, you will find that the planting machine will be unable to keep a good level as it rises and falls over the unseen ridges left by the plough, which will affect the depth of planting. In addition, when you are spraying and mowing you can often feel the tractor rising and falling over the plough ridges as you travel over them. Some operations, subsoiling for instance, need to take place across the site, both ways, and normally up and down as well. This, of course, takes place before ploughing and final cultivations. There is more on subsoiling in the section on *Cultivations* later in this chapter.

Earthmoving

Many sites will have features which it is best that a vineyard does not have: small ponds or low-lying areas, ridges, humps and bumps of all sorts, plus stray pieces of hedgerow and tree stumps. These are all best removed so that the site is as flat and uniform as possible. Although this question is dealt with later under 'frost prevention', if your site is at all prone to spring frost, then it is best to try and fill or re-grade any areas that are low lying, however slightly, as these will hold the frost, even on a gently sloping site. If a substantial area requires filling in, it is best to remove the topsoil first, from both the area to be filled and the area from which the 'fill' is to come, so that subsoil is placed upon subsoil and the topsoil upon topsoil. Always leave the filled area a bit proud to allow for settling.

Whilst trees in the middle of a site can be a lovely feature, they take up a considerable amount of space with both the area they actually occupy, plus the area they shade, and to my way of thinking, are best removed. Before doing so, however, it is always best to make sure there is no TPO (Tree Preservation Order) on them. You will also need to think about access roads to the site, car parking for visitors' and pickers' cars and possibly an area of hardstanding for both storing picking boxes and bins and for loading them on to a lorry.

Water to the site

A supply of water will be required in, or near to, the vineyard for pesticide spraying. Spraying is not a task that most people want to spend any more time doing than is absolutely necessary and the nearer the water supply is to the point of use, the quicker you will be able to fill up the tank and get back to spraying. In order to make tank-filling as quick as possible, it is best to have the water supply as near to the vines as you can and, to make filling up even quicker, erect a supply tank with a ball-cock and valve connected to the mains. This tank, which needs to be bigger than your sprayer's capacity, should be placed on a suitable platform so that it can empty by gravity into your sprayer. Make sure that the outlet from the supply tank is fairly big – at least 37 mm or 50 mm – with a ball-valve to control the supply. In this way you can fill up your spray tank in one quick dump of water and get back to spraying. Don't forget to drain the tank and turn off the water supply after the end of the spraying season and before the winter frosts arrive.

Drainage

The question of drainage – does a site need it or not – is often a difficult one to answer but there are three big things we know: many sites are better drained than not drained; drainage is relatively easy before planting and relatively difficult after planting; and drainage is expensive (although the value of the land will definitely be enhanced). Therefore, the question of whether to drain or not must not be lightly dismissed.

When I am looking at a new site, almost the first thing I am asking myself is whether any of it needs drainage. There are obvious tell-tale signs: visibly damp areas, rutting which tractors and machinery have made in the past; areas where grazing animals have 'poached' the land; and the presence of rushes and other moisture-loving plants. Asking the landowner, unless they have really farmed the land commercially over a period of years, I have found to be almost worse than useless. The stock answer to my question: 'does this field drain well?' is 'yes, this field is never wet and always dries out in the summer' and the merest hint of a gradient will cause them to add 'it's on a slope, you know' as if land on a slope always drains well. Sloping land certainly does NOT always drain well and can often contain springs and wet areas which need to be dealt with.

Before field drainage is considered, one ought to

Field drainage machine preparing a new vineyard site

appreciate the benefits of drained land for vines. The benefits of drained land are that:
- Vines will get a much better start in life, allowing them to establish quickly and evenly and get into cropping by year two or three. This will help reduce stress and lessen the incidence of TDs.
- The soil will dry out faster and more completely and force the vine's roots to dig deeper. Deep rooting is beneficial for many different reasons and it makes for a more stable environment for the vine. Is also allows oxygen to penetrate more deeply, encouraging earthworm and microbial activity at deeper levels.
- Roots that are drying out will produce more abscisic acid (ABA) which vines need to both produce and ripen grapes.
- The soil will warm up more quickly in the spring and the vine's fine root hairs, which actually do all the work of transferring water and nutrients from the soil to the plant, will begin to grow earlier. The earlier the bud-burst, the longer the growing season.
- The vine will develop a larger root system which will help reduce stress in dry seasons; drought is something often given as one of the contributory factors leading to TDs.
- The land will be accessible to machinery sooner in the spring and will be less likely to rut, compact and suffer long-term damage. This can be important in the early weeks of growth when spring and early summer rainfall is most likely, and when the vine needs maximum protection against mildews and other diseases. It is also important at the end of the season at harvest time for the same reasons.

• The land will be more valuable than un-drained land because, for many crops and for livestock, it is more productive and therefore more profitable.

The best way of telling whether land needs draining is to call in the experts. A reputable drainage contractor or soil consultant will carry out a survey of the land and draw up a proposed plan. Costs will vary according to many factors such as drainage layout, distance between laterals (10 m is usual for vines and fruit), whether you have gravel backfill or not (usually recommended) and the overall size of the scheme. The John Nix *Farm Management Pocketbook Edition 50, 2020* gives £2,300–£3,400-ha (£931–£1,376-acre) for an arable scheme with laterals spaced at 20 m apart and using permeable backfill, but on a vineyard, where closer spacing would normally be required, costs would be considerably higher. For a large site (10-ha or more) I would budget nearer £5,500-ha (£2,226-acre) for a drainage scheme with 10 m between laterals using permeable backfill. For a smallish site (under 5-ha) drainage might well cost nearer £6,750-ha (£2,732-acre). Much will depend on the exact layout, number of junctions and headwalls etc.

Whether this cost can be justified is for each grower to decide. Given that land and vineyard establishment costs might easily come to £75,000-ha, and stock costs another £50,000–£100,000 per hectare (depending on whether you are producing still or sparkling wines) a sum of £5,000–£6,000 per hectare added on to a total investment in the enterprise of £125,000–£175,000 per hectare doesn't seem like much of an additional expense in order to have a well-drained vineyard, given that bad drainage often lies at the heart of a poorly performing winegrowing enterprise. It is also possible to do a percolation test on a site using nothing more than a spade and a bucket or two of water to gauge whether your site needs draining. See *Soil drainage* document at www.englishwine.com/wgigb.htm.

The drainage work itself should ideally take place when the soil is dry and as far before planting as possible. Although I have known planting take place within days of the drainage being finished, it is better to get all the drainage done in the late summer or early autumn, followed by subsoiling and ploughing, after which the soil will have a chance to settle over the winter. This will allow the slots where the drains have been laid, especially the larger main drains, to close up.

After the drainage is finished, your contractor will provide you with a detailed plan showing you where all the drains are laid and to what depth. These days, plans are drawn up using GPS so ought to be very accurate. Knowing where the drains are is important for any subsequent work involving digging down – laying water mains for instance – but for vineyards it is especially important as it would be better not to have one's line of end-posts directly over a drain run for obvious reasons. Drains are seldom laid deeper than 800–900 mm and end-posts can sometimes be in the ground to this depth (or even deeper than this).

In addition, posts depend on part of their stability by being driven into undisturbed ground and if they are sited in the immediate area where a drain has been laid, they will lose some of their strength. The Land Drainage Contractors Association (www.ldca.org) has a list of members and some good information on drainage.

Green manuring

If you have the luxury of six months or more between taking possession of the site and planting a vineyard, then you should consider growing a 'green manure' crop which will clean up the soil, add humus and help reduce weeds. If you can get any major site operations – earthmoving, drainage and subsoiling – carried out early enough, then a green manure crop can be planted into cultivated land, left to grow, mown (or grazed) off, and then incorporated into the soil the following spring, prior to planting, but always allowing enough time for the green matter to rot down. The species of plant (or – more probably – plants) to be used will depend on the timing of sowing and the exact task you want it to perform. Green manuring is discussed in more detail in *Chapter 7: Vineyard nutrition* in the section on *Green manuring.*

Humus additions

Additions of humus (organic matter) in the shape of FYM, green waste compost, chicken manure, mushroom compost, hop waste or some other source of organic matter should ideally go on to the site whilst the ground is firm and dry to avoid compaction and wheel ruts. It is also best put on before the ground is ploughed so that tractors and spreaders are not shaken to bits. The amount of organic matter that you put on a site will (almost) entirely depend on how much you are prepared to spend. You could easily put 50 tonnes-ha (20 tonnes-acre) of FYM on a site as a pre-planting dressing without it being excessive. With chicken manure there are rules about the amount that can be applied as the nitrogen content is high, especially if it is fresh manure. Whatever product you apply, an allowance for the amount of nitrogen (N), phosphate (P) and potassium (K) contained in the compost/manure will have to be made when it comes to working out the amount of chemical fertiliser to apply. The ADHB *Nutrient Management Guide* (RB209) has a whole section (Section 2) on 'Organic manures'.[4] Whatever manure you use on your yet-to-be-planted vineyard, a word of warning. Avoid as far as possible stockpiling the product to be spread onto land which is to be planted with vines. Manures of animal origin, especially if they are quite fresh, will leach out all kinds of minerals which can badly taint the soil and prevent vines from thriving. If you are in a Nitrogen Vulnerable Zone (NVZ) – and much of Britain now is

---

4 See next chapter Vineyard Nutrition for details about the ADHB Nutrient Management Guide.

The best type of machine for subsoiling

– there are restrictions on how much N you can put on your land and you need to make sure you comply with the regulations. Further information is available in a Farming Advice Service technical article: 'Nitrate vulnerable zones: Back to basics'.

Cultivations

The extent to which any individual site requires cultivations will depend upon its previous crops and/or current state. Obviously, a field that has been under arable cultivation for many years and is, say, in arable stubble prior to planting with vines will need cultivating differently to one which has been unimproved grassland for decades. In general, most sites will benefit from having any weeds present sprayed off with something like glyphosate (Roundup and other trade names) prior to any work on the site. If the vegetation on the site is not killed off prior to cultivations, not only will the work be harder (especially if the site is old pasture) but also, when the site is ploughed, the old grass will form a rotting layer just where the vine's roots will end up. If the site is not to be sprayed with herbicide, then allow plenty of time (at least three months) between ploughing and planting vines. The next task, apart from any of the other operations already mentioned, is subsoiling.

There appears to be reluctance on the part of some people to accept that deep subsoiling – and when I say deep, I mean at least 600 mm and possibly even deeper – is an essential part of preparing a site for planting vines. In France, Germany, Italy etc, it is quite common to subsoil down to 900 mm prior to planting vines, especially if the site has been a vineyard for many years, so that all old roots are disturbed. The reason that vineyard sites need subsoiling is that it is generally the only chance you have of loosening the soil at the depth where the vines will be putting most of their roots. Once the vines are planted, it is much more difficult to get a tractor of sufficient size and power to carry a subsoiler down the narrow rows. Subsoiling will aid drainage at these levels and will clear

the site of any plough-pan[5] or impermeable rock layers. If the site has had trees on it at some stage, it will also help clear the site of any old roots that are still there.

The action of subsoiling is to drag an implement fitted with at least one, and usually two or more, long vertical blades, often curved, called 'tines', which usually have a 'wing' at the end to help shatter and loosen the soil, through the ground. The depth to which a tine will go and the number of tines that can be dragged at any one time will depend upon two things: the soil condition and the type of vehicle pulling the subsoiler. Obviously a site with tough, heavy soil, with a plough-pan, stones or rock layers and which has not been cultivated for many years will take more work and require more effort than one which has been in arable cultivation for several years and which may already have been subsoiled at some time in the recent past. It is usual to subsoil across the site from both corners and then up and down the site in the direction of planting. In this way, the whole site is thoroughly dealt with.

The best vehicle to pull a multi-tined subsoiler to the right depth will almost always be a tracked vehicle – a bulldozer or tractor fitted with tracks – which has enough power and traction to pull the implement through the soil at the correct depth. The problem with using a wheeled tractor, with or without four-wheel drive and of whatever horsepower, is that most tractors are fitted with sensors so that when the resistance from the implement being towed becomes too great, they automatically lift up the implement in order not to slow down the tractor or damage the implement. This means that when the subsoiler hits the toughest ground, i.e. the ground that really needs subsoiling, the tractor automatically raises the tines up to jump over the very problem it is there to alleviate! Although a tracked vehicle will cost more to hire – it cannot usually be driven on the roads so will usually have to delivered to the site on the back of a low-loader – the job it does is so much better that it really is worth

5 A 'plough-pan' is a layer of hard soil just below the level at which normal arable cultivations – ploughing and power harrowing – take place.

Wine Growing in Great Britain

Very deep, single-furrow plough for preparing vineyards, but not seen in Britain

the cost. One Kent contractor charges £60 per hour for a Caterpillar D6D with haulage to and from the site for £500 and in one day, could easily subsoil 5 ha, therefore a cost of around £180-ha (£73-acre). The *National Association of Agricultural Contractors* rate for subsoiling is currently (2020) £60-ha (£24-acre), but this is for relatively shallow subsoiling (400–500 mm) and is for a 'one-pass' operation. For deeper subsoiling and with three passes across the land, I would allow at least double this rate.

As an alternative to a fixed-tine subsoiler, the McConnel Grassland Shakaerator is an implement worth considering. This is a subsoiler which is powered via the PTO and vibrates as it passes through the soil, making both the passage of the tines easier and achieving a more thorough 'shattering' effect at the ends of the tines. A 2.0 m wide version is available which can carry two to three 750 mm tines, but requires a 35–55 kw (50–75 hp) tractor to pull it. In fruit-growing areas of Britain this type of machine is often available for hire.

The other specific reason why vineyards which are to be machine-planted – and that is the majority of commercial vineyards these days – need subsoiling, is that the planting machine works by opening up a deep and wide slot in the soil, into which is placed the vine to be planted. As the planter moves along, the slot is opened up, the vine goes in, and the machine moves on. What happens next is of great importance if the vine is to settle in well and root properly. Once the machine has passed on, the soil must fall back into the slot and ideally surround the roots of the vine. This would typically be at a depth of 250–300 mm. Although the soil at the very surface will invariably fall back around the rootstock, and visually the vine will appear to be settled, it is what happens below the surface level that is critical. Vines planted in sites where subsoiling has not taken place, and where cultivations have only been at a

relatively shallow depth, may well suffer from having their roots waving around in the air, some way below surface level and out of sight.

After subsoiling, the site will probably look very cut up and raised in places and there may well be tree roots, slabs of rock or stone, or other previously buried debris that is best removed prior to the next activity, which is ploughing. The aim of ploughing is to turn over and bury whatever is on the surface: any manure, humus or fertilisers applied, plus any vegetation that is (or was) growing there. The depth of ploughing will depend upon the plough used and in European vineyard areas it is common to use a very deep single- or two-furrow plough specifically for this purpose. These ploughs, although slow to cover the ground, will turn the soil over to a depth of at least 300 mm, and often much more, leaving the soil in loose, broad furrows which, if left to overwinter, especially if there is enough frosty weather, will break down naturally so that spring cultivations are made easier. This will also allow winter rains to penetrate deeper into the soil. If the vines are to be hand-planted, then really deep ploughing is a great advantage. Most British agricultural contractors will not have this type of deep plough and will typically use six, seven or eight-furrow reversible ploughs, ideal for fast ploughing for arable farming. If this is the only type of plough available, then it is not the end of the world. Ideally all this heavy work should be carried out in the autumn when the land is fairly dry. The site can then be left for the winter to 'weather'.

The next set of cultivations will typically take place

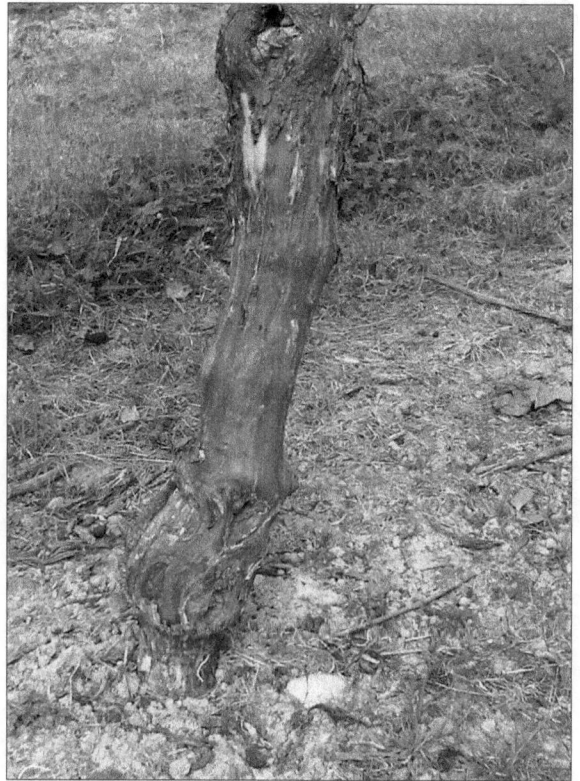

Vine trunks badly gnawed by rabbits

Good rabbit fencing really works

in the spring, perhaps a month or so prior to planting. Assuming that planting takes place in late April or early May, then mid-March is soon enough to get on to the site with machinery, although this will depend on the weather at the time and the state of the site. Most sites, if they have been sprayed off and deeply ploughed, will be carrying very little in the way of vegetation at this stage, so a pass with a disc-harrow, followed by a power-harrow will leave a typical 'seedbed' finish which is what the planting machine requires. On occasions, especially if heavy rain falls after the power-harrowing and the soil caps over, a second pass with the power-harrow immediately prior to planting may be required. Prior to these pre-planting cultivations, the second half of the fertilisers should go on to the site. The aim must be to get a completely clean, flat and even site, ready for planting.

As has been said before in this chapter, as far as is possible, all cultivations should be carried out in direction of the planting line, especially the final disc and power-harrowing, otherwise ridges will tend to form across the planting line which means that the planting machine will have difficulty in keeping its planting level accurate, with the result that some vines will be deeper in the ground than others.

Fencing

Most vineyards will be prone to attack from a range of four-footed pests – rabbits and deer being the most common – but hares, badgers and even wild boar are not that uncommon. In the establishment phase, when the vine's foliage is at, or near, ground level, rabbits can be a real menace and very easily prevent young vines getting

away. Whilst rabbits and hares can be excluded by using individual guards around each vine (fully covered in *Chapter 11: Management – establishment phase* in the section on *Protection against animals* and in *Chapter 14: Weed control*), perimeter fencing is more effective and personally I prefer it.

Rabbit fencing, if done correctly and kept maintained, is almost 100 per cent foolproof (or perhaps that ought to be rabbit-proof) and with luck will keep out badgers and hares as well. It needs monitoring on a regular basis to make sure nothing is trying to get underneath it, through it or over it and you must keep the ground at the base of the fencing sprayed off with herbicide so that it is free of weeds and grasses and you can see what's happening there. The netting used, generally known as chicken wire, must have a tight enough mesh size (don't be a cheapskate and use any that has an aperture greater than 31 mm between diagonals), and make sure the wire used is strong enough – 18 gauge (1.2 mm) is required – as rabbits can bite through 19 gauge (1 mm) netting. The fencing should ideally be buried in the soil – 150 mm vertically down and 150 mm horizontally out, away from the vines – so that rabbits attempting to dig in from outside will meet the netting on the way down. Laying 300 mm of fencing on the ground and pegging it down and/or placing turves on top is often advocated as a way of keeping rabbits out, but it is a short-term measure. Even if rabbits don't work out what to do, badgers, foxes and dogs soon will and push their way underneath, allowing rabbits to get in. The top of the fencing should be at least 900 mm above ground level and fixed to a line wire with an additional top wire (plain wire will do, but barbed wire is better if you also want to discourage people from climbing over your fences) at around 100 mm above the netting. This will also discourage animals attempting to jump over or pushing the netting down with their bodies. I have known rabbits and hares jump fencing of over 1 m high, especially if there is high ground adjacent to the fence, but this is rare and they seldom are able to jump out, so can be hunted down. All gateways need to be properly rabbit-proofed with railway sleeper or poured-concrete thresholds. Also make sure that all diagonal fencing support struts (on corners or on long runs of fencing) are inside the fence as, if they are on the outside, rabbits have been known to run up these and get over the fence.

The cost of rabbit fencing will depend on the size and shape of the site, plus the number of corners, bracing posts and gateways. The per vine cost will also depend very much on the size of the site. However, as a rough estimate and using fencing erected to Forestry Commission standards,[6] the budget cost will be around 60–70 p per vine. This is not far off the cost of buying an individual vine guard, plus the cost of putting it on and securing it to the fruiting wire, plus the cost of undoing it for access, plus the cost

6 The Forestry Commission Bulletin 102 'Forest fencing' is available at www.forestry.gov.uk.

of removing it and storing it or disposing of it. There may be a residual value in a used guard, but it will not be very much. Even if there were a cost saving by using individual guards (which I doubt there is), as I have pointed out above, they don't do as good a job as perimeter fencing and may even contribute to problems with the vine in later years. I accept that perimeter rabbit fencing is not completely a one-off cost and it will require inspection as holes can be made by a variety of animals (man included) and, in time, it will start corroding at ground level. Usually by this time, though, the vines' trunks are sufficiently robust not to be troubled by a little bit of winter rabbit damage, although hares can be a bigger problem and, if they are, wire netting guards should be used.

If deer are a problem, and this is not the case on every site, then perimeter fencing is the only real solution if you want a successful vineyard. Deer fencing needs to be erected to the correct standards for the type of deer to be excluded. Unless you are growing vines where there are the larger, red deer, the most likely deer to cause damage are roe, fallow, muntjac and sika and these can be excluded with a 1.80–2.00 m high fence. Obviously if you are erecting deer fencing, the addition of dug-in rabbit netting at the base makes sense. Because deer damage tends to be when the plants are in their first two to three years and not quite so much when they are established, in situations where deer pressure is light, one could consider a temporary electric deer fence around the site. I have known this to be effective, but I have also known it to be totally ineffective. Electric fencing needs to be switched on to work and if the battery loses its charge or the power goes off or something lands on the fence (a branch perhaps) and shorts it out, deer will soon discover a way over. I have also known growers to erect deer fencing only on the perimeters facing the woodland where the deer appear to come from. This is also generally ineffective as they find their way round the edges. The Forestry Commission and the British Deer Society both have very good advice on deer fencing on their websites. If the deer damage is light, then find a marksman who likes venison.

As for other large animal pests, badgers and wild boar for instance, these tend to go for the grapes rather than the young plants, and therefore are typically a problem in established vineyards, not so much in new ones. Apart from the legal issues surrounding badgers, they are exceptionally difficult to exclude from a vineyard – they seem to see a fence as a challenge rather than a barrier – although two strands of electric fencing at around 100–150 mm off the ground at snout-level will often deter them. One can also install 'badger gates' in rabbit-proof fencing and Natural England[7] has leaflets on these and on electric fencing against badgers. Wild boar are luckily not (at least not yet) a feature in most parts of Britain, although they are found now in parts of East Sussex, Kent and the Forest of Dean.

They can cause a lot of damage to the ground, digging it up whilst they look for grubs and worms and they can push vines about. They also, of course, love ripe grapes. They are almost impossible to keep out with any form of affordable fencing and are best culled, so find yourself a marksman that likes *Civet de Sanglier*.

Windbreaks

As was pointed out in the section about site selection, wind is undoubtedly the curse of the vinegrowing classes and almost all sites will benefit from additional shelter. In top-fruit orchards, hop gardens and in soft fruit plantations, windbreaks have always been a common sight and certainly in Kent and East Sussex, counties that once had many more hop gardens and orchards than they do today, you can still tell where they used to be by the rows of tall poplars and willows that border fields. It always surprises me therefore that today's vineyard owners don't follow the example of their fruit-growing forebears and plant as many windbreaks as I believe they should. The aim of a windbreak is not to stop the wind completely – an impervious barrier would just cause the wind to eddy and jump over the obstacle – but to cause most of the wind to be raised up to the height of the windbreak and continue on its way above the crop. Windbreaks, if planted on the vineyard side of a watercourse (although not if conifers are used), can also be used to reduce your spraying buffer zones. See *Chapter 16: Pest and disease control* in the section on *Buffer zones* for more details.

Windbreaks can be sited both on the windward boundary (or boundaries) of a site – normally in the south of Britain this will be the south-western side – as well as internally. A good, well-established windbreak will protect land up to ten times its height. Therefore, a 7.5 m tall windbreak will protect a 75 m wide vineyard or block of vines. The height of a windbreak will generally be limited by the machine that trims it, so if you intend to allow a windbreak to grow to more than, say, 3 m, make sure your local hedge trimming contractor has a machine that can trim to the height you require. In certain areas, where there are plenty of orchards and plantations with tall hedges, then extra high-reach flail trimmers may well be available which can reach up to 4.5 m high. Remember, when you set out your vineyard, that if your using a tractor-mounted flail to trim your trees, leave enough space on both sides of the windbreak for the machine to work. Alternatively, if you don't trim them with a flail, you can allow the trees to grow to whatever height you like, but know that once they get above a height which can be reached by a trimmer, then cutting them back to around 1 m above ground level and allowing them to regenerate, is the only viable way of keeping the trees from completely shading your site. If the cutting back is limited to, say, every other tree or one tree in three and spaced over several years, then a useful degree of wind cover can still be maintained whilst the windbreak is brought back under control.

---

7 Natural England Technical Information Note TIN 026 for Badger Gates and TIN 027 for Electric Fencing. www.naturalengland.org.uk

The other limit on height is that of shade. Given that most windbreaks will be to the south or south-west of vineyards, its stands to reason that they will, at certain times of the day, shade the vines growing in their lee. Apart from the fact that growing vines too near to trees is to be avoided. The trees will take both water and nutrients from the vines well outside their canopy spread and the shade will lower the vine's fruiting potential and undoubtedly any fruit on them will be more difficult to ripen. However, space is not to be wasted unnecessarily and although the first row or two of vines adjacent to a windbreak may not be as good as the rest of the vineyard, a few 'sacrifice' rows can be tolerated if the vines in their lee are protected.

My preferred windbreak trees are Italian Alders (*Alnus cordata*). They have several attributes: they are not expensive trees, especially if bought as bare-rooted 0.90–1.20 m 'whips'; they grow quickly, attaining a height of 3 m or more after three to four years; they perform very well as windbreaks, being permeable enough to allow some wind through, but dense enough to raise the wind over them; they are not attractive to birds as nesting sites; they leaf up before the vine does and keep their leaves until well after harvest; they do not spread very wide and even when 20-plus years old, spread no more than 0.75–1.00 m either side of the planting line; they can be trimmed mercilessly without being harmed, can be cut back to a low level to regenerate and are more or less indestructible; they do not seem to suffer from canker or other diseases and do not harbour harmful pests; and, lastly, they are attractive trees, with small, mid-green leaves and carry dark green cones in the autumn. After all these attributes there is one thing they do not do well: burn – so do not expect to get much firewood from them. The Italian Alders that I planted in 1977 at Tenterden Vineyards (today's Chapel Down) are still there and still performing well. Other members of the alder family, Common Alder (*Alnus glutinosa*) and Grey Alder (*Alnus incana*) are also suitable, although to my mind not quite as good.

Alnus cordata (Italian Alder) makes and excellent windbreak for vineyards

There are, of course, plenty of other trees which will work well as windbreaks. A hedge of mixed native hedging has a more natural look than a windbreak all of one species, but will take many more years to reach the desired height and will be much denser, and harbour many more birds and their nests, than a windbreak of specialist trees. Poplars and willows used to be common in windbreaks, but one does not see them quite so often these days. Conifers have also been used, but for vineyards and orchards are not to be recommended. They are too dense, causing the wind to eddy and jump over the hedge; harbour all kinds of birds, pests and diseases; suffer from die-back; and are not acceptable if you want to reduce your buffer zone. They are also very ugly. Whatever species of tree you use, make sure it keeps its leaves until after the harvest, which in Britain can be as late as early November.

Windbreaks are best planted as far in advance of the vineyard as possible so that they protect the vines as they establish. If this is not possible, then they should go in at the same time as the vines. If the trees are planted into well-prepared ground, looked after, staked and kept weed-free, then by the time the vines start fully cropping the windbreak should be 3–4 m high and providing some protection. The spacing of windbreak trees will depend to a certain extent upon the species of tree selected and your budget. A row of Italian Alders planted at 0.75 m to 1.0 m apart will give you a good windbreak, although if you have the space, a double staggered row with the trees planted at 1.00 m apart with the rows 1.00 m apart will give you an even better windbreak. The overall width of the double row can be kept down to 2.0 m to 3.0 m, if trimmed. If you are planting internal windbreaks, then plant a double staggered row with the rows 500 mm apart and the trees 1.00 m apart in the row. This will keep the width of the trees as narrow as possible.

I have always used bare-rooted 'whips' which are very easy to plant. A slit cut with a good spade in most soils will normally allow the tree to be pushed in and the ground firmed up afterwards with a boot. These will require staking just to keep them upright as they develop, but a stout bamboo or small tree stake will suffice. Alternatively, put up a single wire with posts supporting it every 6–10 metres and secure the trees to that using rubber ties which will not harm the tender young stems of the trees. It is only larger, pot-grown trees that require properly dug holes, decent stakes and proper tree-ties – a good reason not to use them. Whatever trees are used, they must not be neglected and will require protecting from deer and rabbits – so better inside the vineyard fence than outside – and must be kept weed-free. In other words, they need about as much care and attention as your newly planted vines. They also shouldn't be forgotten when it comes to fertiliser additions. Also note that if you are machine-planting your vines, then the same machine can also plant windbreak trees (assuming they are not too big).

If your vineyard is sufficiently exposed and sufficiently wide enough, then the site ought to be divided into blocks

4 m high Parafence

with internal windbreaks. How wide each block will be will depend upon the height you propose to allow the row of trees on the windward side to grow to and how wide your site is. If you allow your first row of windbreak trees to reach, say, 10 m, then you don't really want any one block wider than 100 m. You also have to factor into your calculations the space that an internal windbreak will take up, both with the spread of its foliage and with the shade it will cast. Because of this it may well be worthwhile keeping internal windbreaks to single rows of trees, topping them at 5–6 m and keeping them sided up so that they are kept thin. Although at least one row of vines and possibly two will be shade-affected, the benefit of the shelter will far outweigh this slight disadvantage.

If you have the luxury of sufficient space to the windward side of your site to plant a small barrier of woodland which can be allowed to grow to its natural full height and provide shelter, then this is ideal. Bear in mind, however, that this may also be a place where birds can nest and roost and hide up prior to launching themselves upon your grapes.

If space really is at a premium, and you don't mind the expense, there are various types of synthetic netting and webbing that can be used. The expense is of both the netting itself, plus the structure needed to hold it up in even the strongest winds. When Lamberhurst Vineyards were trying to improve fruitset and ripening across one windy site, they erected what was then called Paraweb (and is today called Parafence) on every fifth row, which appeared to help slightly. There is no doubt that synthetic windbreaks work – greenhouse owners use them all the time – but they are both expensive and unsightly so not the first choice to my mind. As has already been mentioned in *Chapter 3: Site selection*, Rathfinny have resorted to 4 m high Parafence windbreaks in order to protect the windbreak trees which were not growing fast enough. It will be interesting over time to see how successful these windbreaks, said to have cost £250,000, will be.

Chapter 7

Vineyard nutrition

Vineyard nutrition

The amount of information available about how to address fertiliser requirements in vineyards is vast and, at least to the newcomer, very confusing. In the days when ADAS (the Agricultural Development and Advisory Service) was part of MAFF and offered free advice to farmers and growers, most people just accepted the advice they gave. Today, when competing advisory firms, farm consultants, fertiliser and other agricultural supply companies are all vying for your business, and sometimes selling you their product(s) as well, there is almost an information overload. There are also FACTS – Fertiliser Advisers Certification and Training Scheme – advisers who are trained to offer impartial advice on farm nutrition although, unlike BASIS advisers for pesticide advice, they are not required by law.[1] However, by taking the task one step at a time, working out what nutrients your vineyard needs can be made less daunting.

As my general guide, at least to start with and in the absence of any problems or special situations which might need a different approach, my starting point is the *Nutrient Management Guide RB209* which used to be a DEFRA publication but is now produced by the AHDB (Agriculture and Horticulture Development Board) and has just been updated (2020).[2] This guide covers every aspect of fertiliser and manure use and application and, over the years, I have found its 'Section 7: Fruit, Vines and Hops' very useful and have in general followed its recommendations. However, as with many things, 'junk in – junk out' and therefore before any manures or fertilisers are applied, you will need to have a good idea of the nutrient status of the site. This involves taking samples. For vines, there are two basic systems of sampling: soil sampling, which can be done both pre-planting and in an established vineyard; and what is

often referred to as 'petiole sampling' or 'sap analysis' or even 'tissue analysis' which (in vines) involves taking the petiole (the stalk of the leaf) which connects the blade (the leaf part, as it is more correctly called) to the shoot. Petiole sampling can of course only take place in growing vines and how and when to do this is discussed later in this chapter.

Soil sampling prior to planting

Until relatively recently, the only way to take soil samples was by manually digging holes with a spade, an augur or a soil sampling spear. The accuracy of this type of sampling to tell you about the whole site was limited by two main factors: the number of different soil samples you were prepared to take across a section of a site which could then be amalgamated to be tested as one sample, and the number of different samples so amalgamated that you were prepared to pay for to be analysed. Standard instructions about how to sample a site will tell you to take multiple samples, following a broad 'W' pattern across the site, making sure that you avoid any obvious areas that have been cropped differently, show abnormalities or are obviously of different soil types. RB209 suggest taking '25 individual sub-samples' which will then be amalgamated to form one sample for testing. They also state that: 'the soil sample must be representative of the area sampled. Areas of land known to differ in some important respects (e.g. soil type, previous cropping, and applications of manure, fertiliser or lime) should be sampled separately. Small areas known to differ from the majority of a field should be excluded from the sample'. They also suggest for vines that samples at two different depths should be taken: 0–150 mm and 150–300 mm. This is all a lot easier said than done.

In my experience, this level of sampling is seldom done (I was going to write 'never done' but thought better of it) as the sheer time and labour required to dig or augur this number of holes to take the samples, plus the lugging of buckets across a muddy field, plus the homogenising of

1 BASIS is an organisation that manages the Professional Register for qualified pesticide advisers www.basis-reg.com

2 https://ahdb.org.uk/rb209. The guide is available in whole i.e. all 7 parts for i-Phones, i-Pads and Android devices, but apparently only as separate PDFs for each section for PCs.

the 25 sub-samples and placing into individual bags, plus the labelling, let alone if you are proposing to take samples at the two different levels, is just too great. One might perhaps, assuming the site looks relatively even and there is nothing to indicate great differences in soil type, divide the field up into four quarters, or perhaps on a sloping site take top, middle and bottom, and then for each sub-area take around 10 samples at 150–300 mm, which would then be homogenised to form one sample from each sub-area for testing.

On a small site – say, under 1-ha (2.5-acres) – where there were no obvious peculiarities or problems, I would still advise this type of sampling as, whilst of course there may well be different soil types in even this small area, the variation is not likely to be huge and over the years there will be plenty of time to fine-tune the fertiliser requirements and carry out both soil and petiole analyses once the vines are established. Make sure you know how you are going to sample the site beforehand, take plastic bags capable of holding around 500-g of soil (a medium freezer bag is ideal) plus a marker pen and a bucket, a spade or augur, and some gloves – earth can be very sticky stuff. Then having taken the samples, send them off to a laboratory having first made sure they can handle the type of tests you want. Lancrop,[3] the laboratory I normally use, currently (2020) charge £33.50 + VAT for what they call a 'Broad Spectrum' analysis which covers the basics: phosphorous (P), potash (K), magnesium (Mg), pH, and 'Lime Requirement' (for soils that require a lime addition) plus the minor and trace elements: calcium (Ca), sulphur (S), sodium (Na), manganese (Mn), copper (Cu), iron (Fe), zinc (Zn), molybdenum (Mo), boron (B), and the CEC (Cation Exchange Capacity).[4] Nitrogen (N) is not normally tested for, as this element will have been used up by the crop on the land or will have leached out into the subsoil. They can also carry out test for active $CaCO_3$ which, on high-pH soils (above pH 7.00), will be required so that you can see whether you need a chalk-tolerant rootstock and, if so, just how tolerant it needs to be. Just getting total $CaCO_3$ levels is not enough, as although there is a relationship between the two, it is not linear. This test costs £7.50 + VAT (2020 price).

Soil mapping

In the last two decades, soil mapping using several different techniques has become commonplace in large-scale agriculture as more and more tractors, seed drills, fertiliser spreaders and harvesting equipment are connected to GPS and can be programmed to farm more efficiently. This is often known as 'precision agriculture'. There are currently two main systems in general use. The first is Electrical Conductivity (EC) which works by scanning the field with a GPS-connected, vehicle-towed set of coulters (metal

discs) which send impulses of electricity into the soil to two depths: 300 mm and 900 mm. By measuring the differences in response to the current, caused by differing levels of moisture in the soil, a picture of the different soil types can be constructed, based upon the fact that the water-holding capacity of the soil is closely correlated to its CEC, its clay content and its porosity. Agrii's 'SoilQuest' service is an example of EC. The second type of soil mapping uses a gamma-ray detector mounted on a quad bike which can measure the levels of four naturally emitting isotopes in the soil. Hutchinson's 'TerraMap' is an example of this method. Again, this method detects different types and different levels of soil.

Having been scanned, both systems then draw up maps giving you a view of the relative depths of the soil layers and the varying percentages of clay, sand and silt which are then amalgamated to give you a map of all the different soil types running from sand, via sandy loam and clay loam, to clay, sandy clay and silty clay. Then, having divided the site up into these soil sectors, samples can be taken sector by sector and soil type by soil type for analysis using a quad bike which is GPS-guided and takes multiple mini samples (up to 125 per hectare) from each soil-type area. In this way a much more detailed, soil-type specific, view of the nutrient status can be arrived at. In addition, the maps of soil types give you some very good information about the areas which are likely to require draining, which are likely to be the most vigorous and which could, for instance, do with a heavier dressing of compost prior to planting. They also, of course, give valuable information about which areas might require a stronger (or, more likely) a weaker rootstock and which areas might suit more, or less, vigorous varieties.

Of course, this type of soil mapping and sampling is going to more expensive than hand sampling, but then the information derived is so much more complete. The cost will depend upon the size of the site, the number of different soil types and the number and nature of the analyses required. Current prices (2020) of TerraMap on bare land (i.e. pre-planting) are £24-ha for the standard service and £32 for the premium service, subject to a 30-ha minimum although if they can combine two or more sites on the same day this minimum area will be reduced. (In established vineyards, the charge is £34-ha for the standard service and £43-ha for the premium service). The standard service gives you nine map layers and a basic chemical analysis; the premium service gives you twenty-one map layers, plus the organic matter content, the CEC and the levels of the major, minor and trace elements in the soil.

In my view, the more you know about what's below the surface of your vineyard, the better given that you seldom see much of it (or at least only an occasional glimpse of it when you dig a test pit or have some other reason to expose what's there). After all, the soil is where your vines are rooted and where they obtain almost all of their water and nutrients. Having had several sites mapped, I am constantly surprised by the variability and irregularity of soil types and soil depths even within a relatively short distance. When

---

3 www.lancrop.com

4 The CEC relates to the nutrient holding capacity of the soil and relates to the soil's clay and/or organic matter content.

we see patches of vines that are perhaps more vigorous or less vigorous, more fruitful or less fruitful, if your site has been precision mapped, as the late, great Kenneth Williams (parodying the late, great gardening expert, Percy Thrower) used to say, you will know that 'the answer lies in the soil'.

Petiole analysis in established vineyards

In taking soil samples, the aim is to discover what's in the soil. But, however good your soil sampling, analysis and interpretation, it tells you nothing about what is getting from the soil, via the roots, into the plant. Petiole analysis (also known as 'sap' or 'tissue' analysis), whereby the sap in the petioles (leaf stalks) is extracted and analysed, tells you exactly, at the time you take the samples, what nutrients are in the vine. The leaf stalk is used as this connects the leaf – the part of the vine where the miracle of photosynthesis is taking place – with the rest of the plant. The stalk is relatively narrow and therefore the nutrients running to and from the leaf to the roots via the stem are fairly concentrated. Petiole analysis is not something you would carry out until the vines have started cropping and have an established root structure, but then it should be carried out regularly. Some growers take petiole samples every year, although with a large vineyard and multiple blocks, this can be an expensive process. Lancrop's current price (2020) for 'vine petiole' analysis, which covers nitrogen, phosphorous, potassium, magnesium, calcium, sulphur, manganese, copper, iron, zinc, molybdenum and boron, is £42 + VAT per sample.

In order to get an accurate snapshot which can be compared over time to previous snapshots of the vine's nutritional status, it is important to take petiole samples from the same part of the plant at the same time of the year. Depending on what you are trying to show, you need to select a block of vines that is as uniform as possible, i.e. same variety, same clone, same rootstock, same vineyard management. Of course, you might have one block of vines of the same variety, clone and rootstock, but where one half had been treated one way and the other half another; if this is the case, take samples from each half. Petiole analysis is also valuable when you have one section of what is otherwise a homogenous block of vines showing symptoms of something being not quite right. In this situation you could sample the healthy versus the unhealthy section and can either pinpoint or eliminate a nutritional reason as being the source of the problem.

In vines, samples are taken from the leaf opposite the first flower or bunch of grapes either when flowering is almost over – 80 per cent capfall, as it is known – or at *véraison*, both dates which can be seen visually and at a time when the vine is under a certain amount of stress. Sampling at flowering will give you a chance to address any deficiencies sooner in the growing season rather than later, when any adjustments you make might not have that much effect upon the vines. Make sure your hands are clean and then pick around 75 leaves, remove the blade and place the stalks into a clean paper envelope (and not in a plastic bag as they will sweat). Ideally you should take these before it gets too hot (although I am not sure it ever gets THAT hot in British vineyards in July). You must also make sure that you take leaves from across the block, no more than two per vine and select leaves that are well exposed to the light. Avoid taking leaves from vines at the ends of the rows as these are generally not very representative of the rest of the row (as they have less competition from neighbouring vines and are often sprayed more and fertilised more as the tractor starts off down the row) and avoid any vines that are obviously sick, unhealthy or showing signs of stress. You should also avoid vines that have been recently sprayed (especially if using copper or sulphur sprays) as this might distort the results. Once collected, the petioles should be kept as cool as possible and sent off to a laboratory without delay. If you cannot send them off at once, keep them cool and dry until you can send them off. Obviously, they do not want to dry out completely as otherwise there will be no sap left!

The results from petiole sampling will show the various elements as being 'very low, low, normal, high, or very high' and should be used as a guide to additions of fertilisers and foliar feeds. They should also show nitrogen (N) status, something that soil samples generally do not, and this can be used to see whether a judicious amount of nitrogen ought to be applied, although often a slight leaf yellowing will have told you this already. Fairly small foliar applications of nitrogen are often the best way of perking up vines that are showing signs of N deficiency, but tread carefully as an excess will only result in too much soft growth.

Fertiliser additions prior to planting

Having got your soil sample results, you will see that each of the major nutrients will have been assigned a nutrient 'Index Number' from 0 to 5 (or sometimes 0 to >4) and for each nutrient there is an amount to be applied, depending on the starting index level. Page 16 of Section 7 of RB209 has a table showing the different amounts to be applied in order to bring the level up to index level 3, which is considered ideal. As to the type of fertiliser to be used (because there are – as for most nutrients – several different materials containing the same element) RB209 again has guidance on this, or your local supplier will certainly be able to help, especially if they are used to dealing with fruit growers. Fertilisers come in what are known as 'compounds' or 'straights'. Compounds are named after their constituent nutrients and in the order: nitrogen (N), phosphorous (P) and potassium (K).[5] Thus a 20-20-20 will have 20 per cent of each of N, P_2O_5, and K_2O. There are also compounds with no nitrogen such as a 0-20-20. Often there will be additional elements present such as magnesium and sulphur, in which case the name of the compound might be 9-4-10, 5 MgO, 22.5 SO_3 which

---

5 Potassium is sometimes called potash and always referred to as 'K' after the German for potash which is Kalium.

would have 9 per cent N, 4 per cent P_2O_5, 10 per cent K_2O, 5 per cent MgO and 22.5 per cent SO_3. Straights, on the other hand, are made of one element and would contain a known amount of nutrient. Thus, triple superphosphate (TSP) contains 47 per cent P_2O_5 and muriate of potash (MOP – also known as potassium chloride) contains 60 per cent K_2O. Section 1 of RB209 has a full list of most of the common materials and a conversion table to convert element to oxide and vice versa.

For vines, because they are a very long-lived and deep-rooted crop, considerably higher levels of fertilisers can be applied prior to planting. Page 12 of Section 7 of RB209 suggests that the site be sampled at 0–150 mm and 150–300 mm and, assuming the results of the 150–300 mm show index levels of 0 to 1, then the amount shown on the tables to bring them up to index level 3 should be applied before planting and ploughed down as deeply as possible, ideally the winter before planting. Then, in the spring prior to planting, the amounts shown by the 0–150 mm level results can be applied as a top dressing prior to final cultivations. If only 0–150 mm samples have been taken, then assuming again that the index levels are 0–1, then the amounts required should be applied before deep ploughing, again preferably in the autumn, and the same amount applied prior to final cultivations before planting. If you are also applying compost or manure, the manurial value of these must be known and deducted from the amounts to be applied as fertiliser. Depending on the level of nitrogen already in the soil and on whether any manures or compost containing nitrogen are used, you might consider applying a small amount of nitrogen prior to planting to help the vines establish. Unless the vines were showing symptoms of nitrogen deficiency, I would not apply any more until they were fully cropping. If you are in an NVZ, also be aware of restrictions on amounts of N that can be applied.

When working out amounts of material to be spread, note that all recommendations are given in kilogrammes per hectare of the oxide – P_2O_5, K_2O, MgO etc – and the actual weight of material to be applied must be adjusted according to the 'strength' of the product. Thus, for example, to add 100 kg of K_2O-ha using muriate of potash, which contains 60 per cent of K_2O, one would need 167 kg-ha of product to be applied.

Soils with low pH levels

Soils with pH levels of below 6.0 will require some additions of lime to bring the levels up to 6.5 and maybe a little higher (but certainly not over 7.0) and this is much more easily applied before planting than afterwards. Vines growing in soils of pH 6.0 or less may well have problems in accessing a wide range of nutrients, so if your soil tends towards acidity, then this is something to be aware of. Improving the humus content of the soil and raising the CEC will both raise pH levels and help keep pH levels up. Rates of the material to be applied will vary according to the indices, but 4.0–7.5 tonnes-ha (1.6–3.0 tonnes-acre) is typical. Where amounts

are greater than this (indicating a very acidic soil) it may be better to apply about three-quarters of the total required before deep ploughing and the remainder after ploughing and before cultivations. With soils that tend towards acidity, additional applications may be required after, say, 4–5 years.

Lime suppliers will advise on the best materials to use and if your soil is also deficient in magnesium, then magnesian limestone should be used (which contains around 15 per cent MgO) and the appropriate adjustment made to the amount of magnesium containing fertiliser that is then applied. Lime is a very bulky, messy product and is best applied by a specialist contractor using their own equipment to both load and spread the product. Some lime suppliers now have spreaders that can be tied in via GPS to EC soil maps, so that different rates are applied to different areas. Lime should be applied as early as possible in the land preparation process as it is slow to act and needs to be worked well down into the soil. Therefore, lime should be applied to the site as early as possible, ideally before subsoiling and definitely before deep ploughing, preferably in the winter prior to planting, as you are aiming to get the lime down to 400 mm deep. Once the vines are established you will not be able to get the large equipment typically used by liming contractors down the rows and if, at a later stage, your soil does need any lime, you will have to use 'prilled', i.e. a granulated, products, such as Calciprill and Magprill, which are considerably more expensive than basic lime, although the manufacturers claim they work much faster than lime (which I can believe).

Soils with high pH levels

Soils with pH levels above 7.0 are termed alkaline and pose additional problems to vines growing in them and in such soils trace element deficiencies, especially of iron, manganese and boron, are much more likely to occur. As has been already pointed out, it is important that the level of active $CaCO_3$ is tested so that the correct rootstock can be chosen, otherwise the vines will succumb to chlorosis. Many high-pH soils will also be deficient in humus. The French like to apply copious quantities of ferrous sulphate (iron sulphate) to high-pH soils, up to 10 t-ha (4 t-acre). Whether this is strictly necessary is a matter of some debate amongst viticulturalists and, to date, the jury is out. One thing is certain, though: it is horrendously expensive. At the rates above it costs around £2,100-ha (£850-acre).

Fertiliser additions in established vineyards

Once your vines are established, and as long as they are not showing any signs of mineral deficiencies, then the next time to take soil samples will be after their first or second cropping, i.e. after year three or four. The guidelines in RB209 suggest soil sampling every '3–5 years' but I would be guided by how the vines look, in particular how the leaves look and what stresses they are under, and probably stick

Power harrow preparing a seed bed for sowing with a green manure crop

to soil sampling every three years. Heavy yields and adverse weather conditions (such as the heat in 2003, the cold and damp in 2012 or the very high yields of 2018) can tax a vine's reserves and might mean taking soil samples sooner rather than later. In established vines, take soil samples from near to the vines in the area where the vine's roots are most likely to be feeding and not from the middle of the grass strip. If your site has been previously EC mapped, then samples can be taken from areas which are homogenous from the point of view of soil type and compared to previous soil samples to see what effect the fertilisers and manures being applied are having. Otherwise, samples are probably best taken on the basis of blocks of the same variety-clone-rootstock. As with taking pre-planting samples, you need to take a number of small samples in a broad W pattern across the area and then amalgamate them into one sample for testing. Given that you cannot usually walk across a vineyard from one row to another (although if the fruiting wire is high enough you can duck underneath it) getting a really good soil sample consisting of, say, twenty sub-samples can be time consuming and involve quite a bit of walking. The soil in established vineyards can also be sampled using the same sampling quad bike used for soil mapping. This can drive up and down every row in a vineyard, taking multiple samples which will give you a complete nutrient map of the site.

Having taken soil samples and had them analysed, your laboratory will have made recommendations for fertiliser applications to be applied. The aim should be to maintain the major nutrients (P, K and Mg) at the index level 2 and the trace elements at the appropriate levels, the precise level and nature of which your soil analysis will show. Nitrogen should be applied to vines very conservatively, if at all, and it may well be better to use foliar urea in small amounts. RB209 suggests annual application rates for P, K

and Mg which depend on the index levels. Trace element deficiencies can often be seen by the way the vine's leaves grow and can be dealt with via foliar feeds (or occasionally soil drenches). Levels of boron, manganese and zinc should be kept up to guidelines. Sulphur and copper levels should not be overlooked and even though nearly all vines in Britain are sprayed with sprays containing these elements, additional applications may be necessary. As has been already pointed out, iron may be an issue in soils with high active $CaCO_3$ levels and, if the correct rootstock has not been selected, or if the vines are showing signs of chlorosis, then even though an iron-containing foliar feed (Ferleaf) has been used, a soil-applied drench of chelated iron (Fersoil) will help stabilise the vine and get down to the roots. Molybdenum is a trace element which I believe may be more important in British vineyards than some people realise. In other countries, when vines have low molybdenum levels, they can suffer from poor pollen tube growth and therefore poor fruitset (particularly Merlot) and I see no harm in applying some in British vineyards to keep levels up. It is quite normal to include foliar feeds in the regular spray mixtures, typically both pre- and post-flowering, together with a general plant tonic made from plant extracts and seaweed, Maxicrop being one of the best known and longest established.

If a fertiliser spreader which can strip-spread is available, then the fertiliser can be spread in bands about 500 mm either side of the vine and the rate of fertiliser adjusted pro rata to the amount suggested. Therefore, in a vineyard with 2.0 m wide rows, the fertiliser rate could be adjusted to 50 per cent of the specified amount. In my experience, even when using maximum recommended fertiliser inputs pre-planting, the first post-planting soil tests reveal that very little has changed and it takes up to ten years before much change takes place in the potassium, phosphate and magnesium levels. Trace element levels, which can be dealt with via foliar feeds, can usually be moved more easily.

I have also seen fertiliser spreaders that can 'inject' fertiliser into the soil at a depth of around 500 mm, although not in Britain. Given that many of a vine's roots are at this sort of depth, this might be an interesting technique to use.

Green manuring

Green manuring, the planting of a crop specifically to serve a specific nutritional or functional purpose, is a valuable tool in the vineyard owner's armoury. Whether you are looking at a site before planting, in its establishment phase, or when a vineyard is into its cropping years, the planting of a green manure can help in several areas, the main one of which is the addition of humus to the soil. Growing your own organic matter in situ will also have the additional benefit of improving the soil structure and drainage of the site – Common Chicory (*Cichorium intybus*) for instance has roots that can penetrate to over 1.00 m – and will also provide nutrients to help increase the soil's microbial activity. Depending on the timing of sowing and the species sown, green manures can also provide some nitrogen – either for short-term or long-term use – although in general terms the addition of extra nitrogen in British vineyards must be treated with care. Late autumn short-term plantings will seldom produce much nitrogen as this does not happen if the soil temperature is below 8°C. Some species of legumes – lucerne, sweet clover and fenugreek – which may have been selected to fix nitrogen, will not do this unless the bacterium *Rhizobium meliloti* is present and this will need to be added to the seed prior to planting. Green manures can also help clean up the soil by suppressing weeds and reducing the incidence of certain pests, including nematodes and mustard (and other *Brassicas*) have been shown to act as biofumigants. See *Chapter 17: Trunk diseases* for more details.

As has been discussed in *Chapter 6: Site planning and preparation*, if there is enough time, even if it is only six months, the sowing of green manure before planting is a valuable extra treatment which will really help get your vines established. Once your vines have been planted, unless there is a really pressing reason why you specifically wanted to sow a green manure crop straightaway, I would hold back until the vines are established and entering their cropping phase before doing so. Perhaps if you had a drainage problem that required some very deep-rooting species, or wanted to stabilise a site that was perhaps on a side-slope, then I can see the wisdom of using a green manure at an early stage. In an established vineyard, you could consider planting every other row with a green manure mixture on, say, a two to three-year cycle. In this way you would always have a firm surface to run on for tractor operations and it would allow the sown rows to produce the maximum amount of organic matter.

The actual choice of species to plant will depend on different factors and there are several good seed companies which can provide advice and seed to suit your exact requirements. Cotswold Seeds is one such company which I can recommend and they also publish a very good booklet called 'Sort out your soils. Second Edition. A practical guide to green manures and cover crops'. The cost of the seeds for green manuring will vary according to the species to be planted and will range from £55–£250 per hectare (£22–£101 per acre) plus the cost of the land preparation and seeding.

How to plant the seed will depend on exactly when, in the lifecycle of the vineyard, the green manure is to be sown. If it is prior to planting, then it can be either drilled, or broadcast using a fertiliser spinner (depending on the seed size) and this is something that the contractor preparing the site will handle. A green manure crop is probably best established after all the 'heavy' site preparation operations such as earthmoving, drainage and subsoiling have been carried out, and drilled into a well-cultivated seedbed. If you are considering planting a green manure crop following vine planting, whether it is directly after planting or once the vineyard has started cropping, then you have two options. The first option is if the green-manure crop is to be planted immediately after the vines are planted and the soil is bare and weed-free. If so, you will need to obtain a narrow seed drill (they can be hired) which can plant the area between the weed-free strips, or use a fertiliser spreader which has been fitted with side-baffles to restrict the throw of the spreader. The one thing you really want to avoid is getting seed on to the 600–750 mm weed-free area beneath the vines. After seeding, the land will probably benefit from a light harrowing and consolidating with a roller. The second option is planting into a cropping vineyard, where a grass sward has already been established in the alleyways. You will first need to get rid of the grass by burning it off with herbicide and then cultivating the area to be seeded. If you don't use herbicide and just cultivate then the grass species already growing will quite quickly re-establish and you will find that some of the benefits of the green manure will be lost. Finding a narrow cultivator may well be a challenge, although, as with a seed drill, they can be hired. After seeding, the land is probably best lightly harrowed and rolled, assuming you can find the right sized equipment.

Once the seed has germinated and started to grow, when you start topping – and how hard you top – will depend on what you are trying to achieve and the species being grown. Some species take to early topping, others do not. If you are trying to grow the maximum amount of green matter possible, then later topping will help achieve this, although some plants, if you allow them to grow until they have produced ripe seed heads, may well grow to over 1 m tall (in the case of chicory, a really good pan-busting plant, it can be nearer 2 m) and you will need a powerful flail mower in order to reduce the growth down to a manageable height.

Whatever species of green manure you plant, the crop should be seen as a relatively short-term one as the likelihood is that within two to three years, certain species, especially grasses including any in the seed mix planted, as well as those native to your soil, will start to dominate and eventually will take over. This is especially the case if you start to mow before the green manure plants have had a chance to flower and seed.

In many oversees vineyard regions some form of green manuring is commonly practiced and one often sees vineyards either completely sown or with every other row sown, whether on a short-term or a longer-term basis, with green manures. One of the problems in Britain – and the reason that I suspect green manuring is not more widely practiced – is that most vineyards do not have access to the tools required to do a good job. Ideally one needs to have access to a cultivator, seed drill, harrow, roller and flail mower, all narrow enough to fit down a vineyard row and not trespass on to the under-vine weed-free strip. Together these might cost £15,000 to £20,000 (although some of them can be hired) and this type of capital expense, plus the cost of the seed, the cost of planting and establishing the seed and the extra work involved with the proper management of a green manure crop, is seen as an additional expense with an uncertain payback. However, I believe that many poorly performing vineyards would be greatly helped by the beneficial effects of green manuring, especially in the area of improved drainage, improved soil fertility and increased soil microflora activity.

Chapter 8

Planting your vineyard

Planting techniques

The site has been chosen, the vines ordered and the site prepared. The next task facing the vinegrower is planting. As has already been discussed in *Chapter 6: Site planning and preparation*, I believe the best time to plant vines in Britain is in the first three weeks of May. What follows is based upon this assumption.

By now you will have decided upon the layout of your vineyard, the orientation of the rows, the size of the individual blocks, the row length, the space required for headlands, internal roadways, internal walkways, windbreaks, and areas for parking and loading, and these will have been allowed for in your overall planting plan.

The first decision to take about planting, one that needs taking well in advance of the actual operation, is whether to plant by hand or by machine. Until Denbies planted their first vines in 1986, all vines in Britain (that is ignoring the efforts of the owner of Barnsgate Manor Vineyard at Crowborough who in 1975 used, with very mixed results, a cabbage planter) had been hand-planted. Hand-planting, using a variety of implements, can be very successful. With correctly prepared soil, correctly prepared vines and enough people, vines can be hand-planted quite easily, the rate of planting depending on the soil conditions, the ease of making holes and the implement used. The drawbacks of hand-planting are the time it takes and the problems associated with accurately marking out the row positions and the individual vine positions in the row, and therefore the cost. Although hand-planting will probably cost less than machine-planting, it is by no means a cheap option. The cost of machine-planting will vary according to the number of vines to be planted and what sort of deal you can negotiate with the contractor (and also the rate of the euro, as most planting contractors working in Britain come from France, Germany or Luxembourg), but you can budget on between 40p and 65p per vine. On a small scale, say less than 5,000 vines, you may have difficulty persuading a planting

contractor to turn up for this number, or they may want to charge you such a high price that it is uneconomical. In that instance, hand-planting is your only option. Having been involved with planting hundreds of thousands of vines over the past forty years, both by hand and by machine, I would never now voluntarily opt for hand-planting. The time taken in marking out, the difficulty in getting the vines in the right positions, when compared to the ease of machine-planting makes the decision – as they say – a 'no brainer'. It has been suggested that machine-planting might make TDs worse, but in my experience, this is not the case. Sure, if you plant into badly prepared ground, then vines may not establish properly and TDs might be worse. But prepare the soil correctly and plant the vines correctly, and it is my experience that TDs will be no better or worse for the method of planting.

Hand-planting

If you propose to hand-plant, then the first task is marking out. You will need to decide which is your 'reference' row – the one from which all other measurements are taken – and after that it is just a matter of striking a right[1] angle to this row and making sure all other measurements are taken from it. If the headland where you start planting is not at right angles to your reference row, then you have the job of deciding whether the vines should be in line with your headland or at right angles to your reference row. If you decide the former, then remember that the distance between rows must be measured at right angles to your reference row, not in line with the headland. Although the difference is not great, over many rows it will show. Personally, I always favour planting at right angles to the reference row, but this is mainly because almost all planting I am involved with is by machine. Either way, your end-posts ought to be installed in a nice straight line wherever possible, whichever way you decide to plant. Once you have staked out your reference line and struck your right angle, the rest of the rows can be measured out.

The tools required for hand-planting are some good, narrow-bladed planting spades, string, marker pegs and a decent tape measure, the longer the better. Depending very much on the soil conditions and the organisation of the job, one good hole-digger can keep one planting-person busy and one such 'team' ought to be able to plant up to 300–400 vines a day. Ideally you will want several teams in order to get the job done quickly. This number of vines may not be planted on day one, but once a rhythm is established, in well-prepared and easy-to-dig soil, this number of vines ought to be achievable. The vines for hand-planting need to be trimmed so that their roots are slightly shorter than you need them for machine-planting – 75–100 mm is more than long enough – and each planter should have a wooden box or a bucket in which their vines are kept as they plant. Ideally the vines should be kept under a piece of damp sacking so that they do not dry out. This is especially important if the weather is bright and breezy and it is doubly important to cover the vines up when your planting teams take a break. It is important also for the holes to be dug with a flat bottom of the right size – the hole should be square with one side dug in line with

GPS guided planting machine

the row – and the vine should be placed against the side of the hole, in line with the string marking out the row, with the roots spread out in the bottom of the hole. If the hole is too deep, then some of the soil from the hole needs putting in the bottom so that the vine sits at the right height. After the hole has been filled in (each planter might like to have a small trowel about their person for this task) the vine should be tweaked upwards if needs be, to bring it up to the right level and finally the soil around the vine should be pressed down with a firm footstep. Job done!

In addition to the digging-planting teams, you will want a team of at least two people marking out the ends of the rows, putting out taut strings to mark every row and pegging out every vine position. Although you can attempt to do this by taking a starting point and measuring the distance between each vine using a piece of wood the same length as the inter-vine distance, it is almost impossible, except on a very small area, to keep the vines in the correct position. Not only do you need to worry about the vine's positions at right angles to each other, but it is also good practice to get the diagonals in line. Apart from the visual pleasure of having all the vines in the right place, when it comes to installing the stakes in the vineyard, having the vines in the right place is a great help.

Although hand-planting vines using just a spade is a viable option, I have also very successfully used a water lance. This is a hand-held device which, when attached to a sprayer full of clean water, 'digs' a hole into which the vine can be placed. In this way, one lance-person, plus one person driving the tractor, can dig holes at fairly high speed, the only problem being that you do not want the holes to dry out before the vine is inserted, so you do need a gang of two planters per hole-digger. In this way a three-person team might plant 600–700 vines in a day, or even more. The vine roots should be trimmed back to less than 50 mm – although this looks quite brutal, don't worry, the vine will survive – and as soon as the hole has been drilled, the vine should be placed into the mud-slurry that fills the hole. As the water drains away, the mud will solidify enough to hold on to the vine which can be tweaked upwards to

---

1 Striking a right-angle using just a tape measure and three canes is relatively easy using the Pythagorean 3-4-5 technique. A triangle with the dimensions 3-4-5 (or 6-8-10 or even 12-16-20) will have a right-angle between the 3 and 4 edges. It is easier doing this with 2 people or with strings cut to the right length. Whilst it is not easy to get 100 per cent accuracy this way, especially when the right-angle is extended by eye over a hundred metres or more, it is good enough for most hand-planted vineyards. For machine-planting, a more accurate method is required and if you have your site GPS-surveyed, then make sure you get the surveyor to do this for you.

the right planting level. If I was using a synthetic mulch, such as black plastic film or a woven material (see *Chapter 14: Weed control* for more about mulching) then I would always lay the film or material first and plant with a water lance through it. The lance can also be used for inserting intermediate posts, if not all the way, then at least to give them a start. Successful hand-planting depends on getting the soil conditions right before planting. Well-subsoiled, well-cultivated, dry soil is ideal, and digging a hole with a spade or water lance will be a simple task. Tough, heavy, waterlogged clay is another matter.

As for the cost of hand-planting, I have known contractors quote around 35p–40p per vine, to which might have to be added the cost of marking out (some contractors include this as part of the planting price) and organising the supply of vines to the planting teams. However, in my experience, contractors working on piece-work tend to rush the job and just get the vines into the ground in any way they can and it is better to employ your own labour on an hourly basis. You only plant a vine once, so do it properly.

Machine-planting

On the other hand, given correctly prepared soil, machine-planting is comparatively easy. There are essentially two types of planting machine: laser-guided (although these have almost disappeared) and GPS-guided. When GPS-guided planting machines first appeared (in the late 1990s) they were not totally reliable and not to be recommended. Today, however, there are very few old laser machines working, and very accurate GPS-guided machines are the norm; and they are (almost) fool-proof. Laser-guided machines work by following a laser which gives them a straight planting line (although this depends on the tractor driver following the laser) and the inter-vine distance is set by the machine, which is guided by a steel guide-wire played out from a fixed point as the machine progresses along the row being planted. When the machine gets to the end of the row, the laser is moved by hand to the line of the next row to be planted and the stake holding the end of the steel guide-wire is likewise moved the correct

More vines go in at Domaine Evremond, Kent in 2020

distance at the other end (the starting end) of the vineyard. Laser-guided machines can normally only plant from one end of the field and usually plant best driving down a slope. The rate of planting will depend upon several factors, but I would expect a laser-guided machine to plant 1,000–1,250 vines per hour and a GPS-guided machine nearer to 1,750 vines per hour. Obviously, row length will be a factor in planting speed, with lots of short rows taking longer than a smaller number of longer rows. My best planting rate with a GPS-guided planting machine was 19,000 vines in a 12-hour day (and we stopped for a quick lunch!).

The accuracy of planting with a laser-guided machine depends on three people: the tractor driver, the person moving the laser and the person moving the stake holding the guide-wire. Whilst in my experience mistakes seldom happen, it is always possible and my advice is <u>always</u> to be on hand when a vineyard is being machine-planted (laser or GPS) and have a tape measure (or two) about your person to make frequent checks on both row width and inter-vine distances. It also helps to have a bamboo cane marked with the correct row width so that each time a row is started, the bamboo can be moved across the site as a final check that the row width is correct. Finding that you have one row which is slightly narrower or wider than all the others is not ideal. As with any type of vineyard planting, a nice square site without undulations or strange angles will always be easier to plant accurately than one where the headland runs out at an alarming angle to the reference row and you cannot see the far end of the row from the beginning.

With GPS-guided planting machines, on the other hand, after the GPS receiver has been set up and locked on to the satellites and the planting machine has 'shaken hands' with the receiver, all you need to do is decide where the very first vine is to be planted – this becomes your reference vine against which all others will be aligned – and the direction of your first row. Go to where you want your first row to end, mark it, and the GPS-guided machine will drive to it, take a reference position and this will then locate the line of your reference row. All other rows will then be parallel to this row. The only caveat to this apparent planting perfection is if the vineyard site is in a valley or there are a lot of tall trees close to where vines are being planted. In these situations, I have known GPS machines to lose their satellite connectivity and, whilst they eventually got it back (as different satellites come over the horizon), it did result in delays in planting. This was a few years ago, and there are now more GPS satellites in space, so it hasn't happened recently. On an open, non-valley, tree-free site, however, GPS-guided planters work exceptionally well.

Because GPS-guided machines work with total accuracy, if you have a straight headland to which you would like your rows to be at right angles to, it is important to make sure that your reference row is exactly at right angles to the line of the headland. To do this, the Pythagorean '3-4-5' system (see footnote above for more details) using three pegs and a tape measure is not usually accurate enough and

Roots coming from above the graft. Make sure your vines are not planted too deeply

the right angle needs setting out using a theodolite or other measuring device. If you are having the site GPS surveyed, then the surveyor can strike your right angle for you. Having inputted the row width and inter-vine distance into the computer, there is little else to do apart from keep the machine fed with vines and occasionally make sure everything is in line. The accuracy of a GPS-guided planter is to within around 12–15 mm and, apart from where soil conditions disturb the planter as it travels across the field, the accuracy of planting, both in-line and across the diagonal, is impressive. Although a GPS-guided machine can plant in both directions, it is generally easier for it to always start from one end – the end where you have your stock of vines. On a flat site, starting at each end works well, but on a slope, try and get the machine to work from uphill to downhill. Of course, planting contractors always want to plant from both ends as its quicker.

Again, as with laser planting, you will need to peg out the line of the headland at the far end so that the machine operators know where to stop feeding the machine with vines. In truth, it is easier for them to overplant a vine or two if they are unsure as, immediately after planting, any extra vines can be pulled up and placed back on the machine for replanting. It is also wise to have two spades to hand – one at each end of the vineyard – in case you need to hand-plant the odd vine. Occasionally the machine operators will miss a vine or the vine they put into the planting jaws will fall out. Hopefully they will tell you this as it happens so that you can walk down the row in question with your spade, find the missing vine and replant it.

Both types of planting machine operate in the same way. A large double blade is dragged through the ground, creating a slot and the vines to be planted are placed into the slot by a rotating wheel which has a gripper system which holds onto the vine until the moment of planting.

The machine travels forwards, opens up the slot, the vine is planted and two blades and two wheels follow behind, moving and firming the soil around the vine. The planting height is set by the operator, but depends upon two large wheels upon which the machine runs. See the *Wagner Pflanzen-Technik* website for details and a video of their 'IPS-Drive' planting machine.

Although I pointed this out in my section in *Chapter 6: Site planning and preparation* in the section on *Cultivations*, it is worth repeating what I wrote:

Once the [planting] machine has passed on, the soil must fall back into the slot and around the roots of the vine. This would typically be at a depth of 250–300 mm. Although the soil at the very surface will invariably fall back around the stem of the vine and visually the vine will appear to be settled, it is what happens below the surface level that is critical. Vines planted in sites where subsoiling has not taken place and where cultivations have only been at a relatively shallow depth, may well suffer from having their roots waving around in the air, some way below surface level and out of sight.

After planting, especially if the soil is wet at the time of planting, the planting machine – or any machinery you have had on the vineyard during planting – may have left wheel marks in the rows. Depending on the severity of these, and assuming that you will eventually be mowing in the alleyways, it is sometimes necessary to cultivate the rows immediately after planting, to even things out and level the soil. If, of course, you propose to cultivate your vineyard during the establishment phase, then this will happen anyway. It is also very occasionally necessary to press down the soil either side of the line of planted vines if the soil hasn't properly closed up around the roots of the vine. A quick 'waggle' test can establish if this happened. Grab the newly planted vine by the top and see if you can wave it backwards and forwards in the direction of the planting line or pull it up and see if it moves. If you can, then this means that there is very little soil around and in contact with the roots. With well-loosened soil and perhaps some rain soon after planting, the soil may well firm up around the roots on its own, but if it looks like it is not going to, it may be necessary to run a tractor wheel as near to the vines as possible in order to try and press the soil down into the area where the vine's roots are located. It's not ideal, but it's better than letting the roots dry out. Don't obsess over this and in my experience, it is seldom necessary, although badly prepared clay soils are sometimes a challenge.

Soaking vines before planting

Although you will often read in the literature and sometimes even upon the sides and tops of the boxes containing the vines that they should be 'rehydrated by soaking in water for 24–48 hours prior to planting', I have never done this and have never had a problem with vines suffering from lack of water. I suspect that nurseries state this because: (a) almost all their customers are in warmer climates than Britain's; (b) they are not sure how the vines are stored once they leave their premises; and (c) they are covering their backs. In *Chapter 17: Trunk diseases* you will see reference to a document written by three of the foremost researchers into the subject who conclude: '*Vines, particularly the roots, should not be allowed to dry out during planting. However, standing vines with freshly trimmed roots in a bucket of water is detrimental, spreads disease and should be avoided. It is better to cover bundles with a clean damp cloth.*' I have never had to water vines after planting although I do know growers who have had to in dry years. If the ground is correctly prepared and the vines are correctly planted, then at root level the vine's roots should be firmly surrounded by moist soil and should not suffer from a lack of moisture. To see if your vines have enough water, dig down 200–300 mm to the level where the vine's roots are and see if you can 'ball' the soil in your hand. If you can, then there is enough moisture. If the soil will not ball and collapses into dust, then you might need to water. The other pointer is the weeds: if they are happily growing, then there is almost certainly enough moisture for your vines.

Mycorrhizal fungi dips

In recent years there have been experiments in planting a wide variety of hardwood plants, including vines, using mycorrhizal fungi dips on the roots just prior to planting.

As was pointed out in the *Chapter 3: Site selection* in the section on *Soil type*, mycorrhizal fungi are necessary for plants to access the nutrients contained in the soil, and in soils deficient in these fungi, the growth and development of plants may be slowed. Whilst I have seen experiments using these fungi on hedging plants and fruit trees, which showed much better growth and establishment with their application, my experience with vines is not that positive. Unless vines are to be planted in very humus-deficient soil and a soil that has perhaps been poorly farmed, I doubt whether it is worth the expense and, more importantly, the extra time taken to dip the vines at an already very busy time – immediately pre-planting – and I would not recommend the use of mycorrhizal fungi. However, if you feel you need or would like to use them and it makes you feel better, I do not believe there are any disadvantages.

Products based upon *Trichoderma* fungi, which are naturally occurring beneficial soil fungi, are sold on the basis that they can protect roots from many damaging pathogens present in soil. Products such as RootShield are sold in several different forms – as granules and as dips – and claim to offer such benefits as 'forming a shield around' and 'preventing pathogens from damaging' the roots. Again, as with mycorrhizal fungi dips, I am unsure as to how much benefit these products offer to vines planted into well-prepared soils.

Chapter 9

Trellising systems – construction

Introduction[1]

The aim of all trellising systems in viticulture is to secure the fruiting wood and to allow the annual canes and foliage that comes from that fruiting wood to be contained so that operations, both manual and mechanical, can take place with ease and without damaging the vine and its fruit. Trelliswork has to be constructed with materials and methods that are cost-effective and durable and last as long as possible. What is not required is a trellis that is cheap to install which, although it may perform well initially, requires costly maintenance and/or replacement. Good trelliswork should last the lifetime of the vineyard and its cost can then be spread over that lifetime.

Of course, not all vines require trellising systems, but these, typically bush-trained vines, are only grown in warm to hot climates where growing-season rainfall is low and, because sunshine and heat levels are high, less foliage is required to produce the sugar levels required. This is not the situation in Britain. In Britain, as in all cool growing regions, vines require more foliage in order for there to be sufficient leaf area for enough photosynthesis to take place to ripen grapes to the right sugar level. Although it is not a straight-line equation, in a cool climate, the more foliage your vines have (within reason), the riper the grapes, the higher the sugars and the better the wine. In addition to requiring more leaf area, vines growing in cooler climates, where there is typically more growing-season rainfall, will suffer from more disease and require more canopy management and pesticide spraying (especially fungicides) to keep them healthy and their fruit in good condition.

In Britain, the coolest of all regions where vines are grown on anything approaching a world-scale,[2] the problems faced are those of over-vigorous vines which shade both the canes bearing the buds for the following year's fruiting, as well as the buds, flowers and eventually grapes, whose production is the purpose of the vineyard. This excerpt from Smart and Robinson's *Sunlight into Wine* (page II) is worth quoting in full:

*The buds that contain the shoots and clusters for the following season begin growth early in the spring as shoots develop. Flowering is a critical stage. It is at this stage that the cluster primordia are laid down. Cluster primordia are microscopic pieces of tissue which eventually develop into clusters the following season. Within the young 'compound' bud there are three small shoots which develop – the primary, secondary and tertiary. Whether or not these shoots produce cluster primordia is **very dependent on shade in the canopy**. It is [still] important that shade be minimised on the sections of shoots that will be pruned to in the winter, as the light response is very localized.*

The trellising, therefore, must do all it can to help growers achieve an open, airy, light-inviting canopy which will play its part in the battle to achieve sustainable yield levels and ripe fruit. In *Chapter 5: Pruning and trellising systems for vineyards in Britain*, I explained why I feel that a two-arm flat *Guyot* system is the most appropriate for Britain, and therefore it is the construction of this system that I propose mainly to discuss in this chapter.

Materials

The basic materials for a trellis system are posts – both end-posts and intermediate posts – wire, anchors and the fittings and attachments that aid various operations in the vineyard. A visit to a viticulture equipment show such as Intervitis (in Stuttgart), Vinitech (in Bordeaux) or

---

1 The best book on trellising I have found is Practical Aspects of Grapevine Trellising by Mahabubur Mollah. There is also a good section on Constructing Trellis Systems at the end of Sunlight into Wine (pp 70–84). See bibliography for full details. I would also recommend the Fencing Contractors Association New Zealand www.fencingcontractors.co.nz for excellent advice on fencing in general.

2 My apologies to all those growers in Belgium, Denmark, Ireland, the Netherlands, Sweden and any other 'cool' regions who feel they are also growing vines on a 'world-scale', but with around 3,500-ha and growing, Britain's vineyard area is several times bigger than any of the above.

Sitevi (in Montpelier) is littered with companies showing a very wide range of different materials and different fittings and attachments for growing vines and it is soon obvious that there is no right or wrong way, or right or wrong materials, used in vineyard trellis construction. Personal opinion, custom and practice, and availability of machinery and labour will all play their part in how any trellis is constructed and used.

The main thing to understand about a vineyard trellis is that each component plays its part in the construction of the whole. The aim of the trellis is to hold the wires that carry both the vine's canes[3] and the vine's fruit with the minimum of sag. The fruiting (i.e. lowest) wire carries most of the direct weight of the vine's canes and the fruit and this therefore must have as little sag and be as taut as possible. To do this it needs to be attached to two solid end-posts each attached to an immovable anchor. The fruiting wire is supported by the intermediate posts and it is important to appreciate that the load on the anchors is largely dictated by the distance between the intermediate posts. In any trellis that I am involved with, the fruiting wire is also supported by the individual (galvanised steel) planting stakes to which each vine is trained, and which must be firmly attached to the fruiting wire, usually with (removable) stainless steel clips. If vines have no individual planting stakes then these can play no part in the overall strength of the trellis. I have known growers who plant using individual planting stakes, but then, once the vine's trunk is big enough, remove them to re-use on a new vineyard. To my way of thinking this is a false economy and, once planted, the vine's individual stake stays with it for the lifetime of the vine. The other thing to appreciate is that a row of vines in full leaf and carrying a decent crop will, if exposed to a side-wind, exert a considerable sideways breaking force on the posts. This is particularly so on the outside few rows of the vineyard, especially in Britain on the south-west side, and will cause weak posts to bend or break and strong posts to lean if they are not adequately tensioned by the end-post assembly: end-post, anchor wire and anchor. Once posts start leaning, it is very difficult to stop them leaning even further.

Posts

The choice of materials for posts – end-posts, intermediates and planting stakes – is wide and includes timber of many different types, steel, reinforced concrete and several different types of plastic and composite materials. Until about fifteen years ago, almost all British vineyards were trellised using timber posts. Pressure treated softwood was mainly used, although various hardwoods and semi-hardwoods – oak, acacia, sweet chestnut – have all been used, both with and without some treatment, either completely treated or just to their bases. For decades, the pres-

Timber or steel? Take your pick

sure-treatment of softwood was carried out with what was known as CCA (copper chrome arsenate)[4] preservatives, but between 2003 and 2006 these were banned. Today, there are many different proprietary timber treatments used by companies that sell posts and they all claim that they have the best one. The only sure way to get long-life posts is to order them treated to 'Use Class 4' specification which is a British Standardclass for wood 'in permanent ground contact' (see below for more on this). In some countries, however, the burning of treated timber of all sorts has been banned and this has hastened the use of other materials for vineyard trellising, mainly steel. The main determinant of post material will be cost.

Timber posts

Depending on the length, diameter, actual material and treatment, a softwood post suitable for the intermediates – say, 60–75 mm diameter tops,[5] 2.50 m long and pressure treated – can cost from as little as £3.50 + VAT each to well over double that. Jacksons Fencing, who use what they call their 'Jakcured' process, will charge you nearer £7.35 + VAT each, but they do come with a 25-year guarantee. Softwood treated end-posts that are 2.75 m long (100–125 mm tops) can cost you three times these prices.

In my experience, buying softwood posts is a complete lottery. Whatever the manufacturers of posts tell you and (almost) whatever guarantees they give you, there are too many variables in the production of softwood posts for any to last longer than 10–15 years (and often a lot less). The origin of the timber, the exact type of tree (redwood or whitewood), the time the trees have taken to grow, whether they have been naturally or kiln-dried (and for how long and to what moisture content), how long after cutting the

---

3 I am only discussing VSP cane pruned vines here. For cordon trained and spur-pruned vines, other criteria will apply.

4 Also known in some countries as chromated copper arsenate – the initials are the same.

5 Beware the difference between the diameter at the top and at the bottom. Most suppliers will quote you the top diameter which is almost always smaller than the bottom diameter. Some posts have been machined so that they are parallel but I cannot see the point in paying extra for these for vineyard use.

Timber end-posts – rotten after 8 years

posts have been treated, and, lastly, how well they have been treated, are all factors outside of your control. Even if you can afford posts produced to the Highways Agency's specification,[6] most guarantees that I have seen only cover the replacement of the actual post and not the cost of removing the old one and replacing it – and heaven help you if you have so much as touched the post with a tractor or implement. Some suppliers treat their posts to British Standard 8417 (2011) but you need to be careful as there are five classes of treatment and three levels of Desired Service Life (DSL) – 15, 30 and 60 years – which are not guaranteed periods of use and each class and each level has different specifications. The most commonly found for exterior timber posts in contact with the ground is what is known as UC4 (Use Class 4), but typically only to the 15-year DSL which is too short a time for vineyard use. If they will (and you will undoubtedly have to pay for it), get your supplier to treat posts to UC4 and DSL of 30 years. Also get them to confirm in writing that this is the specification they are using. Given that timber is such an unpredictable medium for long-term use, my advice is: use steel posts.

Having said 'use steel', if you really have your heart set on timber for aesthetic or other reasons, then I would advise using posts pressure treated with creosote rather than other materials. Softwood pressure treated with creosote was always the timber of choice for railway sleepers and telegraph and electricity poles and many of these are still in use after decades. In Kent, and I am sure in many other fruit-growing counties, pressure treated creosote softwood posts are used for supporting apple trees and work very well. They can be messy to work with when they are new and operators must always wear gloves when handling and installing them. They also tend to exude creosote from time to time, especially in warm weather, and it's best not to run your hands up and down

them without protection. However, compared to posts treated with other types of material, creosote-treated posts are not as brittle, in my experience they last longer and, if anything, are a bit cheaper. This means that you can afford to get larger diameter posts for the same money as you would be paying for posts treated with other materials. There are restrictions on the use of creosote for treating timber[7] – basically, the public cannot buy them – but for professionals they are still permitted and in my experience are perfectly safe to use in orchards and vineyards. It has been suggested to me on more than one occasion that growing a vine near to a creosoted post will taint the grapes (yes, really!), but I don't believe this – another old wives' tale – and I have never come across this problem in real life. The last time I ordered a full load of posts pressure treated with creosote (in 2019) they cost €9.72 (ca. £8.70) for a 3.0 m x 120 mm top and €5.73 (ca. £5.12) for a 2.50 m x 100 m top. All prices + VAT. You could also use a 'Postsaver' which is a polythene and bitumen sleeve which is heat-shrunk on to the post. To fit a 100 mm diameter post, these cost around £1.06 + VAT each, and for a 150 mm post, £2.59 + VAT, plus the cost of fitting. I have no experience of these, and have never seen any fitted in a vineyard, but according to their website, they have been in business for twenty-six years, and offer a twenty-year guarantee on their products.

If you do decide to use timber posts, then apart from the obvious task of finding a supplier who you can trust and who will give you some sort of reassurance that their posts have not just been recently hauled out of some Russian forest and dunked in a tank of preservative, the main recommendation is not to skimp on diameter at the base. For intermediates you ought to have posts with a base diameter of at least 75 mm, with 100 mm being better; and for end-posts I would recommend a base diameter of at least 150 mm, with 200 mm being better. The breaking force of a 75–100 mm treated timber post is roughly twice that of a 50–75 mm one. Also, the ratio of the surface area of a post to its internal volume decreases as the diameter increases, meaning that the amount of area where the wood preservative is held, increases with the diameter of the post. A 100 mm diameter post has approximately twice the volume of treated timber protecting it when compared to a 75 mm post. Larger diameter posts will also displace more soil and have a greater surface area in contact with the soil, thus making them firmer and stronger. In most soils, timber posts with pointed ends are preferable and are easier to drive in straight and upright, although I have seen non-pointed posts driven in as well. They just take a little longer to bang in.

With timber (as with all posts) it's what's in the ground that counts – not what is showing – and my advice is to have at least 750 mm in the ground for intermediate posts and 800–900 mm for end-posts, although this will depend on the type of soil you have and how the posts are installed. Ideally posts want to go into undisturbed ground as far

6 Department of Transport's Specification for Highway Works No: 0300.

7 The Creosote (Prohibition on Use and Marketing) (No. 2) Regulations 2003.

as is possible and be hammered, punched or shaken into the ground. Steel posts displace very little soil when they go into the ground, which makes them easier to install, and it is best if they go into undisturbed soil and not into pre-augured holes. Fencing and trellising contractors love to pre-augur, or at least pre-spike, holes as this makes their job easier and faster. In my experience, after time, the soil does close up around the posts, but if the soil is easy (i.e. no stones) then best to install them into undisturbed ground. As has already been said, once a post starts leaning, it is difficult to get it to stop. Although I prefer buried anchors (see section on *Anchors* later in this chapter), if you are using timber end-posts, there is one end-post assembly you could use. This is the 'horizontal stay' assembly beloved of Australians and New Zealanders (who use it for all their long rabbit and deer fences) and, if correctly installed, is immensely strong and gives a nice vertical finish to the ends of the rows without anchor wires which are asking to be snagged.

With treated softwood posts, you will be very lucky if you don't get any breakages (from rot) within 10–15 years of them being installed (and often sooner). Over the lifetime of a vineyard (30–40 years), you must expect to replace softwood posts at least twice, and some of them probably three times. Timber posts tend to break at the worst possible time – just before harvest when there is a full crop and you get an autumn gale blowing through the vineyard – and replacing posts, especially end-posts, is not just a question of 'slipping a new one in' as someone hopefully described this task to me – it's a hateful job and one that can be avoided – by using steel posts. When a timber post breaks, it is always at ground level and the bottom, below ground, section of the post is usually still sound and impossible to remove without digging a huge hole to extract it. The replacement post therefore has to go into the ground next to the base of the old one and (annoyingly) slightly out of line with all the others. If using timber posts, you also need to make sure that the nails and staples used are the right ones (the ones with barbs) and that the staples are installed at an angle across the grain so that they will not pull out as the timber dries and shrinks.

Having re-read the last few paragraphs, readers will probably be thinking I work for the Steel Vineyard Post Council or whatever such a body for the promotion of steel posts might be called – far from it – I speak from the experience of having used local-sourced coppiced sweet chestnut posts, hot-dipped them myself (all of them for 5 acres of vines) and seen them start to fail after less than ten years.

Steel posts

Like timber, steel posts come in many different types (and even colours) and, like timber, you tend to get what you pay for. There are several factors that account for differences in price. Apart from the most obvious ones – length and type – the thickness of the steel and how they are galvanised

account for quite large differences.[8] The thickness of the steel is one factor in a post's strength, the other being the exact shape (the profile) and manner of its construction[9] and you need to make sure you are happy with the manufacturer's recommendations and guarantees. The most telling recommendation is how far apart the posts can be placed, as this has a direct effect on the cost per hectare. I always use posts that can be placed at 6 m apart (i.e. every five vines @ 1.20 m apart) which gives you approximately 800 posts-ha (324 posts-acre) in a vineyard with 2 m wide rows. Given the cost of installing them and the other costs associated with having more posts per hectare, it is better to pay a bit more for posts that can be installed every five vines, than to have weaker ones that have to be installed every four vines, giving you 990 posts-ha (401 posts-acre). I would not recommend that intermediate posts be more than 7 m apart, whatever they are made of, although I do know of one vineyard that spaces them at 10 m for the establishment years and then in-fills so that they end up at every 5 m. If they are wider apart than this, the strain placed upon the wire in an effort to get them tight and sag-free will cause the wires to be tightened beyond their point of elasticity and they will start to elongate and weaken. I accept that with really good anchors, so that forward movement of the end-posts is kept to a minimum, you might be able to stretch the intermediate post interval a little bit, but I think it better to be safe than sorry and keep them to around 6 m, depending on your vine spacing.

The other major factor in steel posts is how they have been galvanised. Cheaper ones tend to be 'pre-galvanised' or 'strip galvanised' i.e. the sheets of steel from which they are made are galvanised before they are cut into post-sized pieces and then formed into their shape. The more expensive posts are always 'post-galvanised' or 'hot-dip galvanised', i.e. the sheets of steel are cut and formed into posts and their notches and/or clips are pressed/cut out before the posts are galvanised. In this way all the cut edges and stress points where the steel sheet has been bent are fully galvanised. Whilst there is undoubtedly a saving to be made by using pre-galvanised posts, they will not last as long and in certain aggressive soils, especially where they are stressed by, say, side-winds or have been spaced too far apart, some of them may well fail. Even if they do not fail, pre-galvanised posts will start to rust along the edges and where the hooks have been formed quite quickly. However, if you don't mind this, then save some money and use them.

Another difference with steel posts is whether the hooks are 'internal' or 'external' and this will depend on such things as whether you propose to use a harvesting

---

8 Whilst I have seen untreated bare steel posts in vineyards, they have always been in drier climates than Britain's and seldom in European vineyards. The only exception to this is bare steel planting stakes made from 6 mm 're-bar' (reinforcing rod) which I have seen and which doesn't appear to harm the vine, although I personally wouldn't recommend it.

9 See Hadley Group's 'Ultra-Steel' process for instance: www.hadleygroup. com

machine (which can damage external hooks), whether your vineyard has a dip in it (which causes the foliage wires to jump out of the hooks) and various other related matters. Most growers in Britain do not have harvesting machines and find external hooks easier to use when manually moving wires up during canopy management. Each manufacturer has their own design differences, most of which are claimed to be 'superior' to other types and it is worth speaking to other growers with similar vineyards to your own before making a final decision.

The final factor in the cost of posts – whether end-posts or intermediate posts – is how long do they need to be? This decision depends on three factors: how high you want your leaf-wall to be; how wide your rows are; and how low are you prepared to bend down?

Leaf-wall height is the distance from the fruiting wire to the top of the foliage, which is typically around 100–200 mm higher than the top of your posts and/or the top wire (which ought to be almost the same height as your posts). The height of leaf-wall is actually a very important factor in vineyards in Britain and I fear not one greatly appreciated by all growers. Although I hadn't realised why, I seemed to know that higher was better than lower and it was only when I heard a talk by Cornelis van Leeuwen[10] (known as Kees van Leeuwen) who is the Professor of Viticulture at Bordeaux Sciences Agro School, part of the *Institut des Sciences de la Vigne et du Vin* (ISVV) based in Bordeaux, that I realised why. In cooler climates, excess vigour is often a problem and whilst soil type and depth and rootstock play their part, the availability of nitrogen and water are of greater importance. Nitrogen can be managed by depriving the vine of access to it and by growing cover crops in the vineyard alleyways which will compete for it. Water, on the other hand, in an un-irrigated vineyard, cannot be controlled: when it rains, it rains. Good drainage can help get rid of water and again, alleyway management – growing competing plants – will help absorb some of the available water. The vine, however, also uses water and the greater the number of leaves, the greater will be the amount of transpiration and water use. This is just one other factor in your fight against over-vigorous vines.

Of course, one has to be realistic about leaf-wall height and there are other factors to consider, but in Britain a leaf-wall height of at least 1.00 m must be the aim, with 1.25 m or 1.35 m better. The height of your trellis must also be related to the width of your rows and the overall height does not want to be much more than the row is wide, otherwise shading of the fruiting zone will occur. Therefore, with a 2.00 m row width you might have posts of 2.50 m long with 750 mm in the ground, 1.75 m out of the ground with the vine's foliage sticking above the top of the posts by around 100–200 mm. Your fruiting wire could then be at around 600–700 mm from the ground, giving you a leaf-wall of 1.25–1.35 m. If you didn't mind your fruiting wire being at, say, 300 mm (which would be far

End-post assemblies at Bolney Vineyard, West Sussex

too low for most people's comfort) and you decided that a 1.00–1.20 m leaf-wall was adequate – neither of which I would recommend – then you could use posts of around 2.25 m long with, say, 650 mm in the ground, bringing the overall height of your trelliswork down to 1.75 m. This would also allow you to have narrower rows,, say, down to 1.65 m wide.

The reasons I do not recommend having a fruiting wire at 300 mm is (a) because it's a back-breaking height to prune at, work at and pick at and (b) it's too near the ground for the fruit with regard to: weed control in our often-damp climate; water-splash from the soil on to the fruit which adds to the disease problems; and proximity to animals that will eat and/or damage the fruit. However, with shorter posts the canopy offers less wind resistance and therefore posts can be made from smaller material, do not have to go so far in the ground and are therefore cheaper.

End-posts

For end-posts, the same holds good as for intermediates except that they need to be longer and made of stouter stuff. This ensures that there is more material in contact with the soil and that the soil they are in contact with is more compact. Together these make for greater strength and stability. Don't think you can get away with using intermediates as end-posts. All steel post manufacturers make end-posts with a different set up of hooks and holes from those in their intermediate posts. According to the boffins, a sloping end-post offers no greater resistance than an upright one, but in my experience, all end-posts tend to move forwards a little bit over time, so it's best to start with a slight backwards tilt.[11] Therefore, insert end-posts at a slight angle to the vertical – 10–15° is about right – and to a depth where the top is level with the top of (or very slightly above) the intermediates. If you are using 2.50 m

---

10 www-ecole.enitab.fr/people/kees.vanleeuwen/english_index.htm

11 According to Practical Aspects of Grapevine Trellising (p. 21) end-posts are no stronger when they lean backwards and there is no advantage in them being installed so that their tops are shorter than the intermediates – both things one often sees.

intermediates, then 2.70 m or 2.75 m end-posts with about 850–900 mm in the ground is about right. The end-post should be installed with its base about 600–700 mm from the last vine, i.e. about half the inter-vine distance, with the anchor installed so that the angle between its wire and the post is 25°. This will give you a distance from the last vine to the anchor wire of about 1.50 m (the exact distance will depend upon post height) and this must be allowed for when setting out the vineyard. The top of the end-post must be attached to a wire which is attached to a ground anchor. Whilst I have seen plenty of end-posts installed without anchors, this is most definitely NOT to be recommended. Over time, the end-posts will start to lean inwards, wires will slacken and as they are tightened, the posts will be drawn ever-inwards.

Coloured steel posts

Coloured steel posts are available and several British growers have used them, although why you would want to is beyond me. The reason always given is that the galvanised posts do not look 'natural' and that green, or brown, or black posts will somehow look better. Given that planting thousands of vines in dead-straight rows, putting in thousands of posts and stringing hundreds of kilometres of wire along the rows isn't very natural, I fail to see how colouring some of those posts alleviates this situation. But, if coloured posts are what you want, then they are available. Made from post-galvanised steel and powder coated in the shade of your choice, they naturally cost more than non-coloured. Hadley Group, probably the only British vineyard post manufacturers, charge around £5.40 for a 2.50 m intermediate post, non-coloured and £6.25 for the same post, coloured. End-posts cost £11.50 for a 2.75 m non-coloured and £12.55 for the same post coloured. Prices depend on quantity and are all + VAT.

Individual planting stakes

In my view, every vine should have an individual planting stake against which it can grow, up which it is trained, to which it will become permanently tied and which will stay with it for the life of the vineyard. In this way, the trunk of the vine will be straight so that in the event of mechanical weed control or water shoot removal, the vines will not be harmed. It is my experience that vines that are tied to a planting stake during their establishment phase develop a better structure and start cropping sooner than ones that are left to their own devices. In some countries (Australia especially) vines are trained up a string tied around the vine and attached to the fruiting wire. Whilst this may be a cheap form of 'support', to me it is barely better than no support at all.

Again, as with the other posts in the vineyard, there are a multitude of materials available: timber of all sorts, plastic, fibreglass, and steel and again, cost will probably dictate the final choice. One other factor to bear in mind is that the thicker the planting stake, the more tying material – longer ties or more tape or cord – will be required to go around the stake and secure the vine's stem. With 4,000-plus vines-ha and two to three ties per vine, this can add up to a lot of extra material and cost. Another consideration is the ease (or difficulty) of getting them into the ground and the length of time they will last.

Probably the cheapest planting stake and one that is relatively easy to knock into most soils is a 1.2 m (4') bamboo cane. Whilst you can get canes from 8 mm diameter, those at 12–14 mm diameter will last longer. However, no bamboo will last indefinitely and after 3–4 years, the end in the ground will probably have rotted and the cane can either be removed (as hopefully the trunk will be strong enough and straight enough to stand up on its own by then) or they can be pushed back into the ground again for another few years' service. The Chinese size their bamboo canes according to their weight per 100 canes in pounds (i.e. lbs), so the thinner ones are known as '10-12s', slightly thicker ones as '20-22s' and even thicker ones as '24-26s'. Confusing, but it works for millions of Chinese! Bamboos are also packed in different bundle sizes. A bundle of the thinner ones usually contains 500; the thicker ones hold 250. A 1.2 m, 20-22 lbs (i.e. around 15 mm diameter) bamboo will cost around 12p–14p + VAT, depending on the numbers involved. Pushing bamboos into the ground by hand can be very wearing on gloves (don't attempt to do too many using bare hands) and hands and wrists will take a battering. Make yourself a 'knocker-in' using a short length of iron pipe with one end closed off. This you can slip over the end of the bamboo and work it up and down to knock the cane in.

The best (but more expensive) planting stakes are made from 6 mm 'wavy' steel rod (ArcelorMittal sell them and call them 'undulated support') which are usually 1.2 m long (although shorter and longer ones are available). These are easy to push into the ground, will last forever and are thin so that they do not consume huge quantities of tying material. In addition, the 'waves' mean that the ties around the vine do not slip down and the stakes can be secured to the fruiting wire with either permanent ties or detachable clips so that the fruiting wire itself will not slip down. However, at around 25p–30 p + VAT each they are two to three times the price of bamboo. Personally, I would always use them as, when secured to the fruiting wire, they become an integral part of the whole structure. There are also similar stakes made from 6 mm galvanised rod which have grooves on them to stop the ties slipping. At the Sitevi vineyard show in Montpellier in 2013, I saw for the first time what ArcelorMittal call a *tuteur trombone* and Ancrest call a *tuteur a décavaillonner*, both of which consist of a hoop of the same 6 mm wavy steel rod as described above, which can be put in the ground over the vine with its 'legs' placed on either side of a vine. In this way, the vine is guarded from the hydraulic arm of an automatic hoe if one is being used. These cost around £1.00 each, so not cheap, but better than causing damage to a young vine each

However big and strong your end-posts, without anchors they will eventually lean inwards

time an automatic hoe is used.

If you are using individual spray guards or rabbit guards around each vine, then you will probably need to remove the guard in order to attend to the vine – shoot selection and side-shooting – so you will need to attach the planting stake with a removable clip. These are made of sprung stainless steel and can be clipped on and off as required. Do not use the ones made from plastic as they become brittle after a few years and fail. However, if the guard you are using is fairly lightweight, flexible, and of a large enough diameter, it can be pushed up the stem of the vine without detaching the clip.

Anchors

For anchors there are, again, a very wide variety of methods and products. If you are using timber end-posts, then there are several end-post anchors which can be used which are not available to those using steel end-posts. The horizontal stay end assembly already mentioned works well if properly installed, and looks neat. The diagonal stay end assembly, where an inward facing strut rests on a buried stay-block is to be avoided as over time it will force the end-post out of the ground. For my money, for both steel and timber end-posts, a decent buried anchor attached to the top of the end-post is the surest solution. When I first installed vineyards, I always used what hop growers call 'deadmen' which are baulks of oak – normally a 3' long piece of 6" x 4" sawn green oak with a hole through the middle. A long slot (the grave) was then dug in the headland on the outside of where the end-post was to go and a long connecting rod, with a loop or hook on the top and a short section of thread on the bottom end, was hammered through the undisturbed soil at an angle, so that it appeared in the slot and could be pushed through the hole in the deadman and finished off with a large washer and nut. The hole in which the deadman lay was then filled in and the anchor wire, attached to the top of the end-post, was connected to the loop on the top of the connecting rod and strained up. Deadmen, correctly installed and tensioned, are impossible to move – the anchor wire will break first. However, the

materials are expensive and they are costly to install and, unless the rows were to be exceptionally long, I probably wouldn't use them again.

The most commonly used types of anchors used in vineyards in Britain are the screw-in disc and the driven-in anchor of which there are two types. The screw-in disc is usually attached to a fixed anchor rod with a hook or a loop at the end – all completely hot-dip galvanised – and the diameters of the disc and the length of the rod vary. I used to think that the larger the disc, the better would be the hold, but a trellising contractor I know well (and use) who has been installing posts, anchors and trellising for most of his life, told me that a smaller disc, which is easier to drive through the soil, is better when attached to a longer rod so that it sits in deeper, undisturbed soil. This seems to make sense. I normally use a 120–150 mm disc with an 850 mm to 900 mm rod which will cost around £3.50–£3.75 + VAT each. These anchor rods can be wound into the ground by hand, with a hand-held hydraulic driver, or with a hydraulic winding device on the end of a mini-digger. For maximum anchorage, screw-in anchors must be wound in until the hook or loop at the end is barely visible so that you know the disc is firmly embedded in undisturbed soil. Ideally, the anchor should go in at an angle so that there is a straight line between it and the top of the end-post, but if you are using a mini-digger this is not always possible; however, a slight angle helps. If you drive the anchor disc in vertically, it will start to pull straight as you tension up the trelliswork and your end-post may move forwards very slightly. A screw-in anchor combined with a correctly sized and installed end-post makes a very good end-post assembly and only in really hard stony and/or chalky soil will you have a problem getting them in.

If your ground is very hard or stony, then drive-in anchors such as those made by Platipus Earth Anchoring Systems[12] and Fenox[13] are a better choice and will drive in more easily and with fewer failures. The Platipus anchors can be driven into the hardest, toughest soil, using the appropriate drive-rod and can be hand-installed with a sledge hammer, a mechanical driver or with a hydraulic driver. There are two sizes of anchor suitable for vineyards: the S4VIS and the S6VIS which cost (2020) around £5.60 and £7.90, although you will get them for less if buying in volume. Whilst Platipus are happy to recommend the smallest one, the S4, for use in vineyards, the S6 (for only £2.30 extra) will give you a much better hold. Platipus offer a complete 'end-post kit' which contains both the anchor (made of aluminium) and the anchor wire (made of stainless steel), which naturally costs a bit more.

The other type of drive-in anchor is made by Fenox and, again, can be driven in by hand or by machine and will work in the hardest soil. They operate on a very different principal and act rather like an umbrella being opened up in the ground after the anchor has been driven in to the

---

12 www.platipus-anchors.com
13 www.faynot.com

correct depth. They come in several sizes: 500 mm, 600 mm and 1.0 m long. The smaller one is probably best avoided, but the 600 mm one works well and costs around £10 + VAT, depending on numbers. Gripple, which makes wire connectors and associated products, produces something called a GPAK (Gripple Plus Anchoring Kit) which contains the wire and the Gripple joiner required for each end-post. Gripple have also started making an anchor very similar to the Platipus (although subtly different I suspect for patent infringement reasons) which they claim works just as well, although I have no experience of them.

If your vineyard has undulations in it, you may need to anchor some of your intermediate stakes to stop them rising up out of the ground when you tension the wires (this may not happen immediately, but will occur over time). The Platipus S4 and Fenox 500 mm are ideal for this as they can be driven in very close to the post with a wire then being secured to the post to hold it down.

Wire – and other similar materials

There are several different types of wire which can be used in vineyard trellising, the most common being purpose-made vineyard wire which has a mild steel core and is galvanised with a zinc and aluminium (and sometimes magnesium) mixture. This is made by several manufacturers: ArcelorMittal with their unfortunately named 'Crapal' range and Bekaert with their 'Bezinal' range being the best known. This wire is lighter than plain galvanised wire, so it gives you more metres per kilo, puts less weight upon your trellis, has a much better resistance to corrosion with its superior galvanising (important when the wire is subjected to sulphur and other corrosive sprays) and, when tightened, will elongate (stretch) by 10 per cent, compared to 20 per cent for plain galvanised wire. This means less work at both installation and later on in the lifetime of the vineyard. If you are using this wire, then 2.4 mm or 2.5 mm wire for the fruiting wire(s) and 2.0 mm for the training wires and top wire are adequate. This is based upon having your intermediate posts at around 6.0 m apart with each individual planting stake secured to the fruiting wire. For joining wires, Gripple connectors are now used very widely and work well. Plain galvanised wire, such as you will find for sale at most agricultural merchants, is to be avoided as much of it is cheaply made (in China) and apart from the other disadvantages listed above, the galvanising will not always last the life of the vineyard. Vineyard wire is also available plastic-coated in different colours – ideal if you want it to match the colour of your posts – but otherwise I cannot see any advantages.

Other types of wire – high tensile steel wire, spring steel wire and stainless steel wire, for instance – are available and sometimes used, but are more difficult to work with, more expensive and don't offer any significant advantage that I can see over the vineyard wire mentioned above. At one time a 'wire' made from nylon and called Bayco was touted as the best thing since sliced bread for vineyard

trellising but, although it is still available, it is seldom seen in British vineyards. There is also a similar product called Deltex which is made in France. Whilst these are rust-proof, light-resistant, light in weight and retain their elasticity once stretched, they are a challenge to install, cut very easily with secateurs or leaf trimmers and, in my experience, because they constantly exert a pull on the end-posts, are prone to pull the end-posts and anchors up (unless they are exceptionally good) over time. However, I do know growers who use this type of 'wire' and manage to get along with it perfectly well. This type of product is sold with another benefit: it does not conduct electricity and is therefore lightning-proof. If this is what you are looking for in a vineyard wire, here is the solution! If you do use it, I would only use it for the training wires and never for the fruiting wire as it will almost certainly get cut at pruning.

Trellis fittings – spreaders, chains, end-hooks and other items

There is a huge range of fittings used in vineyards and here is not the place to go into all the different options. There are two areas however, which are worth mentioning in some detail.

The training wires (or 'catch' wires) are the two (and sometimes three) pairs of wires whose function is to catch the vine's canes and foliage as it grows up from the fruiting wire towards the top of the trelliswork. There are two main options as to how these are to be managed.

Aluminium spreaders attached to wooden posts

Wire spreaders attached to metal posts

Many growers attach these wires to the end-posts by means of chains and hooks. The chains are fastened on to the ends of the wires and the hooks to hold the chains are on the end-posts. For each pair of wires you require one hook assembly for steel posts or two square-headed nails if using timber posts. The chains are there so that the wires can be tightened or loosened according to the state of the foliage in the leaf-wall. The intermediate posts, if they are steel, will already have hooks on them (another advantage of steel posts) and for those using timber posts, hook nails will have been driven in at the appropriate spacing according to the pruning system, height of leaf-wall and even vine variety. Although these wires are moveable, many growers, especially ones with smaller vineyards, will leave them in situ for the whole year, not moving them up or down and merely using them as static wires. Some will even dispense with the hooks in their posts and permanently staple them to the posts.

The 'correct' way, i.e. the traditional way as practiced in European vineyards, of managing these wires is to start the year off with the wires at the bottom of the posts, usually hanging on hook nails provided for this purpose or in the lowest set of hooks on steel posts or sometimes just laid on the ground. Then, as the shoots start growing and turn into canes with foliage, the first pair of wires is moved up to catch as much of the canopy as possible. Then, as the canes grow, the second pair of wires is brought up from the bottom of the posts and placed in a position above the already positioned first pair. Depending on the height of the leaf-wall, these two 'liftings' may be enough, but with some canopies, a third lifting is required, in which case the first pair is detached from where they were initially placed and raised above the second pair. It is for this reason that some growers install a third pair of catch wires, although if you move the first pair, this is not really necessary. If you are using an arched-cane pruning system, where the shoots extend over a larger vertical area, then this may also be a reason for a third set of catch wires, otherwise two pairs are more than adequate. Every trellis should also have a single fixed top wire to which the growing canes will fasten their tendrils.

A refinement to the moveable wires system is to install wire 'spreaders', made from sprung stainless steel, on every other intermediate post at an appropriate position which carry the pairs of catch wires. These spreaders hold the wires away from the centre line of the canopy and allow (most of) the shoots to grow up between them before you come along and close the spreaders up and hook them together. In this way, the foliage is more firmly trapped and directed into the right line and it makes the whole job of 'tucking-in' that much easier. You do need to get the positions of the spreaders correct and this will depend on your pruning system and the variety being grown. These spreaders are available for most makes of steel post (and it is important to get the correct one for the exact post you are using) and are also available as nail-on for timber posts. They are however quite expensive. The current price per spreader is around 75p each, which will add about £600 per hectare (£250-acre)[14] to the cost of materials for your vineyard, plus the cost of their installation. Over the years however, they will repay themselves in the saving in annual work in the vineyard. They can also be tried out on a few rows and then easily retro-fitted, especially with steel posts.

A slight variation on the above is to fit on to each (or every other) intermediate post two permanent cross-bar spreaders which carry the catch wires. These can be made of aluminium, coated steel or plastic and vary in width from around 200 mm to 350 mm. Most are centre-mounted using a hexagon-headed screw or coach-bolt. Some growers use them to keep the wires apart permanently, in which case they can be tightly fixed to the post to stop them rotating; others use them to keep the wires apart initially, then bring the wires together and hook them on to the post, turning the spreader from the horizontal to the vertical so that they are 'hidden' by the post; there are also spreaders that have fold-up arms for much the same purpose. My preference is for the fixed type. Before you fit spreaders like these, you need to make sure they are compatible with any shoot-trimming or deleafing machine you might use.

The benefit of using fixed spreaders is that they give the leaf-wall more width and more volume, allowing air and light to penetrate, drying out both fruit and leaves, and making sure that sprays reach their target. With fixed spreaders (as opposed to the ones that you close up when tucking-in) you will also find that the bunches have much more space to hang in and, except in years when crops are very large and the leaf-wall is crowded with bunches (not often seen in Britain, but it does happen sometimes), most of the bunches will hang down in their own space, not touching other bunches. Again, this helps with both ripening and disease control.

Getting good at using spreaders of whichever type makes canopy management much easier and certainly helps in achieving a quality canopy.

---

14 Based upon a 2 m row width and a 1.2 m inter-vine spacing with intermediate posts every five vines and two spreaders on every other intermediate post.

Left: Daltons Supertags marking the end-post with row number, variety, clone and rootstock. They really help keep track of what's going on. Right: Another new vineyard gets trellised

Marking the ends of rows

Marking the end of rows, whilst not of course essential, is one of those things that make life a little bit easier. Whether it's spraying, weed control, taking samples or just telling the pickers which row to start on, having a number on every end-post (or every other if you want to save a few pounds), makes a lot of sense. Over the years I have probably seen every method going, but by far the most effective is to use animal ear-tags which can be obtained in a multitude of colours and printed with additional information such as variety, clone and rootstock – really useful if you are trying to spot the subtle differences between, say, 41B and 3309C. There are several manufacturers, but I have always used Dalton's Supertags Size 4 which work out at about £1 each, including printing. You need to order the female half only (the one with the hole in it). They fade over time, especially on the ends facing south, but will last a good decade or more.

One of the questions I get asked most often by non-wine people is: 'why do some vineyards have roses at the ends of their rows?', to which I go through the story of how, in the past (I say 'in the past' as most roses are now resistant to mildew) roses would get mildew before the vines succumbed to mildew (not the same one, but similar) and acted as an early warning system. Today, you will still find roses planted at the ends of rows, purely decorative of course, and if this pleases you, plant away. In my experience they look nice for a while, but once they get large and start scratching you as you drive by on the tractor, the bushes get trimmed back.

Cost of establishing a vineyard

As stated in *Chapter 2: Why plant a vineyard in Britain?* in the section on *Establishing a vineyard*, the cost of putting in and establishing a vineyard, i.e. to the end of year two, can vary widely, and each vineyard needs to be costed up individually. However, it would be unwise to budget for less than £30,000-ha (£12,000-acre), to which you would have to add such things as perimeter fencing and land drainage.

Chapter 10

Machinery and equipment

Note: A list of typical vineyard machinery, with prices, can be found in *Appendix IV*

Machinery and equipment

The basic machinery requirements to run a commercial vineyard, ignoring small items and machinery and equipment of a more general nature, are: a tractor, mower, pesticide sprayer, weed control equipment and pruning equipment. These will be dealt with first. Other viticultural equipment, which falls into the 'optional' category and will probably only definitely be required by larger growers, will then be covered and will consist of: pre-pruning machine, prunings pulveriser, leaf trimmer, deleafer, cane tying equipment, stem cleaner, tucking-in machine and subsoiler. Miscellaneous items such as trailers, lifting forks and three-point linkage mounted fork-lifts, all of which would be useful in vineyards, especially at picking time, have not been detailed. Other more general cultivation equipment, such as disc harrows, power harrows, rotary harrows, spring-tine cultivators, and seed drills have also not been dealt with, but are to be seen in vineyards in Britain. In recent years we have seen the emergence of machinery hire companies and specialist contractors catering to vineyards and some of the little-used or once-a-year items are available to hire with or without an operator. If you only need to pulverise your prunings once a year, or deleaf your vines twice a year, why invest money in machinery that sits in the barn when you can hire it? Much will depend, of course, on the size of the vineyard, and your appetite for investment in plant and machinery.

Basic equipment

Tractor
• Whatever tractor you have, the trellising system needs to be matched to the tractor. If the tractor is too wide, the vines will be damaged; if it's too narrow, then space is being wasted. Having said that, do not try and economise by using a tractor you already own and force yourself to plant at, say, 2.50 m row width, and regret it over the next few decades. The vineyard will last much longer than the (old) tractor that's sitting in the barn.

• For a double *Guyot* VSP system, the tractor must be able to fit down the rows with at least 350 mm clearance between the outside of the tractor and the centre line of the row. Therefore, a 2.0 m row width requires a tractor that is no more than 1.3 m wide. Other systems, such as Scott Henry, will require more space for the vine's foliage. Wider systems such as GDC can be managed with much wider tractors.

• It must have a standard three-point linkage to mount implements and machinery and a PTO to power any machinery that might be mounted or towed behind it.

• It must have sufficient power to be able to run a suitably sized air-blast sprayer up any gradients you have in the vineyard and sufficient oil flow and hydraulic take-off points to power any machinery you might think of mounting on the tractor and which would require

Fendt Vario with mid-mounted Braun undervine weeder and mower behind

Left: 'Perfect' rotary vineyard mower keeping the grass short. Right: Berthoud 3-row sprayer for maximum efficiency

oil pressure to work them: leaf trimmers, deleafers, inter-vine weeders and some weed sprayers.

- The tractor should ideally have a 'Q-cab' (quiet cab) for protection against sprays when spraying. Most cabs also have air conditioning, radios etc. Cabs should also have a HEPA-carbon[1] filter to prevent sprays entering the cab. These filters should be renewed annually.

- Tractors are also available with a front-mounted three-point linkage and PTO which can be used for mounting a mower so that operations such as spraying and mowing can be carried out simultaneously in one pass. A front-mounted mower will increase the overall length of the tractor and may require additional headland space. Your mower must be a suitable model for front-mounting as the blades need to run the opposite way to a rear-mounted mower.

Mower

- It is a good idea to have a mower around the same width as the tractor so you know that if the tractor fits, then so will the mower. Do not have it much narrower than the tractor as then you will be running on earth rather than grass, or you will have to make two passes for each row.

- For inter-row grass cutting, a single, twin- or three-bladed rotary mower is best. These are typically mounted on the three-point linkage and powered by the PTO via a gearbox and belts, although some are direct driven.

- Mowers either rest on four wheels (or sometimes only two) or are carried on the three-point linkage with a roller or skids at the rear. The choice depends on various factors, but almost any mower suitable for orchards will work in vineyards.

- There are also mowers which can be rear mounted, but which have a through-PTO so that a sprayer can be towed behind the mower. This will not only require additional power, but also, because of its length, require more headland space. However, they will of course halve the time taken to spray and mow the vineyard.

- Mowers are also available with swing-arms which mow right up to and around the stem of the vine, but these

are not generally suitable for use in British vineyards. It is important to maintain a completely weed-free strip beneath the vines and this type of mower cannot do this. It is also not advisable to use these mowers, especially on young vines (under 5 years old) as the mower (or to be more accurate, the hydraulic arm) will hit the vine in exactly the same place each time which will cause lesions and is a possible entry point for TDs.

- A flail mower, whilst it will tackle tougher conditions and heavier material than a rotary mower, is not as suitable for inter-row mowing as it will not cut the grass as short. This is important as, at certain times of the year, really short grass in the vineyard will help keep moisture levels low. A flail mower is suitable for pulverising prunings, see below for more details.

- There are also mowers with side-mounted strimmers which will, at the same time as the grass is being mown, strim the area under the vines, removing both unwanted weeds and water shoots. If you are using mechanical weed control, then this additional non-chemical weed-removing equipment is a useful tool to help keep the under-vine area clean.

Pesticide sprayer

- Your pesticide sprayer is the most important tool in your vineyard and will help you win the battle to keep your vines and your crop clean and pest- and disease-free. Do not try and economise on it.

- Go to a dealer who is experienced in selling sprayers suitable for vineyards and fruit trees and who can specify the right nozzles for your row width and cover requirements.

- The best type of sprayer for vines is an air-blast sprayer which has an axial fan at the back to direct the spray on to the crop either directly or via 'snakes'. I prefer a sprayer without snakes, but the Berthoud snake-type is becoming popular. For multi-row sprayers, air snakes are standard.

- Sprayers come in various shapes and sizes. There are mounted ones carried on the three-point linkage which, depending on the lift capacity of your tractor's arms, will contain up to 500–600 litres of liquid. There are also trailed ones for the larger vineyard which have a capacity

---

1 HEPA stands for 'high-efficiency particulate absorption'.

Left: Re-circulation sprayer saves sprays and saves time filling up. Right: Rear mounted axial-fan pesticide sprayer with 500 litre tank

of around 1,000–1,500 litres, or even larger.

- There are also so-called 'tunnel' (or 'recovery') sprayers, with and without fans, which have shields either side of the row (or rows) of vines so that as they are sprayed, any spray passing through the canopy does not go to waste by falling on the ground, but is caught by the shields and returned, via a filter, to the tank containing the spray. Reports suggest that between 35 per cent and 50 per cent of material can be saved. The fan-assisted ones are undoubtedly better than the non-fan ones. Some growers using tunnel sprayers use them in conjunction with air-blast sprayers, depending on the growth stage of the vine. Early sprays are done with a tunnel sprayer when the foliage hasn't filled the canopy with shoots and leaves, and an air-blast sprayer is used post-flowering when the canopy is filled and more air movement is required.
- Different types of sprayer are discussed in *Chapter 16: Pest and disease control* in the section on *Machinery requirements.*

Weed control equipment

- Depending on what weed control system you propose to adopt, you will require one or more of the following:
- Herbicide sprayer
- Under-vine cultivation equipment
- Mulching equipment
- The question of weed control and the equipment you need is dealt with fully in *Chapter 14: Weed control*

Pruning equipment

- Whilst many vineyards will be pruned using hand-held and hand-powered secateurs (Felco No. 2 are the best), one increasingly sees powered secateurs in use. Whilst their power and the ease with which they slice through almost anything put in their way is something to be wary of, there is no doubt that, once you have got used to them, they make life a lot easier.
- At one time both air assisted and hydraulic secateurs were quite widely seen in overseas vineyards (which required a compressor or hydraulic pump), but today, battery powered secateurs are almost universal. These are light, completely portable, either with a built-in battery, or a belt-type battery pack, and, whilst not cheap, have got cheaper over the years.
- Powered secateurs also probably help prevent the spread of some TDs as cuts can be made cleanly and evenly in exactly the place where you want them to be and at exactly the right angle. In this way, the cut will drain easily and the surface of the wound can be painted with wound protectant, should this be required.
- Powered secateurs also enable you to make large cuts, larger than possible with hand-powered secateurs, which helps keep the head of the vine cleaner, especially as its gets older and larger.

Optional equipment

Pre-pruning machine

- Pre-pruning machines, although they are generally only used in spur-pruned vines, can be used in cane pruned vineyards to cut up some of the unwanted wood at the top of the leaf-wall, making the job of 'pulling out' that much easier. ('Pulling out' is the task of removing, from the wires, the unwanted canes that are usually tightly held there with tendrils). Owing to their cost, they are in reality only suited to large vineyards or to vineyard contractors and, to date, there are only a handful being used in British vineyards.

Prunings pulveriser

- Whilst a rotary mower (discussed above) will deal with small amounts of pruning wood, especially if it is cut into short lengths and anything large pushed under the vines out of the way of the mower, the full amount of pruning wood from an established vineyard, especially if the wood from two rows is piled into one row (which is the normal practice) will be too much for one. A prunings pulveriser, usually known as a 'flail mower', is required. These mowers are driven via a gearbox and belts and work in the horizontal. The powerful flails cut the prunings into

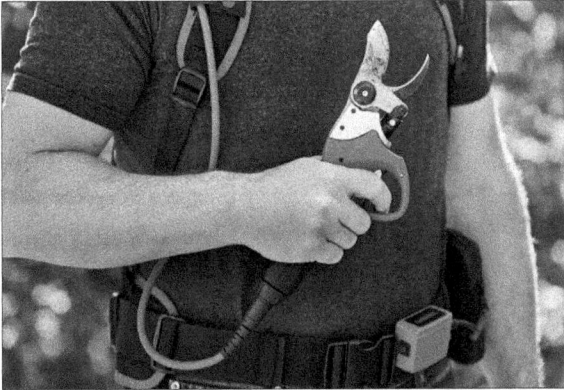

Left: Felco electic secateurs. Right: Front mounted two-sided Ero leaf and shoot trimmer

small pieces and they come out of the back as a mulch which soon disappears into the grass and over time is incorporated into the soil. These mowers are expensive considering one only uses them once a year (on the vines) and it might be better to hire one, given that the timing is not critical.

- Other ways of dealing with prunings – burning in situ with a *chariot de feu*, and removal and burning off-site – are dealt with in *Chapter 12: Management – cropping years* at the end of the section on *Pruning, training and tying down – first cropping year.*

Leaf trimmer

- One of the most important summer tasks is leaf trimming. This is also known as hedging, shoot trimming or summer pruning. Whilst this can be done by hand with secateurs, shears, or powered hedge-trimmers, on anything other than a small scale this is a back-breaking task, especially with tall trellising where the tops of the vines might be 2.0 m or more above ground level. In a typical growing year, vines in Britain would be trimmed at least twice and with some varieties and in wet seasons, three or even four times. Keeping the leaf-wall tight and trimmed is an essential part of canopy management and will help with disease control and improve ripening.
- There are several different types of leaf trimmer but the main two are the reciprocating blade type and the rotating multi-blade type. Both types are usually driven by hydraulic motors although at least one manufacturer is offering one driven by electric motors which has a trac-tor-mounted generator to run it. Some of the multi-blade types have two-armed blades; some of the newer ones have four-armed blades which work a bit like helicopter blades, creating suction which sucks the cut leaves away from the leaf-wall.
- All machines have height and lateral adjustment allowing you to lower and raise them (for different trellis heights and where there are tall posts and other obstructions to be overcome) and to angle the trimmer when on sloping ground (a cross-slope) and to cut the leaf-wall in a gentle pyramid, wider at the base (where the grapes are) than at the top.

- Whilst there are leaf trimmers which will cut both sides of a row in one pass, and even trimmers which will cut two rows at once, most vineyards are not large enough to warrant these and the preference is to use a machine which cuts one side and the top in one pass. Whilst this may be slower, it is much easier to manage and avoids knocking into things and damaging both posts and vines as well as cutting wires.
- Leaf trimming on a site with a side-slope is quite tricky, therefore planting other than up and down a slope is to be avoided wherever possible.

Deleafer

- Together with leaf trimming, deleafing is an essential part of canopy management in most vineyards. As with many tasks in the vineyard, this can be done by hand (and often is), a machine that removes the leaves makes it much more likely that it will be done at the right time and done quickly and effectively. I am convinced that a partial deleafing as flowering finishes, when the bunches are still open and well before bunch-close (and of course well before *véraison* which is often given as the time for first deleafing), helps open up the canopy for light, air and fungicide sprays, keeping this year's crop clean and next year's buds healthy.

Front-mounted de-leafer

- There are several types of machine: some suck the leaves into a void where they are sliced off; others that work with rollers that do much the same thing; and there is also the 'Collard' type that blow the leaves to bits with blasts of compressed air. There is an Italian make of air-blast deleafer called the 'OLMI Vortex' which some growers rate highly. However, some users report that the air-blast type can blow shards of leaf into the centre of the bunch, which doesn't help with disease control. There are also types that use gas jets that singe the leaves so that they desiccate and drop off, but these are not often seen, although – pre-harvest – they might be of interest.
- Whilst no machine is as accurate or as discriminating as a person, the quickness and ease with which leaves can be removed by machine make this an important piece of equipment for any serious vineyard.

Cane tying equipment

- The tying down of annual canes is a task that many growers will carry out by hand using paper- or plastic-covered wire ties, and the size of their vineyard will not really warrant automatic vine tying equipment. However, in larger vineyards these machines, all of which operate in a similar way with wire, covered or uncovered, being dispensed from a roll or cartridge, are common. Whilst they are not suitable for tying up young vines (the wires tend to cut through the canes), for the post-pruning tying down of canes they are ideal.
- For tying canes to the planting stake during the establishment phase the Max Tapener machine is the best device.
- There is more on this topic in *Chapter 11: Management – establishment phase* in the section on *Side-shooting and tying up.*

Stem cleaner

- Most British growers will be removing unwanted shoots from the vine's trunk just after bud-burst by hand or with simple, hand-held tools. Some use an approved chemical defoliant called Shark which works well, although doesn't always completely remove all of the larger shoots or the higher up shoots, and requires some hand-work as well. However, it is the cheapest method.
- Tractor-mounted powered stem brushers are widely used in large overseas vineyards and work simply by brushing the stem from both sides at once with stiff 'road-sweeper' type brushes powered by the tractor's hydraulics. Again, like some of the other 'minor' pieces of machinery listed above, a stem cleaner is not an essential 'must-have' item, but as agricultural wages rise and labour gets both more expensive and difficult to obtain, I expect to see larger vineyards investing in them.

Tucking-in machines

- In many vineyards around the world, tractor-mounted tucking-in machines are widely used, although so far only one vineyard in Britain uses one. There are two basic types: ones that work with the existing foliage wires, lifting the canes and trapping them with the wires; and ones that work by lifting the canes and trapping them with thin plastic twine (similar to baler twine) that requires removing at the end of the season. These would only be affordable to a vineyard of considerable size, at least 20-ha (50-acres), or perhaps to a contractor.

Subsoiler

- A subsoiler is probably unnecessary except for the larger vineyards as they can be hired, although having one to hand when the soil conditions are right and you have time to carry out the work is always better. As has already been discussed in *Chapter 6: Site planning and preparation* in the section on *Cultivations*, there are several different types available and tractor power will be an important factor in which type suits a particular vineyard.

Chapter 11

Vineyard management – establishment phase

Caring for vines after planting

After planting, your vines will sit for a while and take stock of their new surroundings before they wake up and decide to start growing, so don't worry if you don't see any signs of life for two to three weeks. The exact time of bud-burst will depend on the date of planting, the soil temperature and various other factors. However, once they have started to shoot, your work of training them can begin.

Protecting against animals– rabbits, hares, deer, badgers, wild boar

If you have not fenced the site against deer and rabbits, you will need to have decided how you are going to protect your vines against these unwanted animals. Whilst perimeter fencing is to my mind the best way of doing this, against rabbits there are various types of individual guard that can be employed. There is the tree-guard 'grow-tube' type which come in a wide variety of shapes, sizes, colours and materials. As has already been discussed in *Chapter 6: Site planning and preparation* in the section on *Fencing*, I don't really like the small diameter guards (75 mm to 95 mm) for a number of reasons, mainly the long-term health of the vines. Larger diameter round guards (150 mm diameter) or square guards (125 x 125 mm) can be used, but they need to be 750 mm high and will be expensive, probably around 85p + VAT each. However, they should be large enough to get your hand down without having to remove them each time you wish to weed the vine, but they will still have to be opened up or removed in order to carry out shoot selection, side-shooting and tying up the stems if this is to be done properly. Once the vines are established and have grown trunks robust enough to take herbicide spray, the guards can come off, although in vineyards where rabbits are a real menace, such as those planted near woodland or scrubland, then it may well be necessary to provide

protection even after the vines have developed trunks. In the absence of perimeter fencing, this is best done with individual guards made of chicken wire mesh which must be at least 600 mm high and 125 mm in diameter. These are typically used by apple and pear growers to protect established trees against 'barking'. There are also flimsy (and therefore cheap) individual rabbit guards made of plastic mesh, usually coloured blue (don't rabbits like blue?) which are slightly better than no guard at all, but only just. I wouldn't use them unless your rabbits are

Ventilated rabbit and spray guard keeping vines safe from damage

extremely timid or it was a temporary arrangement until your fencing is installed. Electric fencing can be effective in the short term against rabbits, but not as a long-term solution. As for various preparations that can be painted or sprayed on to the stems, these are totally ineffective against rabbits eating the shoots, leaves and canes of a young vine, although might work against winter bark nibbling by rabbits and hares.

Whichever route you choose to protect your vines against rabbits, it needs to be effective as otherwise you will never get your vineyard properly and evenly established.

Training young vines – year one

There are, dear reader, you will not by now be surprised to hear, several different theories about training vines in their youth and, much like child-rearing, theories come and theories go and the world moves on. Like much of this book, I can only speak from my practical experience of growing and advising people how to grow vines in Britain over forty-five years and the methods and techniques I will set out here are the ones that I find work best. Always remember that what you are trying to do when establishing a vineyard, is to get the vine to grow ROOTS and that what you see above ground level is a reflection of what's happening below ground level, not the other way round.

Shoot selection

When most of your vines have started to shoot, the time has come to select a single shoot that will grow away from the ground and encourage the vine to put down as big a root system as it can in its establishment phase. To do this you must select the most central and most upright of the several shoots that have probably come out of each vine – NOT the strongest shoot – and rub off all the others with finger and thumb. If this task has been left too late and some of the shoots have already started to turn woody, then you must use sharp secateurs to take these off. Ragged wounds where shoots have been torn off are not a good idea so close to the graft, and they might possibly be an entry point for various diseases. Get used to making pruning cuts at an angle so that any sap that seeps from the cut drains downwards and seals the wound and any rain drains away. Make sure when you cut any piece of wood, you cut fairly near to the cane so that a long piece of dead wood isn't left behind. Also, when cutting near a bud, make sure you angle the cut away from the bud, otherwise this too will become covered in sap. Whilst some of this is just viticultural housekeeping, it pays to get into good habits.

The reason for selecting the most central shoot and not the strongest is simple. The main reason that vines fail, apart from physical damage and disease, is graft failure and at this stage of the vine's development, when the graft is only around fifteen months old, it is important not to place any additional strain on it. The idea is to create as straight and as true a path from the vine's roots, up through the rootstock and graft and upwards into the shoot that is starting to grow. In this way, there are no kinks and chicanes within the graft which, as the vine expands over the next forty years, could give rise to too much twisting and tension. The aim is a straight sap line from earth to heaven. Once the vine has been 'singled out' and you have a stem, the vine will put all its energies into growing away and climbing its planting stake. Depending on how many of your vines have developed shoots when you first carry this task out, you may need to go through the vineyard one more time to get all of the vines dealt with. Also beware of rootstock shoots emanating from below the graft. These occasionally will grow away strongly and inexperienced shoot-selectors will mistake them for scion shoots.

Side-shooting and tying up

As the vine climbs upwards, it will naturally try and grasp onto the planting stake with its tendrils, but these are not usually strong enough at this stage to support the vine's weight and, in addition, the task of side-shooting – taking out the vine's side-shoots – must take place. Side-shoots are the shoots that will grow out of the angle between every leaf and the stem and, if left to their own devices, will continue to grow and take away energy and vigour from the main shoot. In some cases, side-shoots actually think they are the main stem and attempt a takeover coup. This must not be allowed to happen and early side-shooting is both easy and essential.

In addition to side-shooting the young vine, the growing shoot needs to be tied to the planting stake using something that will hold it in position, but only loosely. The best device for this is a Max Tapener machine which is a self-contained tying and cutting device with a roll of tape in its base and sharp blade and stapler at the other end. Whilst some people seem to take a while to get the hang

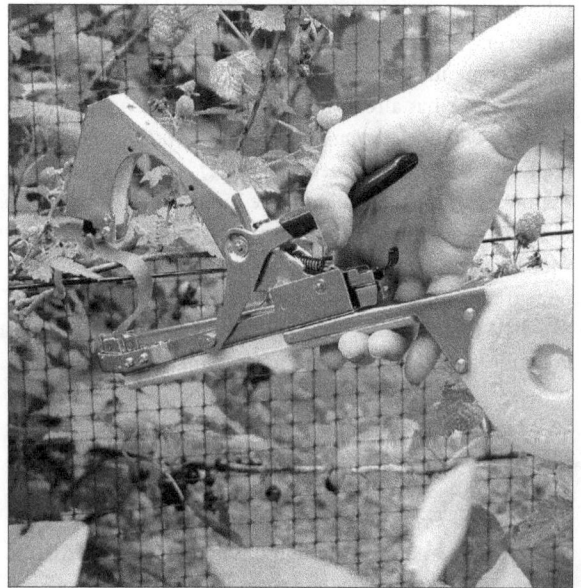

Max Tapener tape machine for tying up young vines

This is how your vines should look at the end of their first year

leaves, many will get blown over. As a temporary measure, if you are not installing the full trelliswork in the first year, consider installing the end-posts and a few intermediates and stretching one wire along every row in order to secure the planting stakes to something.

The end of the first growing season is also the time to walk through the whole vineyard and take stock of the number of vines that didn't grow – hopefully for you and your vine supplier, not a large number – and get replacements on order. The number of losses you can expect will depend on a whole host of factors – site preparation, drainage, planting conditions and post-planting care being very important – but even in very well-managed vineyards some losses are inevitable. Some graft failure is to be expected and there will always be vines that are broken, run-over and otherwise destroyed. I always used to say if the losses after three years were lower than 0.5 per cent then that was a great result, 1–2 per cent about normal, and, if much above that, something has gone wrong. Whatever the number you have lost, get replacements on order before the winter and before your supplier has run out of the varieties, clones and rootstocks you need.

The 'don't do anything in the first year' school of vine training

There is a school of thought that says vines should be left alone in their first year, either completely or partially, in order to toughen them up and teach them about the real world, and one should only start bothering with them in their second year. Whether this 'sink or swim' style of management is borne out of laziness, indifference, or unwillingness to spend money on them, I have no idea, but it exists (and in France it seems to be the norm). I suspect in warmer and drier climates, where weeds and fungal diseases are not such a problem, this may be done. Also, if you already have several hectares of cropping vines and have just replanted a small area, then there is no urgency to get them into cropping, and a long-term approach like this may have some validity. Not however, in Britain. Many is the time I have seen recently planted vines lying on the ground or barely supported and with multiple shoots growing out at all angles and heard the owner say: 'oh, I am letting them find their own way for the first year' or something equally nonsensical and then they wonder why half their vines die and the other half take five years to produce a crop. Sorry, but I cannot be doing with this way of caring for (if that is the correct description) young vines. Single them down to one shoot soon after planting and train them up a stake.

of them (and some never do) I have always found them a pleasure to use. The tape comes in different colours and different strengths. For vines, the lighter-weight white or green tape are quite adequate (40 m per roll), whilst the medium-weight blue and green tape (26 m per roll) and the heavyweight tape in red and green (16 m per roll) are too heavy and a waste of money. What you do not want to use for tying up the vines at this stage is anything that has a solid core, such as a plastic- or paper-covered wire tie, as this will resist the vine's stem as it expands and, if put on too tightly, will cut right through the stem. If you do use this type of tie, make sure you put it on loosely and hope it gets taken off before the stem expands into it.

Side-shooting and tying up must continue for as long as the vine is growing which, in a good season, might be until the end of September and even into October. By then, vines planted in a well-prepared site and correctly looked after, should have grown at least 1.00 m and very often twice that. This is the reason that, ideally, the full trelliswork wants to be installed in the first summer as otherwise your young vines will be flapping about in the breeze and some will, in all probability, break off at the top of the planting stake (which will be at about 1.00 m out of the ground). If the trellising is up, then you will have some wires to tie the canes to for support and some catch wires to hold the canes. If you do not put up your trellising in the first summer, then you might need longer planting stakes for the vines to grow up and these need to be stuck in the ground fairly well, otherwise with the first autumn gale in September, when the vine has its maximum complement of

Weed control

Weed control is probably THE most important single task to get right in newly planted vineyards and is dealt with fully in *Chapter 14: Weed control*.

Vineyard alleyways

There are three basic ways of looking after the ground between the rows of vines, the vineyard alleyways, in the establishment phase: cultivating them, growing something in them or keeping them free of vegetation with herbicides. Each system has its benefits and drawbacks.

Vines will normally be planted into bare, cultivated soil and for a while, all will be peace and quiet. Then, within a week or two of planting, a green mist will appear on the surface of the soil as the weeds and grasses announce their intentions which, unfortunately for you, are not very benign. Their sole purpose in life is to smother or strangle your precious vines. In warmer, drier climates, especially in more traditional winegrowing regions, vineyard alleyways (as well as the under-vine areas) are often cultivated for the first few years, if not for the whole of the vineyard's life. Cultivating the soil is done so that the vine is persuaded to root more deeply in order to look for moisture at deeper levels. It also aerates the soil and, when compost or fertiliser is spread, enables this to be incorporated into the soil. Cultivations will take place at regular intervals, as and when the weed growth requires it, using spring-tine harrows, disc harrows, power harrows or rotary cultivators. In young vineyards in Britain, say in the first and second years when no crop is being taken, cultivating the rows is possible, although not seen very often. The reason why is probably because it requires extra equipment and machinery which will, in all probability, only be used in the establishment phase of the vineyard and because, traditionally, our vineyards have been grassed-down (either naturally or seeded) so that a firm working surface is created as soon as possible. Given the problems we have seen with TDs in some vineyards in Britain, which might be associated with weak root growth, plus the fact that on average we are seeing drier, warmer summers, I would not be against keeping a vineyard cultivated for the establishment phase, although am happier to see mown alleyways in vineyards, especially once they start cropping, as this provides a better working platform and, in addition, helps use up some of the available moisture.

In regions where summer rainfall is low and soils are consequently drier than they are in Britain, keeping the vineyard alleyways cultivated throughout the year doesn't pose a huge problem at harvest time. Given typical conditions during many British vintages though, cultivated alleyways would soon become mud-baths for tractors and pickers alike (as many growers saw in 2019) and are not to be recommended. One way around this, often seen overseas, is to alternate rows, with one row cultivated and one row grassed-down or sown with a green manure. This allows for some vineyard operations – spraying and picking – to stick to the firm, grassed-down rows whilst giving some of the benefits of cultivations to the vines. After two or three years the rows can be alternated, cultivating the grassed-down rows and seeding the previously cultivated ones. The use of green manures is discussed in *Chapter 7: Vineyard nutrition* in the section on *Green manuring*.

My preferred option for vineyard alleyways during the establishment phase is to do nothing, apart from keeping them topped with a rotary mower. Whilst this may look unsightly for the first year, or even the first two years, over time a very satisfactory multi-species grass and weed sward will establish itself and provide a good working surface for both people and tractors. Many different grass and weed species will appear, providing habitat and food for predator insects and, the best benefit of all (apart from it being free), is that the sward will be slower growing than sowing grass seed. The vines will also have a year or two without much competition from the alleyways, giving them a better chance of rooting properly. Often vineyard owners insist (against my advice) on sowing grass seed in the alleyways soon after planting, and finding that within weeks they have what amounts to a silage field which needs cutting every week or ten days. Grass seed manufacturers sell seed on the basis that it will grow and provide lots of green matter – almost the opposite of what you want in a vineyard. Sure, there are 'orchard mixtures' which are often sold as being 'low growing and shade tolerant' but in my experience they still grow too much. And anyway, why pay several hundreds of pounds for something you can get for nothing and which works better? You also want to avoid planting anything that might increase the amount of nitrogen available to the vine which might make it more vigorous than is good for it. Therefore, mixes which contain clover, alfalfa and other legumes which, assuming their roots are populated by the right bacteria, can fix nitrogen from the air and will return some of that nitrogen to the soil, should be avoided until you know the vigour status of your vines. Let nature take its course and provide you with a natural carpet.

The third route, the chemical route, is one not often seen in British vineyards (or for that matter oversees either, now) and I suspect it is not advocated so much these days, although it certainly was. In other fruit crops – apples, pears, hops, blackcurrants and other soft fruit – 'total herbicide' is still sometimes used and, done well, is an effective way of keeping the floor of the orchard, garden or plantation free of weeds. This allows the plants being cropped to benefit from reduced competition and have access to water and nutrients, both of which grasses and weeds also need to survive. However, vines in vineyards in Britain are not usually short of water and their deep roots allow them to access enough nutrients to survive, so some of the benefits of total herbicide are lost. Another benefit is that bare soil, being darker, warms up more quickly and stores its heat and this may be a benefit worth considering for vines. In addition, if the vineyard is one with a spring frost problem, then bare soil, being drier than grass, stays frost-free for longer and, in conjunction with other frost control techniques, may help. Biddenden Vineyards in Kent have used total herbicide in some of their vineyards for decades and manage to both make it work and harvest good grapes.

The disadvantages of total herbicide, especially at this stage where the vines are still young and will have foliage at, and close to, ground level, are the problems involved with getting the material on to the vineyard at the right time and in the right weather conditions (dry and still), plus the cost of both the chemicals and the application. Given that you will almost certainly require at least two applications – one pre-bud-burst and another in the mid-summer after flowering – and possibly a third after harvest as a pre-winter 'clean up', the material costs could be in the region of £300-ha (£121-acre), plus the tractor costs and time. Having the right herbicide spraying equipment will certainly help with both getting the material on at the right time and getting effective coverage. If you are considering total herbicide in your vineyard, a machine such as Micron's Undavina shielded weed control spray head and guard would be ideal.[1] The problem of 'poisoning' your soil through using herbicide for total weed control I personally feel is not one to worry about, although from a PR point of view, this might not go down so well with some people. Glyphosate (aka Roundup), the main weedkiller used in vineyards, is probably going to disappear from use in the next few years, so unless an alternative appears, the days of total herbicide in orchards and vineyards are probably numbered. Herbicide use is discussed in more detail in *Chapter: 14 Weed control* in the section on *Herbicides*.

Training young vines – year two

Assuming most of your vines have made adequate growth by the end of their first season – and some (hopefully, many) will have grown to 1.50 m or more – then the question of how to prune and train them in the second year has to be tackled. I always prefer to get the vines off the ground as quickly as possible and retain whatever good ripe wood of lead-pencil diameter (around 7–8 mm) that has grown. This ensures that as much fruiting wood as possible is retained and can fruit in year two. I know plenty of instances where two-year-old vines have cropped at up to 5 tonnes-ha (2 tonnes-acre) without any short-term or long-term harm to them. However, some growers prefer to prune all vines back to a short one- or two-bud shoot and going through the whole 'year one' process of shoot selection, side-shooting and tying up again. Their aim is to enable the vine to establish a better root system before it starts cropping. Whether there is any science behind this is debatable and, in my experience, if a vine has grown well and produced a fruit bearing shoot long enough and thick enough, then it is able to bear a crop in year two.

If your vines have not made good growth in their first year, then there is no other option but to cut them back to a one- or two-bud shoot and work out why they didn't grow as much as they might have done. As I said at the start of this section, the thing to remember about young vines is that what you see above ground level is a direct reflection of what is happening below ground level and good cane growth equates to good root growth and its good root growth that is the purpose of the establishment phase. At the end of year

two, before the winter sets in, is another time to walk the vineyard and count up the number of replacement vines required and get them on order.

Pest and disease control

In the 'old days', when the main varieties grown in vineyards in Britain were German cross-breeds plus Seyval blanc, it was quite common to start spraying against diseases at the start of the first cropping year (usually year three). This didn't seem to worry the vines and I don't recall it leading to any real disease problems. Today however, it seems short-sighted not to put at least two to three maintenance sprays on to vines in their first and second years – assuming that year two is not a cropping year. Obviously if you are proposing to crop your vines in year two, then a fuller spray programme will be needed. Chardonnay especially seems particularly prone to Downy Mildew in its youth. Whether this is because we now enjoy warmer growing conditions, or (and this is my theory) French *pépiniéristes* are not so rigorous in their disease control as their German counterparts, who can say? The fact is that without some first year sprays it often succumbs to Downy Mildew. As well as protection against Downy Mildew, it would be senseless not to include some sulphur against Powdery Mildew. Botrytis is probably going to be less of an issue without a crop but, again, a few maintenance sprays against this wouldn't do any harm.

Slugs and snails can sometimes be a problem with vines in their establishment phase, especially with growers using unventilated guards. The warm, damp environment seems to be to their liking and young vine leaves certainly are. Some growers routinely sprinkle some slug pellets into the guards immediately after planting which seems to solve the problem.

Conclusion – establishment phase

The aim of the first two (and sometimes even three) years of a vine's life, the period I have called the establishment phase, is to allow the vine to develop a root system that will support a small crop in year two or three and a full crop in year three or four. If your vines are weak in their growth, then they will have a small root system and normally will not have grown sufficient wood of the right diameter to tie down. In this situation, cropping should be delayed. You have no idea what the first cropping year will be like; it could be the driest or wettest season on record, the warmest or the coolest. A young vine, overburdened with fruit, will struggle in adverse conditions and stress is never good for the long-term viability of a vine. The thing to always remember is that what you can see growing above ground is a reflection of what you cannot see below ground – the vine's roots. Good vigorous growth above ground level at this stage will mean the vine is putting down roots, searching for moisture and nutrients and preparing itself for fruiting. The opposite – poor growth above ground level – typically indicates slow root growth and the vine should be allowed to take some time to catch up.

---

1 www.micron.co.uk/viticulture/undavinas

Chapter 12

Vineyard management – cropping years

The vinegrower's year

The timetable of the vinegrower's year will vary according to many factors and no two sites, varieties, pruning, training and trellising systems or years will be the same. All dates and timings are therefore approximate. The timetable that follows is a general guide and more options can be found in the sections that follow it.

- Pruning: can start at any time once vines are dormant and wood is ripe. Typically, January, February and March, are the main pruning months, although December is not unknown in larger vineyards. Must finish before buds start swelling in mid-April.
- Training and tying down: will follow pruning and should be completed before bud-burst.
- Fertiliser: soil and petiole samples will have been taken during the previous growing season and fertilisers can be applied during the dormant season.
- Disposal of prunings: prunings can be pulverised and left to rot down in the vineyard or taken out and burnt. Best done during the dormant season and before alleyways need mowing and before herbicide applications.
- Repairing trelliswork, replanting vines: vineyard maintenance including replacing broken posts and anchor wires and replanting of missing or dead vines. Ideally this should take place prior to bud-burst.
- Alleyway management: depending on the alleyway management system, mowing or cultivating should start once grass and/or weeds start growing and continue until growth stops in the late autumn. Specialist operations, such as subsoiling or root pruning, will ideally take place when the ground is at its driest, i.e. mid-summer to pre-harvest.
- Weed control, under-vine: weeds beneath the vines can be dealt with in several ways: chemical, mechanical or mulching. Each technique will have its own timetable. Weeds must be dealt with before they start growing and

should be kept under control throughout the growing season. Often a post-harvest 'clean-up' is beneficial. Spot treatment of troublesome weeds (nettles and docks for instance) may need to be carried out using a knapsack sprayer at almost any time of the year.
- Pest and disease control: spraying often starts just before bud-burst (mid-April to mid-May, depending on the site, variety and year) and will continue at regular intervals, every 10–14 days, depending on varieties and weather conditions, until 2–3 weeks prior to harvest, depending on products used and relevant harvest intervals (HI). Post-harvesting applications may also be necessary. See more about HIs in *Chapter 16: Pest and disease control* in the section on *Harvest intervals and residue levels*.
- Shoot removal, trunk cleaning: shoots growing from the vine's trunk below the fruiting wire will need removing once they appear, typically in late April to mid-May. Usually this is only carried out once, although a second round might be necessary, particularly in younger vineyards.
- Shoot thinning: excess shoots, especially in the crowded

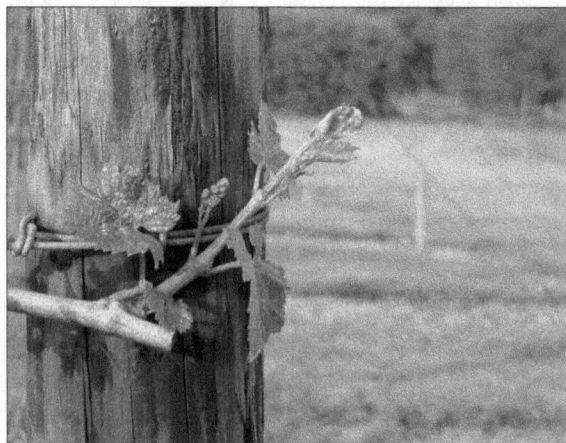

The start of the growing season

A chariot de feufor burning prunings as you prune

Pruning, training and tying down – first cropping year

The first task facing the vinegrower at the beginning of the year is pruning. For traditionalists, 22 January is the date to start as this is the day of the patron saint of vinegrowers, St. Vincent.[1] In reality pruning can start whenever the vine has shed its leaves and the wood has 'ripened', i.e. lignified, turned brown and hardened off and the vine is fully dormant. In most vineyards in Britain and in most years, this would not happen much before mid-November and often later and many growers wouldn't think of pruning until after the New Year. If your vineyard is young and has little wood to cut out and remove, then this is a quick and easy task and can be left until the weather warms up, say, early March. There is no advantage to be gained in pruning earlier. Don't be concerned if your vines start to 'bleed' – exude sap from the cut end of canes – this is quite normal and is purely a sign that the ground is warming up and the roots are starting to stir. A vine can bleed for several days and the ground beneath the end of the cane will get noticeably wet, but this will not harm the vine. Pruning later can also help with frost protection. See *Chapter 13: Frost protection.*

The vines will be ready to crop when they have grown a length of ripe cane of around 7–8 mm diameter. In Germany they say as 'thick as a lead pencil'. Any thinner than that and the cane is less likely to bear viable buds; any thicker than that and – well that's an interesting question. Canes thicker than 10 mm are certainly not unknown and with some varieties, especially if they on a vigorous rootstock and in their first few years, can have canes up to 15 mm diameter, so-called 'bull canes' although this is rare. These thick canes also typically have long internodes, quite possibly a sign that they have been growing in shaded conditions and are therefore likely to be less productive than thinner canes. Vines on rootstock SO4 are especially prone to producing thick canes, especially in the early years of a vine's life. However, if thick canes are all you have to work with, then better thick canes than no canes at all.

Assuming your vine has grown a length of cane above and beyond the fruiting wire, then you have the opportunity to leave a length of fruiting wood for the coming year. The length of wood you leave should not exceed half the distance between the vines, i.e. if your vines are planted at 1.20 m apart, then your single cane should not be longer than 600 mm. A little bit longer is no great crime, but avoid the temptation to leave a cane that stretches almost to the next vine. This is a young child-vine and mustn't be sent up too many chimneys on its first day. Trim the cane slightly longer than the correct length, take off any tendrils (although this is purely cosmetic and leaving tendrils on does no harm)

crown area of the vine, should be removed when they can be easily rubbed off, typically in May–June. Removal once the shoots get woody is to be avoided.

- Tucking-in: once shoots get long enough to be tucked in and depending on variety and rootstock, this will begin in late June–July and continue until vines reach the top of canopy and trimming starts. Most vines will require tucking-in at least twice in the growing season.
- Flowering: the most critical period in the year as it determines crop size. Typically takes place in Britain between the second week of June and mid-July but timings will vary according to the year, the site and the grape variety. In recent years, some vineyards have started flowering towards the beginning of June.
- Shoot trimming (hedging): will start once the shoots have grown above the top wire, typically at around the end of flowering. As the season progresses, both side growth and top growth will require cutting. Tipping of canes during flowering can also be carried out.
- Deleafing: depending on variety and rootstock, most vines are best deleafed twice, once just after flowering, and again around the time of *véraison*. A third pre-harvest deleafing might also be carried out, depending on leaf cover and variety, in order to speed up hand harvesting. White grapes are much harder to see than red grapes beneath a full canopy.
- Crop thinning: when vines have set a heavy crop and for certain varieties and particular types of wine, some crop thinning might be required. Normally this will take place once bunch-close has occurred and before *véraison*, although other timings are possible.
- Picking: once grapes are ripe for the type of wine to be produced. Will depend on variety and weather during the growing season, but for early varieties and in early years it might be the beginning of September (in 2018 it was the end of August for Ortega) and, with late varieties, as late as early November. In 2013, Chardonnay was still being harvested on 20 November.

---

1 St. Vincent is also the patron saint of winemakers and vinegar makers. The Bulgarians have a different saint for this purpose, Trifon Zarezan – St. Trifon the Pruner, whose day is 1 February, although apparently they also worship him on 14 February as well. I am not sure how St. Valentine feels about this though.

and leave the cane pointing upright. Tying down can wait. Any small pieces of excess cane can be cut into two or three and thrown into the row. The mower will deal with those, assuming you are mowing the alleyways. If you are cultivating then it is still probably acceptable to leave any excess wood behind in the rows as such a small amount will not interfere with your cultivations. It may not be the same in future years when the volume of old wood is larger. If the vine has produced some cane above the fruiting wire, but not so much that you are able to leave 600 mm of ripe cane, then you have two options: leave as much ripe cane as there is, so that some fruit will be produced; or trim it back to the height of the fruiting wire and let it spend another year in a non-fruiting mode and hope it produces some ripe canes for the next year. Any vines that have still not produced much in the way of a trunk may well be in trouble and replacing them should be considered.

Traditionally in Britain, almost all growers left the wood from pruning behind in the vineyard – usually putting two row's worth of wood into one row – and smashing it up with a prunings pulveriser. In France one often sees growers using a *chariot de feu* to burn them as they go (and I have even seen one in a British vineyard) whilst others gather the prunings up and take them off-site for burning. With concerns that TDs may be spread by from one vine to another, some growers in Britain have started to remove

Chainlock (above) and Treefix ties

prunings and burn them. Whether this helps or not is open to speculation, although it cannot do any harm. There are also machines which will gather up the prunings and turn them into faggots which can be sold for barbeques, but the machines are too expensive for the scale of most British vineyards. Whilst I can understand the wish to 'clean up' the vineyard by taking out old wood, I cannot quite get my head around the science. All the trimmings from summer pruning, all the leaves that have fallen, some rotten bunches of grapes and other detritus, plus the old wood that forms the permanent framework of the vine remains behind in the vineyard whether or not you remove prunings. Removing the prunings hardly turns the vineyard into an operating theatre. Still, if it makes you happy, take the prunings out.

Once you have been through the whole vineyard, pruning to one cane all those vines that have grown a good length of cane and trimming back to a suitable point all those that have not made the grade, you can think about tying down. There is no need to do this immediately after pruning, although this does often happen. The best time is when the canes have got a little sap rising in them which makes them more supple. They can then be bent over at ninety degrees

without snapping. Take the cane at the point of bending in your left hand (assuming you are right-handed) and gently bend the cane over, trim it to the correct length, twist it once round the wire and attach it with a tie. If your vineyard is on a slope, and you have single-cane pruning, then it is normal to bend the cane downhill as this is said to help even out the bud-burst.

Tying materials are many and varied and they have been partially discussed in *Chapter 10: Machinery and equipment*. Although Max Tapener machines can be used, they are not really suitable as it is difficult to get a really tight tie around both cane and wire and the tape tends to slide along the cane and the cane slips out. Better to use paper-covered wire ties, remembering that if you use these on the trunk of the vine, tie them loosely. If they are too tight, they will cut through the wood as the vine's trunk grows and expands. Plastic-covered wire ties are not to be recommended as they have sharp edges and after a day of tying down, you will have very sore fingers. For bigger vineyards, there are semi-automatic tying machines which have already been discussed. Other proprietary cane clips are also available, but personally I cannot see the use of them. At some stage you will need to tie the trunk of the vine to the planting stake with something more robust than the Max Tapener machine and there is a very wide range of products available. Whatever you use, my advice is to use something that can either be unfastened and opened up as the vine's trunk expands, or something that is elastic enough to stretch as the trunk grows. The best type of tie I can recommend are the 'Tree Fix' rubber ties which are easy to use and will not cut into the vine as it expands. They are made in several sizes so make sure you use the right size according to the age of your vines, although if you get the next size up you can always wind them twice around the trunk and release them at a later stage. I have also used plastic 'Chainlock' ties which are cheap and easy to use. The material comes on a roll and if cut to the appropriate length can be opened up and expanded as the vine ages. It will last almost indefinitely.

After you have finished tying down, if you have a vineyard with all the vines carrying a single cane of around half the distance between your vines, you should have a potential yield of 30–50 per cent of the vineyard. In reality, taking into account vines that have not made it to the fruiting wire and vines with canes shorter than the correct length, it is safer to work on a potential crop of one-third of what one might expect in a mature vineyard, i.e. around 2.5 tonnes-ha (1 tonne-acre).

Pruning, training and tying down – subsequent cropping years

After the first year of cropping, your vines should have grown a substantial amount more wood and be ready for the full pruning experience.

The aim of the vinegrower is to leave each vine with its full complement of fruiting wood, plus a spur (or 'thumb') to provide next year's fruiting canes – the 'replacement' canes which give the system of 'cane replacement pruning' its name. The fruiting wood should more or less fill up the space allowed for each vine – leaving a gap of around 100 mm between the ends of each vine's canes – and both the fruiting canes and the spur must come from below the fruiting wire. Ideally the canes should come from above the spur and the spur should be around 100–250 mm below the fruiting wire. However, as the vine gets older it may be necessary to select a spur position further below the fruiting wire than this and as long as it's not too near the ground (where it might get hit by herbicide) it doesn't really matter how low it is. *In extremis*, a shoot coming from just above the graft could be trained up as the trunk, but this would usually only be in the case of problems with TDs such as Esca where the vine needed rejuvenating. The reasons that the canes and the spur should always come from below the fruiting wire are that if they start appearing above the fruiting wire, tying down becomes problematic, fruiting wood is lost, and the trunk may stop producing shoots from which the spur can be formed. A mental sign saying 'always below the fruiting wire' should be hanging in front of everyone as they prune cane replacement pruned vines.

Many books will also talk about the vine being 'in balance' and in my experience, in relation to vineyards in Britain, this is a difficult concept to understand. It is worth reproducing what appears in the Oxford Companion to Wine (OCW) written by Richard Smart, its viticultural editor:

Vine balance is a viticultural concept little appreciated by wine consumers, yet one which is essential for producing grapes for premium wine-making. A vine is in balance when the leaf to fruit ratio[2] is in the correct range. The amount of early season shoot growth should also be in balance with the vine's reserves of carbohydrates. Vine balance concerns vigour and it can be managed by the viticulturalist, with balanced pruning and water stress the principal tools. One of the best measures of vine balance is the ratio of fruit yield to pruning weight, now often called the Ravaz index, following its promotion by the French researcher of that name. ... A balanced vine has shoots of moderate vigour with no shoot tip growth during fruit ripening. Leaves are of moderate size and number so excess shade is avoided, with both leaves and fruit well exposed to sunlight. Unbalanced vineyards

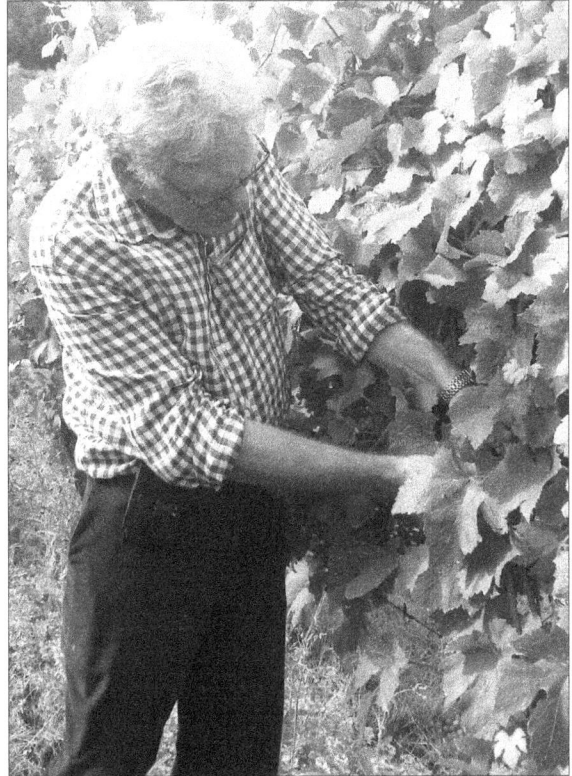

Richard Smart attempting to find some grapes in an English vineyard in 2012

are either too vigorous—in which case poor ripening results from shading and competition between the ripening grapes and shoot tips for carbohydrates—or not vigorous enough—in which case there is insufficient leaf area for proper ripening. Monitoring shoot tip growth is seen as an important management tool for vine balance.

The theories about pruning and how to do it are all very well, but none relate to British conditions and I have found very few of any use in practical terms in vineyards in Britain. Given that your inter-vine spacing is set at planting and your trellising has been designed to carry your fruiting wood on a single wire (or perhaps on two wires in an arched-cane system), then the best you can do is to lay down as much wood as the space allows, leaving a small gap between the canes belonging to the neighbouring vines and definitely not crossing the canes over each other (except in extreme cases – frost prevention being one – of which more later). The 'Ravaz Index', mentioned above, is arrived at by dividing the yield per vine by the weight of pruning wood, giving you a figure of between 0 and around 15 which is an indication of the vine's vigour. This same technique is described on page 26 of *Sunlight into Wine* (although not called the Ravaz Index) and suggests that you categorise your vines into low, medium and high vigour, and prune accordingly. A figure of 30 buds for each kilogramme of pruning wood is also advised by Smart in the OCW, although he also adds: 'experienced vine pruners can achieve a similar effect by looking at each vine, judging how it grew last growing season, and adjusting this year's number of buds accordingly'. We also, of course, do not

---

2 According to Smart in another entry in the Oxford Companion to Wine, the 'leaf to fruit ratio' lies between 5 and 30 sq. cm of leaf area per gramme of fruit, with 10–15 g considered 'adequate' for most varieties although 'higher values are considered by some for Pinot noir'. I am sure someone knows how to measure leaf area, but I have never seen it done.

have the critical element that Smart mentions above: that of 'water stress', i.e. the availability of (and need for) irrigation, with which to control our vines.

My problem with these systems, based upon weight of fruit, is twofold: vines growing in our conditions tend to be relatively vigorous whether they have a high yield or low yield (or even no yield such as in many vineyards in 2012 and 2017); and yields are so dependent on the weather at flowering time that they often bear little relation to the amount of wood that the vine produces. I also do not quite understand how, with a cane pruned system, one is meant to vary the amount of wood left after pruning. Sure, you can shorten the canes to the point where there are no canes at all, but how do you lengthen them without altering the trellis system? This may be fine with spur-pruned vines, but cane pruned? I just don't see how.

Therefore, 'balance' in a vine in Britain is something that you have to work at even before your plant. Given Britain's climate, its small number of days over 30°C, its annual rainfall, especially that which falls during the growing season, and its usually fertile soils, excess vigour is something we have to tackle by using lower-vigour rootstocks, high-density planting and with appropriate vineyard and canopy management techniques. With these we can attempt to tame the vine's vigour – something incidentally more difficult in young vines than in vines of twenty-plus years old – and the rest is down to prayer. Even when we have taken all the above vigour-reducing measures, the level of crop is often due to flowering conditions over which we have no control whatsoever. In addition, I do not think that vines growing in British vineyards respond to yield in the same way that some textbooks think they ought. In most years (years with extremely dry summers are possibly an exception) vines carrying what is, by British standards, a very large crop – say, 20–30 tonnes-ha (8–12 tonnes-acre), as we saw in 2018 – do not show any great reduction in vigour compared to vines carrying a tenth of that yield (or even no yield at all), except perhaps in very dry summers. Perhaps if British vineyards had large yields more often and we routinely had dry summers, then our vines would show a definite vigour difference between low and high yields, but as things stand at the moment, I don't see it. Some varieties are also inherently more vigorous than others and this has to be factored into the equation.

My recommendation is to take each vine on its own, judge how well it has grown in the previous year, and prune it accordingly. If the vine has grown plenty of pencil-thick canes and you have two suitably positioned canes to leave, plus a spur for next year, then prune away and fill up the wires. If your vine has struggled to produce canes and what canes there are, are short and pretty thin, then you have problems! In these cases you prune short, perhaps leaving only one cane, or perhaps only a few short spurs, and see what happens. If next year it has recovered, then fine. However, if next year it has died, then dig it up and plant a new one. If your vine has produced canes that are too short near the head of the vine and there are longer ones further along the cane, avoid the temptation to select the longer ones as this will seriously spoil the shape of the vine and may well exacerbate the problem. Leave the vine with one or two short canes coming from the correct place at the head of the vine and hope, that with a smaller crop, it can sort itself out in the year ahead. Avoid also the option of spur pruning the vine as this almost always results in the vine bearing too many buds and only weakens it, not makes it stronger. Remember, it's not <u>more</u> buds you need, but <u>better</u> buds.

Of course, if you find that your vines are hugely vigorous, producing masses of canes which tend towards 10–12 mm diameter and your crops are poor and intermittent then, Houston, you have a problem. What you can do about it in the short term is debatable. If the rootstock, the variety or the clone is the issue, then these cannot be changed. Your vineyard management might be a contributory factor and maybe you need to look at that. You could try growing green manure in the rows in order to use up more moisture and resorting to more shoot and deleafing to try and improve cropping levels, but this is not a certain path to follow. Root pruning might also help (see section below on *Alleyway management*). Most probably the long-term solution (short of grubbing and replanting) is to re-trellis but this will undoubtedly be an expensive way round the problem, but one that might help. Using a training system, such as Scott Henry, is a textbook solution, but (as I said in the section on *Pruning and trellising systems*) not one that many British growers have used successfully. Taking out every other row and re-trellising to GDC (which is something Lamberhurst Vineyards did to around 10-ha of their vineyards in the 1980s) will probably make looking after the vines cheaper, but is unlikely to reduce the vigour much, nor increase yields.

The mapping of vineyards for vigour using sensors that detect the amount of foliage in the canopy is a technique that is slowly gaining ground (although I have not seen it yet in our vineyards). Sensors are normally mounted on the side of the tractor or sprayer, and each time the tractor passes through the vines, a snapshot is taken and, tied into the position of the vines using GPS. In this way a vigour map can be built up showing which areas of a particular vineyard are less vigorous or more vigorous. The different areas can then be individually treated with respect to a wide range of inputs and operations: fertilisers, irrigation, pruning, canopy management and harvesting. This technique could be very useful in fine tuning a vineyard, especially, say, where a very high-quality crop is required: select the block with the lowest vigour, shoot select to open up the canopy, green harvest to reduce the yield and wait for riper grapes. On the other hand, on more vigorous blocks, leave more buds to try and increase the yield and help tone down the natural vigour in the vines. I have also read about drone operators offering vigour mapping and this may be another way of getting to know which sections of a vineyard need special treatment. I have yet to see this in practice though.

Alleyway management

The management of the vineyard alleyways will, of course, depend on what system you have opted to follow in your vineyard: mowing, cultivations or total herbicide, and this has been discussed fully in *Chapter 11: Management – establishment phase* in the section on *Vineyard alleyways*.

In the case of mowing, the system is quite simple – mow when the grass gets long enough and keep on mowing for as long as the grass is growing. An early 'clean up' mow after pruning and after the prunings have been pulverised, burnt in situ or taken out and burnt, will make things neat and tidy. It is also good practice to keep the grass as short as possible towards harvest. This helps keep the vineyard as dry as possible and maybe helps keep botrytis at a lower level. It will also make a slight contribution to warding off pre-harvest frosts as short grass holds less moisture and doesn't freeze so readily. Those opting for cultivating or total herbicide will have their own timetable dictated by a combination of the weather and soil conditions. The use of green manures in alleyways is discussed in *Chapter 7: Vineyard nutrition* in the section on *Green manuring*. If your vines are suffering from poor drainage there are various remedies available (apart from installing tile drainage which, in an existing vineyard, is a difficult and costly exercise), including subsoiling in the row. What implement you use and to what depth you can subsoil will depend upon the equipment available, especially your tractor. If you have access to a tracked machine (crawler tractor or small bulldozer) which can fit down your rows, then subsoiling, using a two or four-tined winged subsoiler and working at a depth of 500–700 mm, ought to be possible in most soils. With a four-wheel drive tractor, you will probably only manage a two-tine subsoiler and the depth will depend on the soil and surface conditions. With a low-powered two-wheel drive tractor your options are very limited. One way around the lack of tractor power is to use a powered subsoiler such as the McConnel Grassland Shakaerator already mentioned in *Chapter 6: Site planning and preparation* in the section on *Cultivations*. Subsoiling is best done when the soil is as dry as possible, so after harvest is generally not a good time, although if it is dry enough to do it then, the vineyard will have four to five months to settle before you need to get into it again with a tractor. After subsoiling, the earth will have been lifted and it is a good idea to give it as long as possible to settle before driving on it. The best time is actually after you have done the last pre-picking mow and if you do every other row one year and the other rows the next year, you can avoid the subsoiled rows at harvest time and hopefully leave them for the whole winter to settle down.

Root pruning is a technique that has, in other viticultural regions, been shown to be valuable in helping control vigour, but is seldom used in Britain. Classic root pruning consists of dragging a sharp-fronted tine at about a depth of 500 mm down either side of the vines at about 300–400 mm from the line of the vines – the actual distance will depend on your equipment and how close your tractor can get to the vines. In warmer climates, where natural rainfall is less plentiful than in Britain and where in many instances irrigation is used, root pruning might have more effect than I suspect it would in British vineyards. The fact that our vines tend to bear a lighter crop than in many overseas vineyards may also be a factor in its effectiveness. I think that variety and rootstock choice and planting density probably have a bigger effect upon vigour than anything else. However, if you had an over-vigorous vineyard planted with a rampant variety on a high-vigour rootstock, then root pruning might help.

Canopy management

Under this heading are all the post-pruning tasks associated with the vine and its shoots (canes) and leaves – the canopy. They range from spring tasks – shoot removal and shoot thinning – via tucking-in and shoot trimming (hedging), to deleafing and crop thinning. Exactly when this starts will depend upon factors such as when bud-burst occurs and when flowering takes place. This will vary according to the earliness or lateness of the spring, the quality of the site in terms of shelter, exposure and altitude, plus the variety, clone and rootstock in question. Bud-burst is principally controlled by root growth which in turn is stimulated by a rise in temperature in the soil. The books will tell you that this should happen once the soil temperature at around 300 mm deep reaches 8–10°C, but in my experience this is something of a moveable feast. Wet soils will always warm up more slowly than dry soils, so drainage may also be an issue with regard to the timing of bud-burst and vine variety is certainly a factor. The books will also tell you that rootstock is also a factor, although I have not noticed this to be so. In some years bud-burst in Britain can be as early as the start of April; in other years 3–4 weeks later. Much depends on the site and the variety.

Canopy quality

At the heart of canopy management is the concept of what I call 'canopy quality' and it is a concept that needs to be understood in order to grow the best grapes in any climate, let alone a climate as challenging as Britain's. In cool climates, the quality of the canopy can make the difference between a good harvest and a great harvest and between a fair harvest and no harvest.

Given that the condition and quality of the buds on the cane are critical in determining yields, it is important to appreciate that the microclimate within the canopy, especially in the first 0–700 mm of the canopy (measured vertically from the fruiting wire), is critical to future yield as it is in this area that the buds on next year's canes are situated. This is also the area where the vine will mostly carry its grapes. Given that both the quality and condition of this year's crop and the quality and condition of next year's fruiting wood – and therefore of the future yield – is contained in the same area, it is critical that this area should receive as much light as possible.

Within the buds that overwinter and which are left to

Wine Growing in Great Britain

Quality canopy with exposed bunches of grapes, all well spread out

produce fruit in the coming year is all the plant material required to produce everything that the shoot will carry: leaves, inflorescences and tendrils. Before bud-burst, whilst the vine is still dormant, the question of whether the shoot yet to emerge from the bud will produce inflorescences or tendrils is not decided and it is only once the bud bursts and the growing shoot emerges that the changes take place. Although I have already reproduced this excerpt from Smart and Robinson's *Sunlight into Wine*, I think it is worth repeating:

*The buds that contain the shoots and clusters for the following season begin growth early in the spring as shoots develop. Flowering is a critical stage. It is at this stage that the cluster primordia are laid down. Cluster primordia are microscopic pieces of tissue which eventually develop into clusters the following season. Within the young 'compound' bud there are three small shoots which develop – the primary, secondary and tertiary. Whether or not these shoots produce cluster primordia is **very dependent on shade in the canopy**. It is [still] important that shade be minimised on the sections of shoots that will be pruned to in the winter, as the light response is very localized.*

From the above can be seen that, in relation to yield, a factor which I consider to be of great importance to vineyard owners in Britain, is whatever can be done to reduce shade in the critical first half of the canopy should be done.

Cool climates are challenging because the vine is a sugar-producing plant and we judge the ripeness of the vine's fruit – the grapes – by the level of sugar they contain. Within reason (and very dependent on the vine variety in question and the style of wine aimed for) the higher the level of sugar in the grapes the better. We not only judge the level of ripeness by the sugar level in the grapes, we also take acid levels, pH levels and the slightly nebulous concept of 'physiological ripeness' – which includes flavours, tannins, aromas and general 'fruitiness' of the grapes – into account before deciding when to pick. All these are impacted upon by two inter-related natural elements: sunlight and temperature. Again, in general terms, the more sunlight and the higher the temperatures for the vine variety you are growing and for the style of wine you are producing,

the better. In addition, cool growing conditions do not just affect grape ripeness and, therefore, wine quality, they also affect yield levels, and therefore vineyard viability. Without a mixture of both – quality and yield – growing vines becomes an unrewarding enterprise.

In Britain, all of the conventional climate assessment parameters that are used internationally to judge viticultural regions for their ability to grow vines and ripen grapes say we are not just cool, but über-cool. The average temperature during the growing season, the number of degree-days, the Latitude Temperature Index, the mean temperature of the warmest month, the number of days when the temperature rises to 30°C and above – all of these are lower than any other vineyard area growing the same varieties and we cannot change these. Therefore, in growing vines in such a cool climate, we need to attempt, through whatever economically sustainable means possible, to overcome these natural hurdles. Site selection, rootstock, clone and variety, wine style, planting density, pruning and trellising systems, and vineyard management – all of these need to be selected or carried out with two over-riding objectives at the forefront of our decision making process: what can I do to make my vineyard warmer and what can I do to get more of the available sunlight on to the leaves of the vine and, at the right time, on to the grapes. I appreciate that much of the above has been discussed before in earlier sections of this book, but it is worth going back over old ground at this juncture.

In the vineyard, the canopy is where it all happens. You have chosen your site, your rootstock, your clone and your variety. These cannot be changed. You have also chosen your row width and inter-vine spacing and these too cannot be changed in the short term. You will have started out with a pruning and training system in mind, and will have installed your trelliswork accordingly. What happens next is at the heart of canopy quality. The concept of 'balance' has been discussed already in this chapter in the section on *Pruning, training and tying down – subsequent years*, but it is not until your vines get into their stride, probably in cropping year two or three, that you will be able to assess the quality of your canopy and judge whether you have 'balance' or not. In any event, as has already been commented on, 'balance' is a concept with which I have a problem in the context of growing vines in Britain.

The aim of the trelliswork is to hold the fruiting wood in such a way that the annual shoots grow away from the winter buds and, in a cane pruned VSP system, direct them upwards so that a narrow 'hedge' is formed, which can be contained and eventually trimmed. Although a vine's leaves are actually quite efficient at capturing non-direct light – that is, light reflected from other leaves and from other surfaces (the ground for instance) – and in warm to hot countries shade can actually be of benefit as it stops bunches getting sunburn, in a cool climate this is not the case. In cool climates, a vine's leaves need all the direct and indirect sunshine they can get and, therefore, given that the sun is moving from one side of the canopy to the other, the canopy should ideally be no more than three to four leaves thick so that the sunshine will shine directly on to one layer of leaves and partially on to a second layer on each side. Maintaining a thin canopy will not only help sunlight to interact with as much of the leaf area as possible, but will also allow air and light to penetrate in order to dry out the leaves, shoots and grapes – thus aiding disease control – and allow pesticide sprays to reach their target.

How such a canopy is achieved is the subject of the next few sections, but always remember that the canopy is something YOU have created. You have determined the planting density, the inter-vine distance, the number of buds left at pruning, the architecture of the trellising, and whether you have spreaders or not. You have also decided whether you have allowed every bud that bursts to develop into a shoot and whether every shoot and every side-shoot remains throughout the growing season. You can also decide whether to leave every leaf that grows, or whether some need to be removed, and how many and at what time. You can see for yourself whether the canopy is open and whether you can see through to the other side. You can also see whether there are yellowing leaves inside the canopy, a good indicator of a crowded canopy. You can also decide to leave every bunch that is there after flowering, or whether some ought to be removed. In short, you, the grower, have a great deal of control over the canopy and although there may be financial and other resource-based considerations to be taken into account, it is no good looking back at what might have been, when what could have been was in your hands.

Shoot removal - trunks

After pruning, tying down and dealing with the pruning wood when the vine is dormant, the next task is to remove any unwanted shoots (water shoots) that sprout from the trunk of the vines. This will probably be in late April to mid-May. Once the establishment years are over and the vine has developed a permanent trunk, any shoots that appear on the first 300–450 mm of trunk (measuring from the ground) should be rubbed off whilst they are still fresh and young and haven't started to turn woody. The exact height to which you should clean the trunk will depend on the height of your fruiting wire and do not be tempted to remove shoots too near the wire which you might then need to form the spur (thumb) required for the year after. Most growers will carry out this task manually, wearing gloves and running a hand up and down the trunk to the required height. Others will use a shaped hoe which does much the same job, but without the back-bending. Machines also exist which are fitted with stout bristles that brush the shoots off. Some growers use a herbicide (called Shark) which desiccates the shoots so that they wither and drop off. In my experience, using a herbicide to burn off water shoots still leaves a few shoots on the stem, those situated nearer to the fruiting wire, which will need removing by hand. At least though some of the work will have been done, reducing the amount of bending down. Shark is widely used in hop gardens to remove unwanted lower growth.

Timing again is critical and it is not always possible to treat all varieties, all clones, and all rootstocks at the same time and to the same effect. Often a first flush of shoots will emerge and be rubbed off, only to be followed by a second, later flush which will require a second round of removal. With time, the vine's trunk will run out of steam and stop producing many water shoots, although there will always be a few and the task will always need doing. Growers using 'high-graft' vines will not have to undertake this task (although this in itself is not a valid enough reason for using high-graft vines).

Remember also when carrying out this task, that sometimes a water shoot should be retained. Vines suffering from TDs, especially Esca, can sometimes be regenerated by taking a water shoot emerging from just above the graft and training it upwards over a couple of seasons. Then, once the new shoot is robust enough to take over, the trunk can be cut off just above the point where the new shoot emerges and take over as the new trunk. In some regions – I am thinking of Alsace and Burgundy – this is a very widely practiced and it is not unusual to see up to fifty percent of the vines in a vineyard thus treated.

Shoot thinning

The other spring task, one that I believe isn't done enough in vineyards in Britain and one which I also believe would help achieve better flowering and the production of better canes with more viable buds, is shoot removal, especially around the crown of the vine. This area, at the top of the trunk, is often very crowded and is where many water shoots will emerge. Water shoots may not be very noticeable in the first few cropping years, but as the vine gets older and the size of the crown increases, there is more opportunity for these shoots to appear. Left untouched, the canes that develop from these shoots, which often do not bear any inflorescences, will crowd the canes coming from the spur left for next year's replacement canes and will just add additional leaves and cover in exactly the place where they are not required. If in mid-summer (late July–August) you see yellowing leaves in the centre of the canopy it is definitely an indication that your canopy is too crowded as these leaves have not been getting enough light to keep them green. In cool climates it is important that as many leaves as possible are exterior leaves and as few as possible are interior leaves.

The reason that the task of shoot thinning is not done more often in vineyards in Britain is probably one of tradition and practice – in spur-pruned vines this work is absolutely routine – and also the question of risk and reward. Growers probably cannot relate the time taken, and therefore the cost, of carrying out this work with any tangible reward. I am convinced though that they are missing an important element in getting more consistent and better yields. Remember, the more air, light and heat that gets to the canes being left for next year's fruiting, the more likely those canes are to produce viable, flower-rich buds. In a 2009 study[3] of

---

3 The Flowering Process of Vitis vinifera: A Review by Vasconcelos, Greven, Winefield, Trought and Raw. American. Journal Enol. Vitic. 60:4 (2009). This

the flowering process, the following was stated:

Environment and cultural practices influence flowering, either directly or indirectly via their impact on photosynthesis and nutrient availability. Cultural practices encouraging light penetration into the canopy favour flower initiation, while practices resulting in shading have a detrimental impact.

In *The Grapevine* (p. 68), the authors state: 'late spring is the critical period ... shading at this time has a greater effect upon fruitfulness than shading earlier or later in the season' and (p.70) that: 'there is generally unanimous agreement that modifications of the light microclimate in the bud renewal area is mainly responsible for the improvement in bud-burst, fruitfulness and yields'.

The removal of 'blind' shoots, i.e. those shoots coming from buds along the cane which are barren and bear no inflorescences (or possibly one very meagre one), plus shoots which are stunted and which are carrying very little in the way of leaf is, again, something that has not traditionally been done in British vineyards. Blind shoots which are vigorous, grow to full length and which carry a full complement of leaves, will be contributing to sugar production, despite not carrying any grapes. These can be left if they are not crowding other, crop-bearing shoots. But, if they are shading and crowding crop-bearing shoots and bunches within the canopy, they can be removed. Short, stunted shoots are probably best removed in any event. In the normal course of events, I would not expect many growers to go to the trouble of this level of shoot thinning, except perhaps where you were trying to harvest a really ripe crop of grapes for high-quality still wine.

Tucking-in

Once your vineyard is past the establishment phase, and as it develops its full complement of fruiting canes, you will need to start doing some proper canopy management. Essentially this is the process of managing the annual shoots that grow from the buds left at pruning. Once the shoots get to around 300–450 mm long and start impeding the progress of tractors and sprayers up the alleyways, the task of 'tucking-in' can begin.

Exactly what needs doing, and when, will depend on the decisions taken when you planned the trellising in your vineyard. If you opted for moveable training wires (catch wires), these will already have been lowered to the bottom of the posts, in which case you need to wait until the shoots have grown beyond the point to where you propose to move the wires. If using steel posts, you can move the wires to wherever you want on the posts, as the hooks will already be there, although the position of the chain hooks on the end-posts will limit radical movements. If using timber posts, you can only move the wires to where the first pair of hook nails is positioned. As you lift the wires, the foliage will be trapped between them and be guided on its way towards the top of the canopy. In the first cropping year, when your vines

will only have, at most, one fruiting cane and some may not have a cane at all, some of the shoots will refuse to be trained in this way and will obstinately fall down between the wires. This is to be expected and in subsequent seasons, when all your vines have their full complement of canes, this will not happen. In the first cropping year, therefore, you may need a few ties or some clips, of which there are many different makes and types, to hold the wires together. If on the other hand your vineyard posts are fitted with spreaders of one sort or another, many of the canes will hopefully find their way up between the wires, allowing you to close the wires up (assuming they are of the closing up variety) once the canes have passed through. Inevitably there will be some tucking-in required even using spreaders, but less than with moveable wires. Some growers (but not all) use clips – plastic, metal and even wooden – to hold the wires together and trap the foliage. Much depends on how tight you can pull the training wires.

Like so many things in the vineyard, timing is key and your site and its soil, your varieties, clones and rootstocks plus, of course, the season's weather, will all play their part in determining exactly what needs doing when. To start with, you just have to keep your eyes firmly on what is happening to the shoots, work out what labour you have available and how long it might take you to get around all your vines and work backwards to a starting point. With tucking-in you cannot start too early – otherwise there will be nothing to tuck-in – and it is one of the jobs that is better done with, say, four people in one week than one person over four weeks. You will soon learn which blocks need doing first, which blocks grow fastest and which can be left until last. Some varieties, and in the case of Pinot noir, some clones, are very much more upright in their habits and helpfully send their shoots in a more heavenwards direction than others. Some varieties and some Pinot noir clones are decidedly droopy in their habits and do everything they can to escape the confines of the catch wires.

As the vines continue to grow towards the top of the canopy, and by flowering time many varieties will have reached the top wire and above, a second round of tucking-in will be required. Those with a second set of wires to be lifted will bring them up from ground level and into the second set of positions or hooks. Those with wire spreaders will close up the second set. Quite often, and especially in a cool, wet year when vine growth will be more vigorous than in warmer, drier years, a third tucking-in will be required.

Flowering

Not long after bud-burst you will notice that the shoots are (hopefully) carrying some embryo flowers, the inflorescences. These, barring frost and other calamities, will expand and eventually flower and, after pollination, turn into bunches of grapes. As the canes grow towards the top of the canopy and complete the leaf-wall, the flowers will grow larger as they approach flowering. As I have pointed out already in various sections of this book, yield is one of the most important factors in determining the viability of a vineyard and yield is very dependent on how the flowers pollinate. In order to persuade the flowers to turn into as many grapes as possible, getting as much air, light and heat as possible into the centre of the canopy – where the flowers are mainly positioned – is one of the ways of ensuring good pollination. In both *Sunlight into Wine* (p.16) and *The Grapevine* (p.178–182) there are methods of assessing 'canopy quality' which, for the small grower, are quite complicated and long-winded and I doubt there are many growers in Britain who have ever taken the trouble to perform such a test. However, it is good to be able to visually assess the state of your canopy in order to determine whether it is open or crowded. If you look through your canopy at about the point where the majority of the inflorescences are positioned (around 1.00 m from the ground), you should be able to clearly make out the inflorescences (and later the bunches) on the vines, plus the next row of vines. If you cannot, the chances are that your canopy is too thick. Later on in the season, the presence of light green and yellowing leaves inside the canopy is a clear sign of a crowded canopy.

The exact time of flowering is dependent on several factors. Site, variety and rootstock all play their part, but the

East Malling Research Station, Kent. 1971-2000			
Month	Mean 24 hr Min Air temp °C	Mean 24 hr Max Air temp °C	Average temp °C
April	4.5	12.7	8.6
May	7.3	16.4	11.9
June	10.2	19.4	14.8
July	12.4	22.0	17.2
August	12.2	22.1	17.2
September	9.9	18.9	14.4
October	7.1	14.9	11.0
Data source: Met Office www.metoffice.gov.uk			

Table 29

temperature is the most important factor. Again, the books will tell you that the average daily temperature needs to reach at least 15°C before flowering can start, although Peter May, in his book *Flowering and Fruitset in Grapevines* gives details of various studies which challenge this. Britain is so much cooler than almost all other vinegrowing regions that I suspect very few of the 'rules' apply to us. Between the 1970s and '90s, flowering in most years would start in the second week of Wimbledon, i.e. the first or second week of July[4]. As you can see from Table 29, it is July before the average daily temperature reaches 15°C. However, since global warming has cast its magic spell on Britain, flowering has started earlier and earlier, and today, except in a really late season such as 2013 where it didn't really start until almost mid-July, flowering now typically begins for most vineyards in mid to late June, depending on the variety. As I said above, times of flowering are dependent on many factors and in an early season and with an early variety such as Rondo, flowering can start a full two weeks ahead of Pinot noir. However, in recent years, flowering has been getting earlier and 2020 saw Chardonnay, which is earlier than the Pinots, at 25 per cent open flowers on 2 June in one vineyard, a full eight days earlier than in the same vineyard in 2018.

Unlike some fruit crops, insects play almost no part in the pollination of vine flowers, so encouraging bees and other insects into the vineyard will not help. All commercial vine varieties are hermaphrodite and therefore have both male and female parts on the same flower. The distance the pollen has to travel from the ends of the stamens to the ovaries is only a few millimetres and there is research to suggest that with some vine varieties, much of the pollination takes place before the flower caps (which cover the whole flower) fall off. Research also suggests that in years with favourable flowering weather, pollen can become airborne and wind pollination between flowers can occur. Certainly, in my experience, in years when the flowering conditions are ideal, the flowers give off a rich, citrus and honey-toned perfume, whereas in cool, wet years, no scent at all is given off during flowering. Given warm dry weather and a light breeze, flowering can take place within a few days, but in years with

Shoot with inflorescences showing well

4 Since 2015, Wimbledon has started a week later than previously and it now starts on the last Monday of June or first Monday of July.

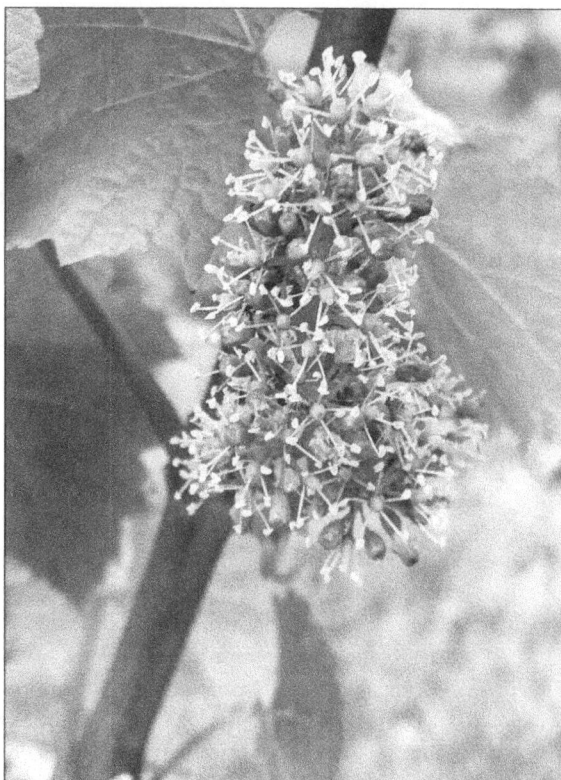

Mid-flower with some caps still on

Classic millerandage after a poor, cool flowering

and will ripen, but remain very small. These 'chicks' will get very sweet quite early in the ripening period, becoming attractive to wasps and birds, and will often succumb to botrytis or rot of one type or another and simply fade away as harvest approaches. In any event, they seldom amount to much in terms of weight.

If you notice that some of the inflorescences have withered and turned brown just before flowering, this is what is known as Early Bunch Stem Necrosis (EBSN) and whilst it can be due to a shortage of nutrients or trace elements, the most common causes are low temperatures and shading in the canopy. EBSN seems to affect some varieties more than others and in a British context, Seyval blanc (in some vineyards but by no means all) can suffer badly in some years. Apart from keeping your canopy as open as possible and encouraging as much light as possible to penetrate the canopy, there is little you can do to influence flowering (short of some sort of covering which is discussed in *Chapter 15: Protected vinegrowing*). It goes without saying that keeping the vines free of disease and nutritionally satisfied is a pre-requisite to good flowering. As has already been mentioned, a shortage of molybdenum has been shown, with certain varieties (most notably Merlot and in countries other than Britain), to make flowering more difficult and some British growers are now applying molybdenum in pre-flowering foliar feeds. Whether this helps flowering is an unresolved question, but the cost is low and it appears to do no harm.

Shoot trimming

The second aspect of canopy management is shoot trimming – often known as 'hedging' – which is the removal of excess growth from both the top and the sides of the canopy. In smaller vineyards, this can be done with secateurs, shears, or a hand-held powered hedge cutter, whilst in larger vineyards a tractor-mounted powered leaf trimmer will be needed. These are described in *Chapter 10: Machinery and equipment*. The exact timing of the first and subsequent trimmings will depend on the growth of the vine, but the aim is to keep the leaf-wall as thin as possible (within reason) and the vines trimmed to around 100–150 mm above the top wire of the trellising. The time to carry out the first trimming is when the shoots grow above the top wire and before they get to the point where they start to flop over. In a normal season, trimming would take place at least twice and possibly in a 'growy' year, a third time. Shoot trimming will encourage the development of side-shoots which will contribute to the supply of carbohydrates, but will also contribute to canopy density, so this also needs taking into account when deciding when to trim. As with tucking-in, you will soon get to know which blocks require trimming first and which require additional trimmings. However, with a tractor-mounted trimmer (especially with a double-sided one) this task is relatively quick to carry out.

poor conditions, flowering can take place over two or three weeks, or in the case of <u>really</u> poor years (such as 2012) even longer.

The dangers of a protracted flowering are that crops are likely to be much smaller and the grapes, because they have not been formed at the same time, will not ripen at the same time, thus making crop sampling and the prediction of a harvest date difficult. When the flowering weather is cool and wet, vines will often suffer from what the French call *coulure* where many flowers fail to set and are shed at or after flowering or *millerandage* (also commonly known as 'hen and chick' setting[5]) where many of the berries stay on the flower

---

5 You often see this referred to as 'hen and chicken' setting which makes no sense as a hen and a chicken are the same size!

In some countries, a first trimming takes place during flowering (generally once the first flower on the cane is fully open and some of the caps have fallen) in the belief that it helps pollination. The aim is to divert some of the carbohydrates being produced by photosynthesis away from the shoot tips, which need all they can to keep growing, into the inflorescences. This view is supported by May in *Flowering and Fruitset in Grapevines* who states that: 'tipping of the shoot is perhaps the only manipulative operation that has consistently resulted in improving fruitset under non-restrictive environmental conditions'. May then quotes a researcher who says that: 'it has to be done when about half the caps have fallen' and follows this by stating that: 'tipping done before the start of flowering or after it has ended has no effect'.

However, tipping at this stage of the season is seldom carried out in vineyards in Britain and I have often wondered why. It may be that it is not seen as having much influence over flowering, given that environmental factors are much more likely to be in control. I also suspect that it is also because quite often, with our tall leaf-walls, the shoots have often not reached the top of the trellis by the time the first flowers are open and therefore, in order to tip the canes, as this could not easily be mechanised, one would have to use a hand-held trimmer in order to get below the top wire and avoid the posts. This is possible in a small vineyard, but not really practical in a larger one. In *The Grapevine* (p. 40), the authors state that: 'it is better to trim relatively early, e.g. close to or at fruitset, thereby allowing a large number of lateral leaves to reach maturity (and thus maximising photosynthetic capacity) around the time of *véraison*'. They then add: 'the presence of active shoot tips after *véraison* is an indicator of vine imbalance, where soil conditions are favouring continued shoot growth'. To which I would add: which is an indicator of the unusual growing conditions that exist in British vineyards. Given that Britain typically has its warmest weeks after fruitset (mid-July), and that typically we get rainfall during the summer, 'active shoot tips after *véraison*' are a very common sight.

A relatively new technique (although it may in fact be an old technique that has been re-discovered), is shoot 'wrapping' or 'tucking', called *palissage* (or maybe *tressage* or even *enroulement*) in France. This is where, instead of trimming the tops of the shoots, they are either wrapped around a wire, the top wire if they have grown that much (known as 'wrapping'), or bent and tucked down in between the catch wires (known as 'tucking'). The aim is to stop the production of auxins, naturally occurring hormones which promote cell division and shoot and root growth, which increase when you cut the tips of the shoots. By wrapping or tucking instead of trimming, auxins are not produced which helps control the vigour of the vine. It is also claimed that this technique lengthens the bunch rachis, thus making for looser, and therefore less botrytis-prone, bunches. However, the jury is still out on this one and it has yet to be proven. *Palissage* is quite a time-con-

suming (and therefore expensive) technique, cannot be mechanised and, from various reports I have seen, makes cane selection at pruning time a bit more difficult. In the north-eastern USA, a two-year trial in 2015–16, produced mixed results and the (interim) report on it concluded that 'the potential long-term impacts of *palissage* on vine vigour, yield, and fruit quality need to be further explored'.[6] I know of three vineyards in Britain that have tried this technique, but two gave up citing the time (and therefore cost) of the operation and the fact that it didn't stop side-shoots growing (one of the claims) and the vines still needed trimming. They also said that the disease situation was worse as the canopy was much more crowded. The other (quite small) vineyard claimed better results but it didn't stop the vineyard being grubbed up. The technique appears to be popular amongst the organic and biodynamic sector in other countries (Olivier Humbrecht at Domaine Zind-Humbrecht in Alsace is a great fan). However, given the extra cost (over mechanical trimming) and the problem of increased disease pressure, I cannot see it taking off in many vineyards in Britain. However, it could be a technique to consider in very vigorous blocks of vines just to see what happens.

Flowering can also be influenced by at least two other techniques: girdling and chemicals, neither of which have much place in commercial vineyards growing grapes for winemaking. Girdling, also known as cincturing, is an age-old practice whereby a small band of bark (containing the vascular cambium) is removed from the cane bearing the shoots which have the inflorescences on them. This disrupts the flow of carbohydrates from the leaves to the trunk and roots of the vine and makes them available to fuel the flowering. This is done just prior to flowering using a special double-bladed girdling knife and, if done correctly, the layer of bark soon grows back. It is a technique routinely used in table grape production where size of grape and looseness of bunch is important. It undoubtedly works with wine grapes, but it is time consuming. The late David Jackson (who was then head of viticulture at Lincoln University at Christchurch in New Zealand), showed me how to do it in the small vineyard I had at Scott's Hall. It appeared to help, but was so labour-intensive that I did not see it as a practical technique to improve fruitset.

Chemical thinning is widespread in orchards of almost all types of fruit – apples, pears, citrus, olives, plums, and nuts – and also in grapes, although again, usually table grapes and not wine grapes. The most widely used product in Britain is called Cerone, which contains an 'ethylene regulator' called ethephon, and has a SOLA (Specific Off Label Approval) for use on apples (including cider apples). In France a product called Sierra which also contains ethephon (or *éthéphon* as it is called in French) is permitted on grapes of all types. The website of the *Institut Français de la Vigne et du Vin* (IFV) has some interesting data on its use and also provides the useful

---

6 Experiment carried out by Professor Justine Vanden Heuval, Cornell University, New York.

information that hand *éclaircissage* (bunch thinning) takes 40 hours-ha (16 hours-acre) at a cost of €600-ha (£545-ha, £221-acre), whereas thinning with Sierra costs €181-ha (£165-ha, £67-acre).

Deleafing

The leaves of the vine are its solar collectors and their task is to collect sunshine and turn it, via the miracle of photosynthesis, into carbohydrates (starches and sugars). These power the plant, end up in the grapes, which we harvest, and are stored in the woody parts of the vine which enable it to overwinter and burst into life in the following spring before any leaves appear. Up until a shoot grows about eight fully expanded leaves, it is dependent upon stored reserves within the vine and until a leaf grows to about one-third to a half of its maximum size, it is a net consumer of carbohydrates, drawing on reserves stored in the plant or being created by other leaves. However, once a leaf reaches this size, it then becomes a net exporter, sending carbohydrates back to the plant and its fruit until such time as it approaches its full size which is around five to six weeks after the leaf first appears. Once a leaf reaches its full size, the rate of photosynthesis is maintained for a further two to three weeks, but then starts to decrease and, as younger and still-expanding leaves, particularly those situated on side-shoots, take over, its contribution to the overall sugar production machine reduces significantly. Eventually a leaf gets to a point (about ten weeks after it first appeared) where it is down to around ten per cent of its peak rate of photosynthesis.[7]

The oldest leaves on the vine are by definition those nearest the winter cane. This is the very area where many of the vine's flowers are situated and where many of the buds required for next year's fruiting are growing and deciding whether to lay down the potential to produce inflorescences for next year's crop. As these oldest leaves are no longer major contributors to the vine's sugar factory, some of them (some viticulturalists say all of them) can be removed at the appropriate time without causing any detrimental effect upon either this year's or next year's crop.

As I have pointed out in several previous sections of this book, air, light and heat are the most important elements in both stimulating and securing crop production and therefore deleafing in the fruiting zone plays its part in allowing these elements to reach the parts that matter. It does this in two ways: by opening up the canopy to air, light and heat; and by opening up the canopy to pesticides, particularly for the control of botrytis. Botrytis can be very damaging to recently pollinated inflorescences when there is flowering residue – spent pollen stamens and flower caps – left on the bunch and also to the buds on next year's canes. In the post-*véraison* period, when ripening gathers pace, opening up the bunches to air and light will help reduce both malic and total acid levels, improve fruit flavours and, with red varieties, help to give better colour and riper tannins. Some growers are concerned that exposing bunches to direct sunshine soon after flowering will lead to sunburn and possible splitting of berries, but there is plenty of evidence that an early exposure to sunshine allows immature berries to 'toughen up' and actually prevents them from damage at a later stage. If you find through experience that some of your varieties seem to suffer more damage from sunshine than others, then deleafing fully on the east-facing side (assuming predominantly north-south orientated rows) and less fully (if at all) on the sunny side, should prevent this happening. Most regions where sunburn on grapes is a regular occurrence have far hotter temperatures than we typically see and far less summer rainfall, and except in exceptional years (such as 2003 or 2018) sunburn (of grapes) is a rare problem in Britain.

In a 1988 paper *Viticultural and oenological implications of leaf removal for New Zealand vineyards* written by Steve Smith (now an MW), Richard Smart, Ian Coddrington and Mark Robertson, and presented at the 2nd Cool Climate Wine Symposium, it was demonstrated how leaf removal improved yields, helped with botrytis control, had no significant affect upon sugar levels and reduced total acidity levels, particularly malic acid levels. This ground-breaking work proved beyond doubt that the practice has huge benefits in cooler climates. This paper is available at www.englishwine.com/wgigb.htm.

The question of when to start deleafing is an interesting one. In past times (pre-1990) very little in the way of deleafing was done at all in British vineyards or, if it was done, it only took place in the few weeks prior to harvesting so that the final anti-botrytis spray could be applied so that the crop would be easier to see and access at harvest time. Today, the date of deleafing has got much earlier and my experiences over the past two decades, but especially in years with difficult flowering conditions such as 2011, 2012 and 2017, is that deleafing soon after flowering was very positive. Some growers that I respect have even taken to a light deleafing pre-flowering in order to get air and light onto the inflorescences without apparent detriment to yields.

By 'a light deleafing' I mean the removal of around one-third to one half of the leaves in the fruiting zone, that is to say in the first 400–500 mm above the fruiting cane (assuming flat *Guyot* pruning). These leaves will almost certainly be the oldest ones on the vine and, as I have pointed out above, are contributing far less to carbohydrate accumulation than their younger neighbours and the emerging side-shoots. Research, all done overseas in climates warmer and drier than Britain's and in vineyards where yields are typically higher than ours, show that an early deleafing may not increase yields, as the flowering process is negatively affected by the loss of photosynthetic capability. With this I have no disagreement. However, I believe that Britain's growing environment is so unusual that the benefits of an early post-flowering deleafing far outweigh any negatives.

A 2011 Luxembourg study[8] on methods of controlling

---

7 Kliewer 1981. The information about leaf sizes and timings is based upon USA data, so it may be slightly different in British conditions.

8 Crop cultural and chemical methods to control grey mould on grapes by Molitor, Rothmeier, Behr, Fisher, Hoffmann and Evers (2011). This document

botrytis stated: 'Manual leaf removal in the cluster-zone just after bloom [flowering] provided a significant reduction of bunch rot infestation without any input of chemical substances. Subsequently, this treatment can be recommended as an important tool in any bunch rot protection strategy for integrated as well as organic viticulture.' It suggested removing two leaves per bunch per cane, i.e. a cane with two bunches would have four leaves removed, the nearest ones to the bunch in question. Whilst some research has shown a reduction in yield by deleafing at this stage (as has been mentioned above), the benefits of a clean crop and the ability of being able to leave grapes to ripen fully, rather than having to pick them because of deterioration in quality due to botrytis, easily outweighed any reduction in crop.

The best practice is to perform one deleafing when most of the flowering is finished and well before bunch-close. Some people even start deleafing when flowering is around 50 per cent caps off with no detriment to yield. The reason for this early deleafing is to be able to get a good fungicide spray applied which will penetrate into the middle of the bunch. For the control of botrytis, this is the most important spray as, if not protected, disease will start from inside the bunch and work its way outwards, only appearing when sugars reach around 5–6 per cent potential alcohol (44–50°OE). The next round of deleafing should be carried out around the time of *véraison* so that the fruit is exposed, thus allowing it to dry out more easily and be more visible, which will make picking easier and quicker. Less vigorous varieties and vines on less vigorous rootstocks may require less deleafing to achieve the same end – clean, healthy fruit for the pickers. Around *véraison* is also the best time to bunch-thin, removing those bunches which are the slowest to change colour. In some vineyards, another pre-harvest deleafing will also be helpful and speed up picking.

How you perform your deleafing will of course play a part in how many times it is carried out. Hand deleafing is a time-consuming, and therefore costly, procedure and in a vineyard planted with 2.00 m wide rows can take up to 120 person-hours per hectare (48 hours-acre). Performing this task by machine, while it may not be so accurate and some bunches may get partially sliced in the process, is so much quicker it means it's much more likely to get done at the right time. Some growers perform their first deleafing by machine, following behind with a hand deleafing team to remove leaves left behind by the machine.

Some growers have experimented with using sheep penned in the vineyard to remove leaves. I have tried this myself, but you need to make sure that you pen the sheep into a relatively small section of vines (although, of course, it does depend on how many sheep you have) and wait until they have taken the available grass down to the point where they are hungry, and look for the next best thing – vine leaves. They will start nibbling these before they take an interest in the grapes (which at this time will be hard and acidic) and will

Two good bunches per shoot is the aim. Bunches just before bunch-close

carry out a partial deleafing. The problem I found was that the sheep will also rub up against individual vines and break or bend planting stakes, plus they will put their feet up on the wires in an attempt to reach higher leaves. I also found that the constant patrolling of the electric fencing used to pen the sheep in, plus the fact that you needed to move the sheep (and the fencing) before spraying could take place, made this a time-consuming method of deleafing. Much easier to employ labour or get a machine.

In some vineyards overseas, I have also seen the removal of side-shoots in the grape zone which is certainly a time-consuming practice and one that cannot be mechanised. If you had a variety that was ultra-susceptible to botrytis or perhaps a block of grapes that you were attempting to get super-ripe, then this is also something to consider. However, I would not remove every last side-shoot on the vines unless I was sure that this did not negatively impact upon sugar production.

Crop thinning

The question of crop thinning may appear to be an academic one, given what I have said about the low yields generally seen in British vineyards, but there have been years – 1983, 1989, 1992, 1996, 2006, 2014, 2018 and 2019 – when crops have been heavy and national yields have been between 33 and 46 hl-ha level (compared to the 2010–19 average of 25 hl-ha). In these 'good' years, many growers will have achieved 25 tonnes-ha (10 tonnes-acre) – even with Chardonnay and Pinot noir – and some high-yielding varieties (Reichensteiner and Seyval blanc for instance) will have produced as much as 37 tonnes-ha (15 tonnes-acre), especially in exceptionally heavy-cropping years such as 2018. Again, as I have pointed out, how an individual vine copes with a large crop is interesting and there is a lot of scientific evidence to show that, in fact, a vine responds to a large crop by upping its rate of photosynthesis and therefore the amount of carbohydrates it is producing. This means that a vine carrying, say, 20 tonnes-ha (8 tonnes-acre) will not necessarily be at a disadvantage when compared to one carrying half that figure and, given good growing conditions and in a well-managed vineyard where both canopy and leaf quality are up to the mark, sugar and acid levels may remain broadly similar. However, much will also depend on the lateness of the year and the weather in the two months before picking, something one cannot predict with any certainty, especially given Britain's fickle climate. In 2018, an exceptional year in that it was both very early and yields were very heavy, many growers said that they 'should have thinned' but mainly complained because the crops were so large that they didn't have nearly enough pickers and then ran out of tanks! The quality of the crop in 2018 was very good, even at the very high yield levels (very high by British standards) that many vineyards achieved.

How much crop thinning is carried out will depend on what type and style of wine is going to be produced. As I showed in *Chapter 3: Site selection*, over-cropping may not be a word in the sparkling winemaker's handbook, considering that in Champagne 20-plus tonnes-ha (8-plus tonnes-acre) is not considered out of the ordinary. However, in vineyards for still white wines, especially those made from aromatic varieties such as Bacchus, Ortega and Schönburger or for hard-to-ripen varieties such as Chardonnay, Pinot blanc, and Pinot gris for still wine, thinning so that the yield is brought down to 7.50-10.00 tonnes-ha (3-4 tonnes-acre) would almost certainly help quality. I would also say that with almost any red variety, but particularly with Pinot noir, thinning down to around 5 tonnes-ha (2 tonnes-acres) would improve both quality and colour. Of course, exactly how much to thin to achieve this level of yield is always going to be problematical and trial and (hopefully not too much) error is the only way. Once you get to know what a light, medium and heavy crop looks like and at what levels your vineyard yields, then working out how much to remove to arrive at a meaningful reduction becomes easier, although it is never a straight-line equation.

Thinning can take place at different times. Inflorescences can be removed prior to flowering; bunches can be halved after they have set; bunch shoulders can be removed on bunches bearing them; whole bunches can be cut off prior to *véraison*; whole bunches can be removed in mid-*véraison* when the difference between early and late flowering can be seen more clearly, at least on red varieties (this is the most common timing); and bunches can be taken off at almost any stage up until harvest.

Removing inflorescences prior to flowering will do the job, but given that one never knows how flowering is going to proceed, it is probably the riskiest technique. Finding the inflorescences may be a problem if left too late, but removing the third inflorescence (i.e. the last) on the shoot may well give the other two a better chance of flowering. Removing bunches well before *véraison* is probably the best time to bunch-thin in order to have a real impact on the yield, and each vine should o be thinned down to an average of one bunch per shoot. It's quite a shock to see all those lovely bunches on the ground, but there is no other way. Thinning to this level will certainly raise the ripeness levels and probably reduce the crop by around 30 per cent. The best time to do this is when the primary bunches turn from the (more or less) vertical position to the horizontal and before they start drooping down. It at this stage that the heaviest, most advanced bunches can be seen easily and the less-developed ones removed.

Cutting bunches in half is a technique beloved of some German growers of Pinot noir, which under their conditions appears to be especially susceptible to botrytis. This involves doing exactly what it says: taking a pair of secateurs and slicing the bunch in half. They call this *traubenteilen*. This has the effect of reducing the size of the bunch, but also of causing the shoulders to droop, thus loosening the bunch and making access for air, light and sprays easier. Removing bunches in mid-*véraison* –known as *vendange en vert* (green harvesting) in France – is certainly the easiest, as there is a visual difference between early and late bunches and if it takes place just after a trimming and a deleafing, the bunches will be much easier to find. Removing bunches in the four weeks prior to harvest probably has little influence over the final quality of the grapes harvested, although it will, of course, reduce the yield and make picking quicker. The cost of bunch thinning at *véraison* will depend upon many factors – vine density, pruning system and crop level – but as a very general guide, a practiced hand, working in deleafed vines, ought to be able to bunch thin 100 vines per hour, so around 40 hours-ha (16 hours-acre) in a 2.00 m x 1.20 m vineyard.

Exactly what effect thinning has upon the quality and quantity of the harvest is something over which opinions differ. In *appellations* with yield restrictions, *vendange en vert* is quite routinely done, although how close anyone can get to the actual yield required is questionable. I suspect that very few growers are prepared to thin sufficiently to reduce the yield by a significant amount and that – at least in hand-harvested vineyards – instructions to 'only pick the ripest' will be used to reduce the level of the crop if this is required. In vineyards where the grapes are machine harvested no such instruction can of course be given, although I suspect there are ways around this. My experience in Britain is with removing 50 per cent of all the bunches from some Reichensteiner vines just after flowering, in order to raise the sugar level, as I wished to make some home-made *suss-reserve*. The yield was cut by around 25 per cent and the sugar level raised by around 2 per cent potential alcohol. Whether this paid for the cost of the work is debatable.

As I have already mentioned under the section on *Shoot trimming* in this chapter, grapes can be thinned chemically, although this is not permitted in Britain. In some countries, harvesting machines are used to thin, by knocking off bunches after *véraison*, but I cannot see this technique catching on in British vineyards just yet, given that there are so few harvesting machines, and in any event, yields are seldom that large.

Reducing the crop is one of those things in viticulture

Date	°OE	Potential alcohol %	Total acidity in g/l tartaric	pH	Bunch weight grammes
8-Sep-14	43	4.8	23.0	2.53	120
17-Sep-14	53	6.6	17.5	2.66	167
23-Sep-14	63	8.1	13.0	2.73	174
2-Oct-14	66	8.6	14.1	2.71	170
8-Oct-14	68	8.9	12.4	2.71	166
14-Oct-14	76	10.2	12.2	2.81	178
16-Oct-14	72	9.5	10.2	2.91	188

Table 30

which is hard to get to grips with. Growers are faced with a dilemma: reducing the crop costs money and lowers income – what sort of sense does that make? However, will the grapes and the wine that comes from them really be that much better and sell for more? Very difficult to tell after the event and many growers take the view that it's really not worth it.

Deciding when to harvest

Deciding when to harvest is always a testing process and to get it right requires experience, patience and a degree of luck. In an ideal world your grapes would slowly ripen to the point where they arrived in the winery, unblemished and undamaged by birds and weather, and with sugar and acid levels perfect for the style of wine required. In practice, this is seldom the case. Except in exceptional years – and there have been a few in the last two decades – when sugar levels were rising too rapidly and those wanting to make top quality sparkling wine were forced to pick Pinot noir before the potential alcohol level got too high, determining a picking date is usually a question of waiting for the acids to fall to the wished-for level. Sampling your grapes prior to harvest will be an on-going process, with the first few plucked from the vines, tasted and rejected as being 'far too acidic'. This will be followed by some random sugar tests using a hand-held refractometer, to the point where you can begin to collect samples and test them for sugar and acidity. This is done easily using a temperature-corrected hydrometer for the relative density (what used to be called 'specific gravity') which gives you the sugar level (measured in degrees Oechsle, or °OE, which is the relative density minus 1,000), and by carrying out an acid titration using indicators and/or a pH meter to determine the total acidity. Whilst the floating of a hydrometer in a 250–500 ml glass or clear plastic measuring cylinder is a relatively simple job, an acid titration requires some simple laboratory equipment and fresh reagents to get an accurate figure. If you have your own winery, then you will require this equipment anyway; if using a contract winemaker, then they will usually offer this service to their clients.

Exactly how many different samples you take, and of what, will depend on your vineyard, the varieties, clones and rootstocks being grown. I prefer to collect whole-bunch samples from each block. Six bunches in total, three from one side of the canopy, three from the other and make sure you take some first node bunches and some second and third node bunches. The aim is to collect a representative sample; not to astound your winemaker with how ripe your best bunches are. Remember that the only sample that really counts is the one you will collect when you pick all the grapes. Sampling ought to start at least three weeks before harvest so that you build up a picture of the time it takes for your different blocks to gain sugar and lose acidity. The sugar and total acidity levels, as well as perhaps pH and bunch-weight, should be put into a spread-sheet and plotted on to a graph so that you can see how the grapes are ripening over time. Many factors will get in the way of a gradual rise in sugars and lowering of acid levels – something that would only happen in an ideal world – and rain in the four to six weeks prior to harvest may well cause both your sugar and acid levels to fall at certain times during the ripening process as the grapes get diluted. Table 30 shows Pinot noir in a typical progression of ripening from around 5–6 weeks before harvest until picking on 16 October 2014. Sugars rise until just before harvest when they fall due to rain dilution, whereas acids drop in a fairly straight line. The average rise in the potential alcohol level in this instance is 0.87 per cent a week. All vineyards, all varieties, and all clones differ in the rate they ripen, with the quality of the year and the yield level probably playing the biggest part in the speed of ripening.

Peter Hall in his Breaky Bottom Vineyard taking some sugar readings with a refractometer

Whilst a lot of growers tend to get excited about sugar levels, to my way of thinking it is the acid levels that really tells you about ripening. Sugar levels, taking different varieties and different yield levels into account, are much more a reflection of the weather during that year, and if you decide that you are always going to pick when the grapes reach 10 per cent potential alcohol, you will not get the best wine possible. What makes the difference between a partially ripe crop and a fully ripe crop is the acid level and, in particular, the proportions of malic and tartaric acidity. At the start of the ripening period, generally taken to be at *véraison*, malic acid levels are high, but start to fall as sugar levels in the grapes rise. Whilst the level of tartaric acid in the grapes – the flesh and the skin – remains fairly constant throughout the ripening process, the level of malic acid in the flesh can fall by between 60 and 80 per cent (Iland *et al.* 2011). The main driver in the decline in malic acid levels is the temperature after *véraison*: the higher the temperature, the faster malic acid levels decline. In terms of taste, the higher the proportion of tartaric to malic, the 'riper' the grapes and juice will taste. Although, in general, only total acid levels (i.e. malic, tartaric and other acids) are measured pre-harvest to determine a picking date, in fact it is the proportion of malic to tartaric that is just as important and can differ from year to year. If you compared ripeness levels in 2012 and 2013, even though total acid levels were similar, the much warmer post- *véraison* period in 2013 helped deplete malic acid levels much more than in 2012, giving riper tasting grapes. Acid levels of course will differ from variety to variety and will differ according to the type of wine you are making, but the difference between a wine made with a properly ripe acidity and one made from a not quite ripe acidity is the difference between an award winner and an also-ran. Be patient and wait for those acids to ripen. I always tell novices awaiting their first harvest: it is better to pick 90 per cent of the crop, 100 per cent ripe, than 100 per cent of the crop, 90 per cent ripe i.e. you have to be prepared to lose a few bunches if you want really ripe and balanced grapes.

In Champagne, growers often refer to *l'indice de maturité* (IM) levels which are arrived at by dividing the grammes per litre of sugar in the grapes by the acidity expressed as sulphuric. They consider an IM of 20 to be an ideal level for harvesting grapes for Champagne. Thus, a juice at 75°OE (170 g/l of sugar or 10% potential alcohol) divided by an acidity of 8.5 g/l as sulphuric (12.9 g/l as tartaric) has an IM of 20. This used to be considered a typical level in Champagne up to the 1990s, but since then their average IM levels have risen. For the years 2000–12 their average alcohol levels were 10.1% and acidity levels 7.5 g/l as sulphuric (11.38 g/l as tartaric) giving an IM of almost 24. In 2018, the all-Champagne-all variety IM was 24.30 in 2018, and 26.40 in 2019. My spies in the Côte des Bar tell me that picking started in 2020 on 13 August, four days earlier than ever before (even earlier than 2003) as acid levels were dropping fast. That's global warming for you!

The one thing you do not want to happen is for any of the following to decide your picking date: a rapidly deteriorating crop; bird damage; frost; a lack of pickers; the children's half-term; and because you decided way back in August that mid-October was the weekend you were going to pick on and have laid on lunch and asked all your friends to help!

Picking

Machine picking

Picking can be done by hand or by machine. Denbies have used a picking machine (a Braud), since their early days, and Buzzards Valley Vineyard in Staffordshire also have one. As has already been mentioned in *Chapter 2: Why plant a vineyard in Great Britain* under *Picking costs* there are several picking machines working in Britain now and getting good results on both still wine and sparkling wine grapes. All grape picking machines work in the same way: by shaking the grapes off the bunch and leaving (most of) the stalks behind on the vine. As they scoop the grapes up, they inevitably collect leaves, bits of stalk and other plant debris (plus the odd nail, if timber posts are used), which is collectively known as MOG (material other than grape) and which in many grape purchase contracts has to be kept below a certain level. Most modern machines have sophisticated on-board sorting devices which pick a much cleaner crop and reduce the MOG level to practically zero. However, given that grapes in Britain tend to be less ripe than their overseas counterparts, I would be concerned about the small bits of (unripe) stalk that might get through any sorting system and end up in the press, and the effect they would have upon the herbaceousness of the resulting wine. I would certainly want to do some trials before I went ahead and machine picked my entire crop. Time will tell whether machine pickers are here to stay. If our post-Brexit trade talks result in a shortage of pickers, then picking machines will definitely become part of Britain's viticultural scene.

In 2013, Gusbourne had a demonstration of a brand new Pellenc harvester on some Chardonnay, but the consensus was that, for sparkling wine, the juice was not as good as hand-harvested and whole-bunch pressed fruit. Gusbourne also had to go through the section of vineyard to be machine harvested prior to the machine going through in order to remove any diseased bunches, which rather negates the cost and speed benefits of machine harvesting. However, in 2019 one well-known large sparkling wine producer test-picked a plot of sparkling wine grapes and I gather the results are favourable. Overseas, especially in countries where labour is either expensive or hard to find (or both), machine harvesting for both still and sparkling wines is the norm and, given the increasing size of British vineyards, I am sure that in years to come we will see more picking machines. For sparkling wines, which are typically whole-bunch pressed, machine harvesting is still frowned upon in some quality sparkling wine regions. I suspect it will be many decades (probably never) before harvesting machines are allowed into Champagne.

However, picking machines get more sophisticated with each generation and their on-board cleaning and selection systems are today very effective. The discrimination against machine harvesting in some regions is understandable and old habits and traditions die harder in winegrowing than in many other spheres. I am sure that good grapes can be machine harvested and used for making very good sparkling wines, although I accept that 'quickly and cheaply picked by machine' doesn't look quite so good on a back label as 'hand-harvested with love and attention to detail'.

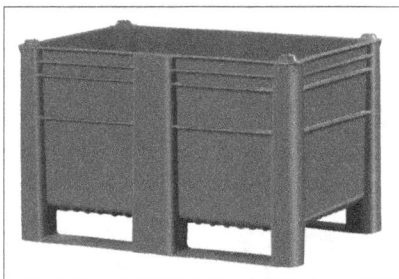

Dolav Box Pallet Type 800. Holds around 300-325 kg of grapes

Hand picking

There are several different ways of picking grapes by hand and each has its advantages and disadvantages. The first thing to decide is what to pick into and this will be determined by one of several factors. If you are picking grapes under contract, there may be a clause stipulating what you have to pick into and, with some contracts, the purchaser will supply picking bins or crates, the type of which will depend on their handling equipment and/or their winemaking philosophy. When it was mainly still wine that was being produced, many vineyards used bulk bins of one sort or another, either plastic or fibreglass bins that could be carried by tractor-mounted forks or a fork-lift and which could be stacked on top of each other for transport or storage. and could often be tipped directly into a receiving hopper, crusher or press. The picking bin used by many wineries is a Dolav Box Pallet Type 800 Solid which has a base the same size as a standard 1.00 m x 1.20 m pallet, has a cubic capacity of 500 litres and will hold around 275-325 kg of fresh grapes (weights will depend on variety and condition and whether the bins have been well filled or loosely filled). These are solid-bottomed bins which are very robust and with care will last decades. They cost (new) around £180-200 + VAT each, depending on number required, but can often be found second-hand for much less. Some sparkling wine producers (Domaine Evremond for instance) have opted for a half-size bulk bin which holds around 200 kg of grapes. Because the depth of the grapes is less in these smaller bins, it is hoped that there will be less crushing of the grapes before they get into the press.

The sparkling wine revolution in British vineyards has meant that many producers now pick into small trays, as careful handling from vineyard to press is all part of the *méthode Champenoise* and enables unbroken whole bunches to be tipped – usually by hand – straight into the press. There are many different types of trays used by wineries and the exact type will depend on several factors – budget probably being the most important. The best, to my way of thinking, are the type that both nest and stack (depending on which way round they are placed) so that when they are full they sit one on top of another, but when they are empty they take up around half the space for

storage. The ones illustrated are the Allibert 11032 which cost new around £15 + VAT each (depending on the number required). They are 32 litres in capacity, have solid bottoms, hold around 12–15 kg of fresh grapes and stack five per layer on a standard 1.00 m x 1.20 m pallet.

Some growers prefer trays with perforated bottoms, but apart from the fact that these allow juice to leak out if any is released by the grapes (which of course shouldn't happen as they ought to be in perfect condition), the perforations will pick up mud and dirt in the vineyard and dried juice from the grapes and they are more difficult to keep clean with normal cleaning techniques. Granted, if you have a purpose-built tunnel pressure washer for your boxes this may not be an issue, but most growers do not have this type of equipment. The solid-bottomed trays are also more robust and less likely to get broken if stood upon or otherwise mishandled. If your budget will not run to new trays, there are usually plenty of second-hand fruit or vegetable trays available at around £3–5 each, but these are usually fairly flimsy and almost always perforated.

In the vineyard, if you are using bulk bins, you will need something to move them about with. Fixed forks or a fork-lift attachment mounted on the front and/or rear of the tractor is ideal. Pickers can then pick grapes into

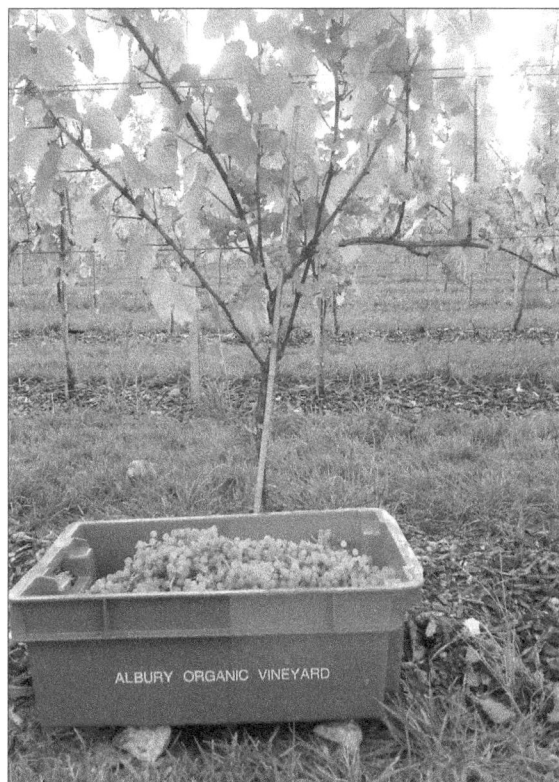

Allibert 11032 stack and nest trays. Each holds around 13-15 kg of grapes

Full grape trays awaiting collection

individual buckets or smaller bins and tip them into the larger bins. If you are using trays, assuming they hold a maximum of around 15 kg (larger ones are not advised as some pickers will only partially fill them as they will not want to lift more than 15 kg), they can be used to pick directly into, although most growers use buckets which can be filled and then tipped into the trays.

Depending on your vineyard layout and/or labour availability, full trays can be walked out of the vineyard to the headland to be loaded on to a pallet, trailer or other carrying device. Or, as is more likely, if your rows are too long for trays to be manually carried out, you will also need some form of pallet lifting and carrying device that will fit down the rows. Again, a tractor with front and rear forks or a purpose-made pallet trailer is ideal. Remember also, if you are picking into trays, that when these are loaded on to a pallet, four, five, or even six high, they will sway from side to side and unless well restrained with straps of some sort, will easily tip over. Scrabbling around in the mud for grapes that have already been picked and were clean and pristine, but now are now muddy and damaged, is not a great experience.

I have always found it quickest to pick four rows at once – two either each side of the alleyway the tractor is in – with two pickers per row, one on either side of the leaf-wall, making a picking gang of eight. Pickers can then hand their full buckets or push the filled trays under the bottom wires into the row where the tractor is. You will need a tractor driver and one 'bucket-person' to handle the buckets as they are passed under the wires and to manage the pickers, keep them up to speed, instruct them in what to pick and what to leave and to check up on the picked rows to make sure all the grapes have been harvested. The bucket-person will also need to load the filled trays onto the trailer and strap the trays down before they begin their journey to the winery or loading area. If you have more than two pickers on each row, you will find that their work rate slows, they spend time overtaking each other, chatting, missing sections and generally slowing down. Of course, if work rate is not a priority and chatting to each other part of the fun, then have as many as you want per row! If you want to pick six rows at once, this is possible, but you will find that three rows is a long way to pass trays or buckets to and fro. If you need to pick more grapes than eight pickers can pick in a day (around 2.5–4 tonnes), then you will need two gangs, each with their own tractor (and driver and bucket-person) and bin or pallet carrying facilities.

In most vineyards in Britain (and indeed in all non-British vineyards that I have ever visited) picking is done by hourly-paid employees and they are kept up to the mark by a desire to get the job done. Some British growers pick on piece-work with each picker or team of pickers being paid according to the weight of grapes they pick but unless you are a really large-scale grower and time and/ or cost are a really important factor in your calculations, I think the extra work involved in keeping tabs on who has picked what, checking to make sure that all the grapes have been harvested and not just the sections with the heaviest crop and managing the payments, far outweighs the benefits. Harvest time ought to be a pleasant occasion and many vineyard owners will provide lunch for their pickers, plus tea, coffee and snacks at the appropriate moments.

Whatever system of payment you opt for and whatever picking containers you use, all pickers will need a pair of picking secateurs and often they will also be provided with protective gloves. You will also need a first aid kit, as a finger often get snipped as well as bunches.

Costs of picking have already been covered in some detail in *Chapter 2: Why plant a vineyard in Great Britain?* in the section on *Picking costs*.

Chapter 13

Frost protection

Introduction[1]

Frost in the vineyard can be a danger at three different times of the year: winter, spring and harvest time. There are several passive measures, ones that require no action at the time of the frost, which can be taken to reduce the level of frost damage, as well as several active methods, those requiring inputs of materials, machinery and labour during a frost event.

Winter frosts

Winter frosts usually occur in late December, January and February, and are seldom a problem in Britain as temperatures do not usually go down low enough. The trunk of a mature vine can withstand -20°C to -25°C when the vine is completely dormant although it is possible that annual canes might be damaged at these levels. If these sorts of temperatures are experienced in Britain (as they have been in the past) the incidence of such things as Crown Gall (caused by the bacteria *Agrobacterium tumefaciens)* and *Eutypa lata* might well increase. However, it is impractical to protect a vine against this degree of frost.

Spring frosts

Spring frosts are by far the most important and damaging type of frosts in Britain, as well as in other northern European vineyards. They occur in April and May (and even very occasionally in June) and come in two types: advection and radiation (or inversion) frosts. An advection frost occurs when very cold, frost-carrying air blows into a vineyard region (sometimes from a considerable distance away) and displaces warmer air. This type of frost is

1 I acknowledge the paper: Practical considerations for reducing frost damage in vineyards by Trought, Howell and Cherry (New Zealand Winegrowers 1999) in helping me write this chapter. This document is available from www. englishwine.com/wgigb.htm.

unusual in the spring and is more associated with winter conditions. The only technique to counter an advection frost is to avoid planting in a frost pocket. The second type, a radiation frost, usually occurs on a clear, dry night when there is little, or no wind, and when long-wave radiation from the ground to the air is not impeded by cloud cover or high concentrations of water vapor in the air. When all these factors occur together, rapid cooling of the air immediately around the vines takes place. This layer of cold air, which can fall to as low as -8°C (although in the south of Britain -2°C to -4°C is more usual), is then trapped beneath a layer of warmer air (hence the term inversion) and cannot escape. Typically, this cold, damaging layer will start almost at ground level and rise to 10–30 m above the vineyard floor. This type of frost is the one most likely to damage vines and is the one that can be more easily, although expensively, controlled.

As has been said above and stated in *Chapter 3: Site selection* in the section on *Aspect*, the best method of frost protection is to select a site that is not in a frost pocket. These 'pockets' can range in size from a relatively small area of land such as a low depression in the middle of a

Frost damage in early May can take most of the crop

flat or gently sloping site, to one low-lying corner of a site where the frost cannot escape, to complete valleys covering a considerable area of land. If you do have a slope on the site and think your vineyard might be frost-affected, plant with the rows in line with the slope and with plenty of space beyond the vines so that the frost can drain away. Make sure that any windbreaks, shelter belts or hedges likely to trap a frost have enough gaps in them for the same reason. However, finding a site that fits all these requirements is not always possible and, in looking for a site that is both well sheltered and not exposed to prevailing winds, you may find that your site is prone – at least in some areas of it – to frost.

Spring frosts have always been an issue in British vineyards – not only vineyards, but in fruit crops in general – and in most years, frosts of some sort will occur. The question of how damaging they are is one of timing: an early bud-burst and a late frost and vines can suffer huge damage. A classic frost year occurred in 1997, when an early spring was followed by a -6°C on 6 May. National yields were cut to 6,460 hl, when they had been over 26,000 hl the year before. At Chapel Down, where I was then winemaking, less than 1-tonne of grapes was picked from the whole 4-ha (10-acres) whereas the year before, from the same vines, 30-tonnes of grapes had been harvested. Ridgeview, who do not have what I would call an especially frost-prone site, say they have used frost protection in all but four years out of the last twenty-three (and three of these were between 2006 and 2008). In some years (2010) they lit their *bougies* (frost candles) nine times. They estimate the cost per night using only 200 candles per hectare at £2,000 for around 2.50-ha (6.18-acres) of vines. To this one has to add labour costs of, say, £500 a night which works out at £1,000-ha. If you say this is the value of 500–750 kg of grapes and a hectare might be carrying 8 tonnes of grapes, then this looks like a reasonable cost for frost insurance.

Ridgeview also have an experiment running with electric warming cables on five rows, (and no frost protection in the two rows either side as a control) and in 2017, a year with really bad frosts (as has already been mentioned), the control rows lost 70 per cent of their shoots and the cabled rows only 20 per cent. The upside of electric cables, of course, is that they can be turned on and off automatically; the downside is the high cost of installation (around £7 per metre of row or £35,0000-ha on 2.00 m row width) plus high running costs and the need for a very good power supply, or a large generator hired for the 6 week frost period. This is probably why this method of frost protection is used very little, either in Britain or overseas. There is more on electric cables below.

In recent years (since 2009) frosts seem to have been more of an issue in British vineyards and more and more growers are now using active techniques to prevent frost damage. Many British vineyards saw considerable damage in 2017 when an up to -3°C inversion frost on 19–20 April, was followed on 26–27 April by an advection frost which went down as low as -7°C in some places. Growers who used their anti-frost devices against the first frost and saved some of their crop were devastated by the second, much deeper frost, against which there was no defence. Some varieties came through relatively unscathed and produced good crops from secondary buds, but overall the national crop was considerably reduced. My harvest report for 2017 recorded the yield of all varieties for 'not-frosted' as 5.06 t-ha (2.05 t-acre) and 'frosted' as 3.45 t-ha (1.05 t-acre), a yield reduction of almost 60 per cent.

In 2020 I instigated a 'frost survey' and, using the services of Mapman Ltd, who put together a website where vineyards could log on and complete a simple form which asked: who they are, where they are, what level of frost they had experienced, and whether they had used frost protection or not. The results were illuminating with responses from 211 vineyards showing 58 per cent of the vineyards suffered damage, but used no protection. Of those that used no protection, 34 per cent suffered damage of 50 per cent or greater with 12 per cent saying the damage was 90 per cent or more. It shows that frost can be a widespread problem in British vineyards and many would benefit from using one of the many methods of protection. Of course, saying you have suffered '50 per cent damage' can only be based upon the visual damage to your buds and shoots. Having half your shoots damaged to a greater or lesser extent doesn't necessarily mean that those shoots will lose half of the anticipated crop. Much depends on timing and the variety, and on the flowering conditions yet to come.

The extent of the damage to vines will depend not only on the temperature down to which the frost falls, the length of time it stays there and the state of the growth of the vine, but also on such factors as the orientation of the site, the vine variety, the rootstock, the height of the vine above the ground, the condition of the floor of the vineyard, the slope on the site, the amount of space allowed for the frost to collect and/or disperse and several other factors. Each frost event on each site will have its own peculiarities. The actual temperature required to damage a vine depends on the state of growth at the time of freezing and the relative humidity of the buds; the higher the humidity, the worse will be the damage. A paper from Michigan State University says that at woolly bud stage, -3.3 °C will kill 10 per cent of the buds and -6.1 °C will kill 50 per cent of the buds; at second leaf, -1.7 °C will kill 10 per cent of the shoots, -2.2 °C, 50 per cent and -5.6 °C, 90 per cent.

Of course, if your vineyard looks likely to suffer from frost, you will need some form of sensing equipment to tell you when a frost is about to occur and/or is occurring. A standard weather station is probably not the best way of obtaining the data as they are (a) usually sited in the wrong place and (b) have their temperature sensors at the wrong height. To predict the likelihood of frost you need sensors in the part (or parts) of the vineyard most likely to be frosted first, with the sensors somewhere near to ground level, plus perhaps at 500 mm and 1.00 m above ground level. The system needs to be able to send text messages to multiple mobile telephones so that everyone involved with frost-fighting can be woken from their slumbers. With experience you will work out where in the vineyard the best place is to site sensors, at what height to have them and at what temperature they are programmed to send out text messages. You can also site additional temperature data-loggers in vulnerable parts of the vineyard to record what actually happens. Being able to predict a frost event accurately will keep the cost of preventing frost damage down to a minimum and some growers subscribe to a frost prediction service which will alert them to the likelihood of a frost in their immediate area so that staff can be warned that they might be needed to help with frost prevention measures.

On a 'frost night' the temperature will normally get towards freezing at between midnight and 2 am (although it can be earlier) and get progressively colder as dawn approaches, reaching its lowest point at around daybreak. During this period the frost will slowly roll from high ground to low, settling in even quite small dips and hollows in the vineyard – you often only notice that there is a slightly lower area in the vineyard when you see how the vegetation and vines have been damaged by frost – and will scorch any green tissue in its path. Having said that, frost can be remarkably capricious, damaging some buds but not others, touching those on the top of the cane but not the bottom and singeing one side of a shoot but not the other. A really hard frost on a dormant or barely open bud may well kill it completely and there will often be no regrowth at all. The damage to buds from a light frost is more difficult to spot and it will often only manifest itself after bud-burst when shoots and leaves will be deformed as they grow. Frost damage to leaves will show up as soon as the sun comes up and the frost lifts. Frost damage can also seriously deplete stored reserves in the vine, affecting the development of subsequent shoots and accounting for lower yields the following year.

Depending on the severity of the frost (depth of temperature and length of freeze) there is usually some recovery afterwards, with many vines sending out new shoots, especially from around the crown of the vine where the buds are better protected by the mass of older wood found here. Typically, what happens is that some buds will be unscathed and continue as normal whilst others will shoot later and will always be behind. This gives rise to different flowering dates, different ripening dates and, consequently, different harvesting dates. Inevitably, frost damaged vines will yield less, often considerably less and, in some instances, nothing at all. Frost damaged vines may also show signs of damage in the following year: more blind buds, deformed buds and, of course, reduced yields.

Passive methods of preventing frost damage

In order to reduce frost damage, there are a number of passive measures that can be taken. If you have a site where frost is likely to be a problem, then planting varieties that are naturally late in breaking out of dormancy might be advantageous, although there is some evidence to suggest that in years where bud-burst is late, spring frosts are also late. However, from my experience, Seyval blanc is less prone to damage than, say, Reichensteiner or Rondo – both early varieties – but this may be because it is a variety that is generally less susceptible to frost damage. Of the Champagne varieties, Chardonnay is the most susceptible because it breaks bud the earliest (of the three) and Meunier the least susceptible as it is the last to break dormancy. If you are only growing Champagne varieties, then plant the Meunier in the area most likely to be hit by frost. Some varieties show a marked tendency to produce viable secondary flowers after the primary ones have been frosted (Pinot noir and Chardonnay for instance) and growing these in frost-prone areas might be considered. You might also consider using a high-wire training system – GDC, Sylvoz or Blondin – as temperatures can be up to 4°C higher at 2.00 m above ground level than at 700 mm. Raising your fruiting wire, say just from 700 mm to 900 mm, will make a difference of around 0.5°C – so, helpful, but probably not critical and generally not enough to avoid frost damage. Spur pruned vines will suffer less damage than cane pruned vines as the buds on spurs will gain some benefit from the body of older wood near them. Spur pruned vines are also much quicker to prune and therefore could be left until the very last, thus delaying bud-burst. Although these are not good enough reasons to adopt this form of pruning over the whole vineyard, if you have one area that you know is always frosted, then this might be part of the solution.

The state of the vineyard alleyways is also important. Clean, firm, bare earth will hold on to heat during the day, even more so if the earth is damp, when compared to rough cultivated soil or alleyways with a covering of grass or other vegetation. The heat held in the soil will be released to the atmosphere during the evening and night and can help ward off frost. Some reports suggest a rise in the temperature by between 0.3°C and 0.6°C. Dark loamy soils are at less risk than light, sandy soils and vines with mulch beneath them will also be at greater risk than those with bare soil underneath the vines. Long, un-mown grass, which holds moisture, is the worst alleyway cover and much more likely to make frost damage worse. In some regions, growers routinely burn off the vegetation on the vineyard floor with a contact weedkiller prior to bud-burst, allowing it to grow back after the frost-dangerous weeks are over. Even if you were to take no other frost prevention measures, this one might be worth considering.

Pruning later in the spring – say, in late March – will

delay bud-burst slightly and could be considered if the area is liable to be frosted. Leaving the canes long and in a vertical position until the danger of frost is over is often stated as being a way to avoid frost damage. Vertical canes are less likely to be damaged than horizontal ones, as the frost cannot settle on them, and in addition, the first buds to grow shoots will be those at the ends of the canes, which you will be cutting off in any event, assuming that you have left the canes over-length. However, given that in Britain it is usually the third to fourth week of May before it is safe to say the danger of frost is over, leaving the canes upright means that when you do come to tie them down, they will have shoots down their whole length – and in most seasons quite long shoots – many of which will get knocked off as the canes are handled whilst being tied down. It also makes the job of tying down much slower and is generally not seen as a practical solution to the frost problem.

Leaving additional 'sacrifice' canes on your vines – a third or a fourth cane if you are normally pruning to two canes – is also a textbook way of coping with frost. The idea is that if you don't get frost, they can be cut off and discarded; if you do get frost, then they can be tied down. It does at least allow you to tie down two canes to start with and then it is only the sacrifice canes that have to be dealt with. What often happens if you do this is that all the buds on the sacrifice canes burst and produce long shoots (often showing multiple inflorescences) whereas the canes tied down are delayed in their bud-burst and seem to lack flowers. However, in an area known to be frost-prone, this is a relatively simple and cheap method of frost insurance.

Active methods of preventing frost damage

Heating vineyards is probably the most used technique (both in Britain and overseas) of frost prevention and there are several different systems and materials used.

Bougies

The most widely used active frost prevention measure in Britain are *bougies* (the French for 'candles') and are made by a company in France called Stopgel. They consist of 6 litre cans of solid wax with a cardboard wick that can be placed out in the vineyard and lit progressively as temperatures fall. They cost around €8.50 (£7.50) + VAT each including transport. The benefit of these is that they require no fixed capital equipment, can be placed out in the vineyard beneath the vines until they are used (meaning that you can still drive up the alleyways to spray and mow if you need to) and then brought into the centre of the row when they are needed. They are also relatively simple to handle and operate and can be stored indefinitely until used. They also meet the recommendation that a larger number of small heaters is more effective than fewer large ones and the number used can be expanded or contracted at will. Stopgel have modified their candles in recent years

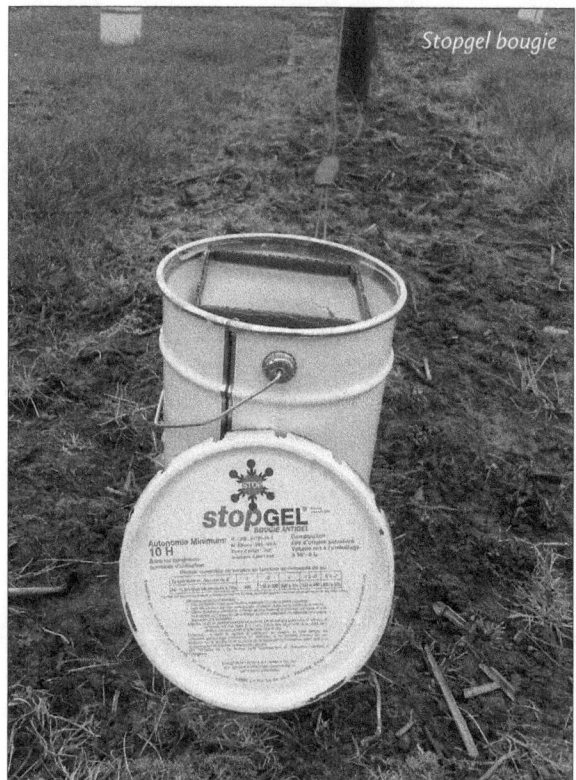

Stopgel bougie

so that they are now made from *stearin* which is derived from natural vegetable fats and they are no longer use oil-based paraffin wax. The new 'green eco' candles burn with almost no smoke. After the 2020 frosts, several British growers are looking at importing a cheaper candle from China. In 2020 a new supplier emerged, The Crop Candle Company, offering '8-hour burn' candles from £6.95 each and 12-hour burn candles from £8.10. Both prices plus VAT and a discount for four pallets (480 candles) and more.

The number of candles required in a given area is something that will depend on several factors. The manufacturers state that a maximum of 500-ha (200-acre) are required to protect against a -7°C frost in a vineyard 'with no wind' and suggest that around 200–250 ha (80–100 acre) are lit to start with and the number progressively increased as the temperature falls. You need to work out, based upon your row width and vine spacing, how many vines per candle you need, i.e. if you are spaced at 2.00 x 1.20 m then for 500-ha you need one candle for every 8.33 vines and space them out accordingly. You need to make sure that the perimeters are well heated as, if you just place them only in the centre, cold air will be drawn in from outside. You also need to place more on the windward side than the leeward side, assuming there is a breeze.

After you receive a frost warning, typically this will be when the temperature falls to around +1°C, you will need to get up and assemble your troops and, assuming the temperature continues to drop, start lighting the candles. Most growers find that a plumber's gas torch works very well for this task. The manufacturers state that each candle burns for ten hours, but users report 8 hours absolute maximum, with 6 hours a more practical average. You can budget on getting around two burns (i.e. two nights), per candle before they need replacing. The actual number used per season will depend upon the number and severity of the frosts, but in some years in Britain you might have to light them around six times, making the cost for the year, including the candles and labour, at least £5,000-ha protected (£2,000-acre) and possibly up to 50 per cent more. Whilst this is a high sum (although unlikely to be this high in most years), hopefully you would be protecting a relatively small proportion of your whole site and with a good crop of grapes valued at £15,000–£20,000-ha (£6,000–£8,000-acre) as a minimum, well worth doing.

Heaters - static

In addition to wax candles, there are several other types of static heaters available, although not used much (if at all) in Britain. Burners using a mixture of waste sump oil and paraffin (known as 'smudge pots') were once popular on fruit farms, but now seldom seen as this mixture of fuel is not very environmentally friendly (and probably illegal today). Added to which the burners were heavy and difficult to move and required filling up with fuel in the vineyard where there is always a danger of spillages and soil contamination. Lighting piles of old car tyres or damp straw bales soaked with a waste oil and paraffin mixture used to be advocated, but are now definitely not permitted, although the smoke was useful in showing where the inversion layer stopped and started. One grower I know has suggested that baling up the prunings and using these in suitably positioned cauldrons might work. It would also be a win-win situation: frost protection and getting rid of the prunings in one move. In Chablis - a region noted for its spring frosts - I have seen gas burners connected with flexible pipelines to a bulk gas tank in the corner of the vineyard, which is probably quite efficient and cost-effective. Fairly new to frost protection are mobile burners which use wood pellets as fuel, but on the current information I have, you need 200–250 burners per hectare, putting the capital cost at €45,000-€50,000-ha,

Old style 'smudge pots' for vineyard heating

plus of course the running costs, and they only managed to reduce the temperature from -4.2°C to -2.5°C, so not very effective: neither cost-effective, nor anti-frost effective.

Fans – with heaters, static and mobile

One of the most widely used techniques is to use tractor-mounted fans that blow hot air, heated by oil or gas, into the vineyard and these can be static or towed around the vineyard. The most widely used in Britain are the FrostGuard (the static version) and the FrostBuster (the trailed version), both made by Agrofrost NV. The units, of which there are several different models, cost between £10,000 and £20,000 and from around £20 per hour to run for the smaller units, up to nearer £80 per hour for the larger ones. Nyetimber have around forty-five static heaters and ten to fifteen trailed ones in order to protect their vineyards from frost. They take a bit of getting used to and, of course, each site will have its own peculiarities, but if frost is a real problem and you have a large area of vines, they are undoubtedly easier than frost candles. The static ones can be automated to start up when the temperature gets low enough, and can then be augmented by a trailed unit which can, naturally, cover more area. They are, however, quite noisy – it's difficult to shift large volumes of air without sounding like a small helicopter – and expect complaints if you are near housing. The trailed ones are probably better if you have neighbour issues, as the noise moves around, and the fixed ones might attract the attentions of your local council's noise abatement officer. Remember this all takes place in the early hours when most sensible people are asleep. There is also a towed machine called a Frostdragon which emits a moisture-laden 'fog' produced by burning more or less anything combustible. They give as examples straw bales, bundles of prunings, or wood. This produces smoke which, when made heavier by having water injected into it, produces a 'fog' which insulates the vineyard as the tractor drives around it. How much noise and smell this device makes is not known, or even whether it is effective

Fans – static, without heaters

There are also frost protection units that rely on the 'selective inverted sink' principal (SIS), also known as the 'forced cold air displacement systems', which draw in the cold air and blast it upwards into the warm layer creating air circulation. Whilst these appear to be popular in some parts of the world, they are not universally liked and have, so far, not made much of an impact in Britain. These fans cost around £20,000–£25,000 each, depending on size and area to be protected. Before buying one (or more) I would want to make sure that no planning permission was required, and I would also want to make sure what the noise level was at the nearest house, especially if it belonged to a grumpy neighbour. There also frost fans, made in New Zealand, which are called 'Tow and Blow'. Their name is actually a bit of a misnomer, as they are only towable to the place where you want them to operate,

Frost windmill in a Coonawarra vineyard

and then they are jacked up and become static! The fan is mounted on a large hydraulic arm that can be raised and lowered, angled and rotated, so that the air can be directed to where it is wanted. They can also be left in position and can be automated so that they come on when the temperature falls to freezing (or just above it). Because they are moveable, they do not require planning permission, although the same comments as above apply about noise. 'T&B' machines cost around £30,000–£35,000 depending on size and control equipment.

Frost windmills – static, without heaters
In many parts of the world, wind machines – commonly known as 'frost mills' – which mix up the cold and warm layers of air in the vineyard, are the preferred form of frost prevention and, when used in conjunction with vineyard heaters, probably offer the best solution. This is the system widely used in citrus orchards. One manufacturer states that one frost mill will protect around 4-ha (10-acres) with a 5.5°C inversion difference at 15 m above ground level. These figures are approximate as each installation and each frost event will be specific to that site. Each machine needs power – electricity, gas, diesel or even tractor PTO – and a control system. A basic gas-powered unit can cost around USA$50,000. Denbies Vineyard installed several Californian frost mills in the mid-1990s at a cost said to be around £250,000 only to find that as soon as they turned them on for the first time, some very irate neighbours, annoyed at having been woken up in the middle of the night by the noise and commotion, successfully appealed to the local authority to prevent their use. A new blade was designed which, although quieter, failed to shift enough air, and eventually they were sold to a New Zealand vineyard.

Helicopters
Helicopters are used quite widely in New Zealand (often in conjunction with vineyard heaters and frost mills) and these are presumably effective, as they continue to be used there. When I ran Lamberhurst Vineyards, its owner, Kenneth McAlpine, who both owned a helicopter and a had a pilot's licence, was interested in seeing if it worked, but both the Civil Aviation Authority and the local residents thought differently and it was never tried. A helicopter needs to be able to fly slowly over the vineyard, re-passing the affected area every 5–6 minutes, to make sure that the inversion layer remains stirred up. According to one New Zealand helicopter company's website, one helicopter ought to be able to protect 20-ha (50-acres) at a time.

Electric heating cables
Electric heating cables are also being used for frost control in a few vineyards and the German company Hemstedt have developed their patented Hem System® Wine Frost Control which uses plastic-covered copper cables strung along the fruiting wire. Their website says that this system was first used at Weinsberg Viticulture School in an experiment in 2017. They state that it reduced damage from 100 per cent to 10 per cent at a cost of €25-hectare for the electricity (using mains power) but say that for a larger area you would need a substantial generator. Capital costs of installing the cables are said to be about €13,000-ha. However, the advantage of electric cables is that they are an 'install and forget' system which can be turned on and off automatically and in theory requires no human intervention. Apart from Ridgeview's experiment with electric cables already mentioned, there are a couple of commercial vineyards in Britain using heating cables. Goodworth Clatford Vineyard, a 0.40-ha (1-acre) vineyard in Hampshire, spent £9,000 (2017 prices) installing 2,000 metres of heating cables and their associated controls, and spend around £25 per night on electricity when they are switched on. Winding Wood Vineyard, near Hungerford in Berkshire, have laid cables on a 0.50-ha (1.24-acre) vineyard and in 2020 turned them on around 5 times. They estimate they suffered only 5 per cent frost damage, whereas other vineyards in the same area suffered 50 per cent damage and more. However, whilst electric cables work well for buds that are dormant or only just bursting, if there are shoots of any length, they appear not to work quite so well. Hemstedt, in their sales brochure, say that their system works on shoots 100 mm long, but Goodworth Clatford Vineyard report that their 150 mm shoots were 'totally wiped out' at -3.5°C. They are experimenting with draping a fleece over the young shoots and around the heating cable, thus trapping the heat. Of course, this means not only buying and putting the fleece up in the first place, but also lowering it and securing it when a frost is due, and then lifting it up after a frost event so that the shoots can continue to grow away and vineyard operations (e.g. spraying) can also take place. This makes this method of frost prevention not quite so user-friendly as it first seems.

Spraying with water
Another technique, known as water aspersion, is widely used in citrus orchards and other soft fruit plantations

(and often in blackcurrant fields in Britain), and involves spraying the vines with a fine spray of water as the temperature falls to and beyond freezing. Unlikely as this technique might sound, where enough water is available (as it must be sprayed all the time while the temperature is at or below freezing), it is a fail-safe method. As water freezes, it releases a small amount of heat (known in scientific circles as the latent heat of freezing) and this heat, coupled with the igloo-effect of the layers of ice that form around the canes and buds, protects them from any damage. Once the temperature has risen above freezing, the water sprays can be turned off and the ice slowly melts as the day warms up.

When the water aspersion system is used, it is important that enough water is available to keep the sprays running for the maximum number of days and the maximum depth of frost that is likely to occur. It is no good running out of water in the last few hours of the last day of frost. It is also essential to make sure that the soil does not become waterlogged and poorly drained soils may well suffer in this respect. There is a risk of breakage to the trelliswork as the weight of ice formed can be very substantial and if this type of frost protection is to be used, the design and construction of the trelliswork will have to take account of this. However, it is generally agreed that where enough water is available, the aspersion system will provide guaranteed protection against frost. The costs of installing an overhead irrigation system will not be small and a considerable volume of water will have to be available prior to the frost period. A vineyard using this system might need to have at least 2.5 million litres of water per hectare of vines in store at the start of the frost period, unless it is possible to top up the water source in between frost events. The big benefit of water aspersion systems, of course, is that they can be automated. Strategically sited thermostats can both start and stop the system's valves and/or pumps, thus removing the task of finding people to get up and work in the middle of the night.

In the 1980s there was quite a large (4.15-ha, 10.25-acres) GDC vineyard at Headcorn in Kent, owned by the decidedly eccentric Major Grant, whose main business was growing potted chrysanthemums ('pot-mums' as they are known) and alstroemerias (a type of South American lily) under around 3.2-ha (8-acres) of glass. In order to provide the greenhouses with water, there was a large lagoon, filled with water from the nearby River Teise. The Grants' abstraction licence allowed them to keep pumping out of the river until the end of May, so an overhead irrigation system was installed in the vineyard. At the time I was their winemaker and, in 1988, when everyone else in Kent was hit by an extremely hard frost, their vineyard was protected by their frost irrigation system and yielded a huge crop. I recall that the icicles formed extended from the high wires of the GDC, about 1.80 m from the ground, right down to ground level and almost succeeded in bringing the trelliswork down with them. I also recall that the site, on heavy Wealden clay, had been land-drained at 5 m spacings (the normal spacing for vines is 10 m) as otherwise the huge amount of water used would never have drained away. Such was the frequency of the drainage runs that they all had to be marked out on the surface before any vineyard posts could be installed as Grant was worried that otherwise the pipes would be perforated. At least two small vineyards are successfully using water aspersion against frost, Mill Lane Vineyard in Lincolnshire and Congham Vineyard in Norfolk. Congham take their water from a well and with their high water table are able to extract for the 6 hours or so they need to keep irrigating. They report that for their 0.36-ha (0.89-acre) they use £5 of water a night and in 2020 applied about 2 mm a night in an 800 mm strip along the rows of the vines.

A traditional water aspersion system uses prodigious amounts of water – 25 mm in a six-hour period would not be unusual – but there are now 'flipper' systems using 'flow-regulated micro-sprinklers' which only distribute water along the row of vines, i.e. not into the alleyways, and water use can be reduced from around 50,000 litres per hectare per hour down to around 17,000. The cost of installing this system is said to be around £3,000-ha (£1,200-acre), plus the cost of a good supply of water. If you have a frost problem and have access to good water, then this system is definitely worth looking at. East Malling Research station has an experiment on vines running with this system which the irrigation experts NaanDanJain Ltd installed.

Covering vines

Using covers of some sort – netting, shade material or fleece – to place over the vines during a frost event may well be a practical method of avoiding some frost damage, although on its own it is unlikely to prevent damage in the event of a really hard frost, say below -4°C. To erect netting so that it can be made secure – problems with wind damage are always possible with any netting or similar product attached to vineyard trelliswork – and so that it can be raised and lowered so that normal vineyard operations can take place (e.g. spraying) requires both considerable capital

Frost protection with water aspersion in a New Zealand vineyard

investment, plus labour inputs. And, when all is said and done, it can only be part of the solution and some other technique will also be required. See *Chapter 15: Protected vinegrowing* for more details about covering vines.

Spraying before a frost event

Sprays of one type or another (various polymers and copper) have for many years been touted as preventing frost damage and every year seems to bring a new product. However, I have never met anyone who said they work. They might have a marginal effect upon delaying bud-burst or they might give the buds a greater degree of resistance to frost damage, but cannot be relied upon to provide total protection. Most of the sprays have to be applied between 24 and 48 hours prior to a frost event, so this implies (a) knowing with some certainty when a frost event is likely to happen and (b) being able to get on to the ground with a sprayer. The chemicals do not last for that long – maybe a week – and are not cheap (£150-ha per spray for materials was a recent quote) and, given that you will probably need to use other methods as well, they do not seem (yet) to be part of a practical solution to the problem of frost damage.

Harvest frost

Although in many years Britain's grape harvest is over by the end of the third week of October, there have been occasions both in the past and in more recent years when autumn frosts have damaged grapes. In some years, a light frost, perhaps around -1°C to -2°C, will do nothing more than singe the top of the canopy and perhaps cause some of the older leaves to fall off, thus exposing the fruit and make it easier to harvest. In years when temperatures have fallen further – down to around -4°C or lower – damage has been done to grapes, although generally only the less ripe ones as grapes with a high sugar content freeze at lower temperatures than those with lower sugar content. In many cases this is a good indicator of which bunches are fully ripe, making picking easier. However, if grapes are frozen, then they must be picked immediately and pressed. In addition, wine made from them may well acquire a 'frost tone' on the palate which will need work in the winery to remove.

As to whether frost prevention measures can be used at harvest time is another matter. Most growers are too tied up with thinking about picking dates and picking arrangements to start putting out and lighting frost candles. Much easier to keep an eye on the weather and hope to harvest before a damaging frost arrives, picking earlier than planned if necessary. I guess if growers had static systems that could be turned on and off at will, they might be able to use them, but at present, this is not the case.

Frost insurance

At one time, frost insurance was available to top and soft fruit growers in Britain (as well as insurance against hail) but this does not appear to be available any more. In some European countries, both frost and hail insurance are available.

Note: The UN Food and Agriculture Organisation (FAO) has produced a huge work on the subject called: *Frost protection; fundamentals, practice and economics*. It can be downloaded from: www.fao.org

Chapter 14

Weed control

Introduction

The adequate control of weeds – which includes weed species, grasses and any other unwanted plants – in the 600–750 mm strip directly beneath the vines, is one of the major tasks in the vineyard. For the 'vineyard virgin', a grower completely new to the myriad of tasks that face him or her as soon as the vines are planted, it poses one of the greatest challenges. Such is the importance of weed control that it warrants a separate chapter to itself.

Without good weed control in new vineyards, vines will soon become choked and smothered and the weeds will compete for both nutrients and water to the detriment of the vine's root and shoot growth. In established vines, weeds will also compete for water and nutrients, will interfere with the crop, if allowed to grow too tall, and will make the disease situation worse by creating a damp and humid atmosphere directly beneath the fruit. Keeping the soil weed-free beneath the vines will also help raise the soil temperature, leading to an earlier bud-burst and better cytokinin production, the all-important element in getting good yields. In *The Grapevine* (p.37), the authors state:

it appears that 'warm' soils are just as important in cool maritime climates as in cool continental climates – in the latter, an early start of the growth cycle is desirable, whereas in the former [cool maritime climates, i.e. Britain], *warm soils are needed to offset the slow rise of air temperature during the spring in order to increase cytokinin production. Warmer soils will usually have more root growth and potentially more cytokinin production.*

There are in reality only three practical methods of keeping the under-vine area free of weeds: cultivating, herbicides and mulching. Other techniques such as flaming the weeds or grazing them with animals (short-legged sheep, large rodents?) may have their places in certain situations and at certain times of the year, but they do not in themselves constitute a year-round solution to the problem. An additional technique to control weeds in the under-vine area – that of the use of another plant species

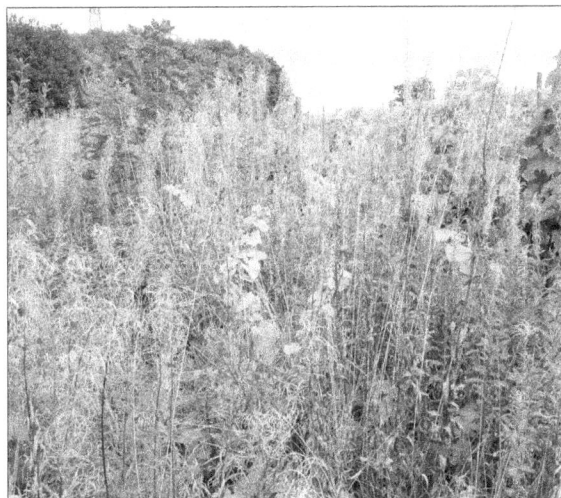

There are vines in there somewhere!

to suppress weeds – does not, to my way of thinking, offer a realistic solution to the problem in British vineyards. The competition from the other plant species is just too great and is to the detriment of the vine's development and long-term well-being. In warmer regions where weed growth is slower (or even non-existent) in the summer, and vines are irrigated, under-vine weed suppression with another species might be possible, but it is certainly not a widely used technique. In organic and biodynamic vineyards, where of course chemical means of weed control are not permitted, some of these less commonly seen techniques might well have their place. However, whether organic, biodynamic, or conventional, vines will still be stressed if weed competition becomes too great.

Cultivations

The control of weeds in vineyards by cultivating, both under the vines and in the middle of the rows, is the oldest method of weed control, and hand-hoeing and the use of horses and oxen to pull ploughs and harrows is as

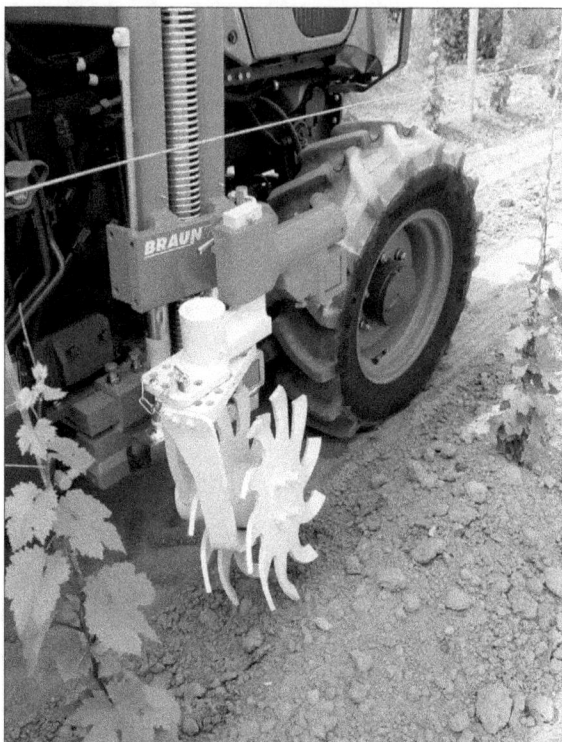

Mid-mounted Braun Rollhacke for speedy weed cultivation

Boisselet inter-vine mechanical weed hoe

In regions warmer than Britain's, and where there is often a shortage of summer rainfall, cultivating between the rows, rather than allowing weeds and grasses to establish and use up moisture, is seen as a way of conserving moisture. The practice of cultivating is therefore part and parcel of managing a vineyard and extending it to cultivating beneath the vines is second nature. In Britain it has not been traditional to cultivate either under the vines or in the middle of the rows, mainly because vineyard owners have tended to follow the established practices of most top and soft fruit growers for whom herbicides around the trees, bushes and plants are the norm. However, the practice of cultivation is becoming more common in British vineyards, although I suspect it will be some time before it becomes universally adopted.

Most under-vine cultivators (and also under-vine mowers) work in a similar way. They have a sensor arm which detects the presence of a stake or a vine which activates a hydraulic ram which swings the cultivator blade away from the vine and then returns it after the vine has been passed by. This leads to two problems. One is that unless the vine is protected by planting stakes placed either side of the vine (or only on one side if you can be sure to drive down the row in the same direction each time), the sensor arm will hit the vine on the trunk at about the same place each time the weeder is used. There are fears, especially with vines which are newly planted and, until their stems have grown and hardened off, might well be in year three or four, that this regular knocking might result in long-term damage and may well be a contributory factor in TDs. As has already been mentioned in *Chapter 9: Trellising systems – construction* in the section on *Individual planting stakes,* the use of a stake such as a *tuteur trombone* would provide protection to a young vine against the damage that might be caused by the arm of an automatic weeder. Using individual rabbit guards around each vine might also help in this respect (although I am generally against their use) but you would have to make sure that the guard was both well staked and firmly attached to the stake.

The other seemingly unavoidable problem is that there is always a small mound of soil around the vine which the cultivator cannot reach and which, if it is piled up against the graft of the vine, will lead to roots growing out from above the graft. Quite why this seems to be more of a problem in British vineyards than overseas is interesting, and I suspect that it is because our soils are typically wetter in the summer than in drier regions and tend not to crumble and fall away from around the trunk quite so readily. The only way to avoid the scion root problem is to go through the whole vineyard on a regular basis, perhaps every two years, and manually remove any soil that has built up with a hoe. If scion roots are left to grow and establish, then the vine will become susceptible to *Phylloxera* and the benefits of the rootstock will slowly ebb away.

There are a wide variety of different types and models of under-vine cultivator available for use in vineyards. Some have fixed blades and no swing-arm, others have

old as vinegrowing itself. In the 21st century, under-vine cultivating to control weeds is very common in Europe and also increasingly so in New World regions. This is partly due to tradition (in the case of the Old World) and also because, in many instances, one grower's vineyard is very often right next door to another's, so there is a strong disincentive to use herbicides that might drift and cause damage to a neighbour's vines. In addition, over the last ten to twenty years, there has been much more emphasis upon vineyard 'sustainability' and 'conservation' and these, coupled with question marks over the use of glyphosate (Roundup and other trade names) has made many growers convert to under-vine cultivating for weed control. Whether using less herbicide but many more tractor hours and diesel is really helping sustainability and conservation is another matter.

powered rotary cultivators that work in the horizontal and yet others with cultivator blades that work in the vertical. However, all cultivators seem to have difficulty in Britain's relatively mild climate and in our often-damp soils where weed growth happens for much of the year. As I have pointed out above, the soil doesn't always shatter and crumble away as it is turned or moved by the cultivator. Much depends on the type of soil you are working with and the timing of when you carry out the work. Some growers report that it is necessary to have two cultivators, each with a different mode of action so that the soil beneath the vine gets attacked in a different way each time. With under-vine cultivators costing from £5,000 for a single-sided, non-powered blade type to £18,000 for a double-sided, powered cultivator type, investing in two makes this an expensive system of weed control.

Whilst mechanical under-vine cultivation might be seen as a more ecological way of managing weeds in vineyards, the amount of tractor time, and therefore fuel used, can mount up. Using under-vine cultivators requires a steady hand and, at least to begin with, quite a slow speed. With a single-sided machine (meaning that each row requires two passes) you would be unlikely to travel faster than 3 km-hr which means three to four hours per hectare in a vineyard with 2.00 m wide rows. Given that you would probably need to cultivate five to six times (and sometimes even more) during the growing season, then this becomes an expensive operation, especially when you add on to that the biennial scion-root clearing, plus the wear and tear of blades on the machine. There is also some evidence that the weeds and grasses that grow under the vine and are incorporated into the soil by the cultivating, increase the nitrogen content immediately around the vine, thus contributing to soil fertility and excess vine vigour. If you were looking to decrease vigour in your vines, maybe another method of weed control might help.

Having said that under-vine cultivators are difficult to use in young vineyards, what is perhaps more difficult is to start using them in vineyards where weeds have been controlled with herbicides for a few years. Soils here tend to be quite hard and whilst perhaps not compacted, take

Front mounted herbicide spraying booms

a lot of work with a cultivator to get them to move and crumble and, in addition, there are bound to be weeds and grasses that have put down established roots which add to the problem. If I were determined to use an under-vine cultivator, I would use herbicides for the first three years and then, with the soil beneath the vines as clean as possible, start with a cultivator that had a powered blade. I think that a static-bladed cultivator would merely bounce along the compacted soil.

However, if you can manage to get to grips with cultivating beneath your vines, and maintaining a weed-free strip, the benefits of earlier soil warming, as was pointed out at the start of this chapter, are probably worth having. Some growers I know use a clean-up herbicide in the spring to remove any overwintering weeds, and then, when the weeds start growing again, use a cultivator for the rest of the year. I am sure that as more and more growers buy, and get used to using, under-vine cultivators, their presence in British vineyards will become much more common.

Herbicides

Herbicide spraying is a widely used and long-established method of controlling weeds in crops of all sorts, both in Britain and worldwide, and with the right materials, right mode of application and good timing, it is generally considered to be a safe and effective method of weed control. To my way of thinking there is a lot of nonsense written about the dangers of using herbicides in vineyards and whilst I fully understand the wish, surely shared by all growers, not to 'poison' the soil by using unnecessary chemicals, the fact of the matter is that every chemical used has been through a testing procedure for safety and as long as the materials are used correctly, then there ought not to be a problem. I also have a real issue with those who are fully signed up to the organic and biodynamic route who believe that any use of herbicides is wrong on the grounds that the soil is being irretrievably damaged; I cannot quite get my head around this view.

If the herbicides are confined to the immediate area beneath the vines in a strip of, say, 600 mm wide, then the volume of soil that the herbicide impacts upon per hectare in a vineyard planted at 2.00 m row width and assuming that herbicides penetrate to a depth of 75 mm, is 5,000 m x 0.6 m x 0.075 m which equals 225 m³ of soil per hectare. Given that a vine will root, let's say, to 2.00 m (and it has of course unlimited access to much deeper regions) and the roots can spread where they want in the vineyard, then the volume of soil the vines have to root in per hectare is 100 m x 100 m x 2.00 m which equals 20,000 m³ or a volume 89 times greater than the volume 'occupied' by the herbicide. Given also that the soil is regularly watered by rain, that other plants will also absorb any herbicides, and that most permitted herbicides break down upon contact with soil or over a relatively short period of time then I find it illogical that, if correctly and sensitively used, any harm will come

161

to either the vines, the soil or the environment by using herbicides. However, I digress.

The first requirement for successful weed control using herbicides is to have the right equipment. You will need a dedicated weed sprayer with tank, pump and spray booms of the right width (or adjustable) and with appropriate nozzles for the area you want to cover. Do not attempt to use the same tank for both pesticide and herbicide spraying. A weed sprayer would normally fit behind the tractor and be mounted on the three-point linkage, although if you have access to a quad bike or similar, it is worth looking at weed sprayers mounted on these. Nozzles for strip spraying are best mid-mounted so that they can be seen without having to keep looking behind. Most weed sprayers will have nozzles fitted to both sides so that two strips can be sprayed at once, but with each nozzle independently switched on and off so that single-side operation is possible. A tank of 250 litres will give you sufficient material to spray 1-ha (2.5-acres) of land, so if you are spraying a 600 mm wide strip in a 2.00 m wide-rowed vineyard, this will give you enough liquid to spray 3.33-ha (8.2-acres) of vineyard. The exact rate of liquid applied per hectare will depend on several factors – the herbicide used, type of nozzle, type of weed to be sprayed, timing of application – so it is difficult to be too specific about the exact size of tank required, but 200–300 litres is typical. There are also a number of herbicide spray systems which use low-volume technology including CDA (controlled droplet application) and firms such as the Micron Group, with their 'Undavina' range, are worth looking at. With their low pressure tyres, they can get on the land when a tractor might not and can be very useful in vineyards. For spot-treating weeds and for places where you cannot get a tractor, a decent hand-powered 20–25 litre knapsack weed sprayer will also be required. You will require different nozzles for different spray widths, plus, if you intend to strip-spray or spot-treat near young vines, a spray hood.

If the vines are correctly protected, the ground around them can be sprayed as soon after planting as required and whilst it is important to be nervous about harming the vines, if the job is done with the right equipment and the right herbicide, the lack of weeds will benefit the establishment of the vines hugely. My preferred method of weed control in young vines is for each vine to be protected with a 500–600 mm high x 125 mm diameter spray-guard which has a ventilated top section, but is closed at the bottom to allow for careful spraying of the ground around the vines. Yes, some weeds will grow up the inside of the spray-guard, but these can be removed when the vines are shoot-singled and side-shooted, as they are large enough to get a hand down inside. At first pruning, depending on how the vine has grown and what your policy is about first year pruning, the guard can either be left in situ or lifted up over the planting stake which should be attached with a removable clip.

I used to advise using a residual (also called a 'pre-emergent') herbicide around the vines as soon as the soil has settled after planting and before any seedlings start to appear but now that there is only one residual herbicide allowed to be used on vines (Kerb Flo) and that has to be used between 1 October and 31 January, that advice is redundant (although in some countries, pre-emergent herbicides are still legally used). In established vineyards, most growers will apply one pre-bud-burst application of herbicide and then be prepared to apply a further one (or maybe two) sprays during the season. In newly planted vineyards, the first spray can wait until sufficient weeds have appeared to make the job worthwhile, which may well be not until mid-July. After leaf-fall the situation needs reviewing and, if there is enough weed growth to warrant it, maybe put on a pre-winter spray. Next spring aim to get a residual on before any weeds start growing (and by 31 January if using Kerb Flo), then spot-treat any clumps of nettles (which typically start growing fairly early in the year) and then put a contact spray on once the first flush of thistles are no higher than 100–150 mm. Let them get higher than that and it is too late. Once you follow a regime like this, weed control becomes easier and simpler as fewer weed seeds are present and perennial weeds get dealt with. Of course, there will always be problem patches of ground, and weed species which may become resistant to the herbicide you are using. If this becomes the case, you need to spot-treat and probably change the herbicide you are using or increase the rate for a spray or two.

Whilst you must consult your chemical supplier about the choice of herbicide, I have used both residual and contact herbicides on newly planted vines without any damage being done. Translocated (i.e. systemic) herbicides must be used in young vineyards with extreme care and it is best done by someone experienced in their use, when conditions are ideal and your vines are fully shielded with spray guards. Different materials will be appropriate at different times of the year and will depend on the weeds to be controlled. In established vineyards, where the trunks are hardened and woody and there are no shoots or leaves at ground level, you can use a much wider range of materials than in young vineyards.

For effective herbicide spraying you need the right weather conditions: without too much of a breeze and dry weather after spraying. Rainfastness times will differ for different products, with some products rainfast in one hour, but up to six hours required for some systemic herbicides. Residual herbicides work by sinking into the top layer of the soil and locking into it so that when seedlings start germinating they are killed as their roots come into contact with the herbicide. For this the soil needs to be damp, so a period of gentle rain after application will do no harm. In dry years residual herbicides sometimes fail to work satisfactorily as they haven't been 'watered in' sufficiently after application.[1]

The question of timing, however, is THE most

---

[1] There is currently (2020) only one residual herbicide permitted to be used on vines (Kerb Flo) and that has to be used between 1 October and 31 January.

important aspect of weed control, whatever method you are using. The first thing to realise is that weeds have their own timetable and will not work to yours. It matters not to them that it is too wet to get on to the vineyard, that it is too windy to go weed spraying and that it will be raining for the next week – in fact the weeds know this and will use it to their advantage to outwit you. For effective weed spraying you have to keep your eyes firmly on what is happening on the ground and be prepared to get out the sprayer at a moment's notice. Once the weeds get away, get too tall and too numerous to get a decent cover of spray on to them, then the battle is lost. Of course, this is a difficult task. Some weeds arrive early, others arrive late and some appear almost all year round. Using a residual at least gets you a head start as a proportion of the weeds will be controlled, leaving only the more perennial weeds that do not come from seeds, but from roots – thistle being a prime example.

Once the majority of the vines have formed a trunk which can take herbicides – and with good luck this will be at the end of year two – the spray guards can come off all those vines that have formed trunks and weed spraying can take place with less concern about harm to vines. Each vineyard will be different and indeed each year will be different. In dry years, weeds tend to grow less for obvious reasons; in wet years, residual herbicides may get washed out of the soil and be less effective and, with good soil moisture levels, weed seeds may germinate more freely. Either way, the battle against weeds is one that requires constant vigilance. Herbicides may sound like a lot of work to the novice grower, but with the right equipment, the correct materials and by spraying at the right time for the weeds to be controlled, I am sure that they are a more certain and more effective way of controlling weeds than any other. Whether they are less environmentally sustainable than other means, I will leave to others.

Mulching

When I studied at Geisenheim in 1976–7, before I planted my first vineyard at Tenterden in Kent (now Chapel Down), I saw the technique of mulching with black plastic 'film' and decided that this would be the technique I would use. The vineyards which Kenneth McAlpine had recently planted at Lamberhurst had also used plastic mulch, so I felt sure that the technique would work. A plastic-laying machine was found and adapted – it had been built to lay 2.0 m wide mulch for strawberry beds – and 750 mm black plastic film sourced and purchased. Planting commenced, by hand, on Easter Monday, 11 April 1977, and after every day's planting, I got on my tractor and, with the help of someone who generally had to sit on the back of the plastic layer and guide the wayward film as it unpeeled off the roll, laid it over the planted vines. I clearly remember the first few attempts at laying the plastic – the unburied edges, the plastic so tightly that it broke the grafts – but after a few attempts we worked out how deep to plant the vines

Black plastic mulch at Camel Valley Vineyard - still working after ten years

and how tightly to lay the plastic. With a few hiccups, we successfully managed to plant 9,300 vines and covered them with almost 12 km (7.5 miles) of film.

As far as weed control went, the black plastic worked exceptionally well. The vines were not hampered by weed growth, save the few weeds that sprouted in the same hole that the vine was planted in, and in the holes which the crows made as they stomped up and down on the plastic looking for food. The soil beneath the film was always moist and undoubtedly warmer than the uncovered soil, as the black film sucked up the heat. The only problem was the edges where the film was buried – of the 750 mm wide film, around half was exposed and half buried – and given that I was cultivating the centres of the rows, there was a 300 mm strip on either side of the plastic that had to be kept weed-free with herbicides (which is about the same amount as would be sprayed if you were just using herbicides).

What I didn't know at the time was that the film I had naïvely ordered from a local supplier had not been made to 'plastic mulch standard' – I was unaware at the time that there was such a standard – and it only had 1.50 per cent of carbon black, the additive which both colours the film and prevents degradation by ultra-violet light. The consequence was that, after less than a year, the mulch started to deteriorate. After numerous letters to the supplier, the manufacturer and to both ICI and British Cellophane, the leading manufacturers of plastic film in Britain at the time, plus tests of the film that had been supplied and used, it was agreed that the film was not suitable for long-term outdoor use. The suppliers eventually accepted that whilst I had told them what the film was for, they had not told the manufacturers this, the consequence being the low level of carbon black. I forget now whether we received any financial compensation or not, but the whole experience was an eye-opener for me and I quickly learnt that you need to spell everything out, preferably in writing, when ordering products whose failure might have long-term consequences and costs!

163

Wine Growing in Great Britain

PAS 100 compost spread around vines to help establishment

The result of the deterioration of the film was that some of the soil around the vines was covered and remained weed-free, but in some places the film disintegrated into ribbons and the weeds soon poked through, especially in areas where the film's colour bleached out and light could get underneath. This made weed control a haphazard affair and after a further twelve months we decided that the plastic had to come up. This was fine – even if hard, dirty work – where the film was exposed, but where it was buried – and completely undamaged and sitting comfortably beneath a solid covering of soil, weeds and grass – the plastic proved impossible to move. We therefore just cut it along the edge and left it there where it remained for many years and all the while those vines remained there, one could still find remnants of the wayward film. The last of those vines were removed in 2015, thirty-eight years after being planted.

For my next vineyard, three years later, I ordered film directly from British Cellophane and specified 4.5 per cent of carbon black and with a thickness of 125 µm rather than the 80 µm of the earlier product. For this vineyard we had decided to use ungrafted, unrooted cuttings from our 1977 planted Seyval blanc and, after laying the mulch, merely stuck the lengths of cane, dipped on the end with hormone rooting powder, through the plastic and waited for them to root. This they did with a great degree of success and this became one of our best – and certainly our cheapest – vineyard. This vineyard has also now been grubbed up, but there are definitely traces of this almost indestructible film in the field!

Today, however, unless there were exceptional circumstances, I wouldn't recommend using plastic film for mulching. There are various reasons. Apart from the cost of buying and laying the film, plus the cost of removing and disposing of it once it starts to deteriorate, there is the problem of keeping the buried edges free of weeds. To do this you end up with having to spray around 300–350 mm on either side of the film (as I have pointed out above, this is about the same amount as you would spray around the vines without mulch). I am also worried that having such a moist, warm, weed-free environment around the vines makes them lazy and doesn't force them to root deeply in search of moisture and nutrients, something I feel a young vine should be encouraged to do at all costs. You might also find that vine weevils, snails and slugs are also attracted to the continually moist soil beneath the mulch, all of which can damage the tender shoots and leaves of a young vine.

I have seen growers use MyPex, a woven weed suppressant which, although significantly more expensive than plastic film, is much stronger and can withstand the occasional tractor wheel and heavy foot traffic, and comes with a five-year guarantee. This is a permeable product and, as such, lets the soil breathe, but I am not sure it offers many other benefits and you still have the 'edge' problem to resolve as the weeds and grasses very soon encroach and root through the fabric. However, Camel Valley have used it on one of their vineyards (planted in 2010) and it is still going strong after 10-plus years. The trick, they say, is to lay it into well prepared soil so that it ends up flush with the surface. In this way the mower can be adjusted so that it skims over the top of the fabric and doesn't damage it, and the edges (if you want it to look extra neat and tidy) can be sprayed off occasionally with herbicide.

Organic and biodynamic vineyard owners face a considerable challenge in the area of weed control and one of the ways to tackle weeds is with mulches. Whilst the problem of how to keep the buried edges of the plastic film or MyPex-type fabric clear of weeds is even greater in the absence of herbicides, if this can be solved – hand mowing and strimming works on small areas – then this does answer the problem in the early years when mechanical methods of weed control cannot safely be used or in situations where tractor-mounted mechanical weeders are inappropriate, i.e. in small vineyards where the cost cannot be warranted.

Mulching with loose materials, such as compost or wood chippings (or a combination of both) can be effective in controlling the weeds, although there are drawbacks. On anything other than a very small scale, you need a dedicated side-delivery spreader which can take sufficient material to cover a reasonable length of vineyard row with each filling. These can cost around £14,000 so are not for the smaller vineyard owner. They can also be hired (at around £200 per day for spreader only, no operator), but getting one that can fit down narrow rows and which can turn in the average vineyard headland can be a challenge. You also need quite a large area of hardstanding which can be reached by road-going tipper lorry (that is unless you are close enough to the source of the mulch material to have it delivered or collect it yourself by tractor and trailer) and ideally some machinery there at the time of delivery to heap up the compost as it arrives. Otherwise, the area you need to store the material before starting to spread it will be huge. For effective weed control, enough material needs to be applied to fully cover the soil to a depth of at least 75 mm, although the accuracy with which this can be judged is difficult and you need to do a test row or two to get the rate of delivery correct for your needs. Spread any less thickly than this and the weeds and grasses will start to poke through quite quickly and, especially at the edges (assuming you have grassed-down alleyways), the cover will deteriorate.

Compost is readily available from waste sites that process green waste and a lot of it is PAS 100 compost meaning it has been produced to BSI PAS 100 specification. This covers every aspect of the compost's production and content, including the amount of 'non-organic' material such as glass, stone and plastic that is allowed in it. There will be some evidence of plastic in it, particularly bits of the blue plastic bags that every corner shop uses, plus bottle tops and other hard plastics that people include in the 'green' waste that they put out for the dustmen, but these should be below the specified threshold. If you are considering buying non-PAS 100 compost, make sure you know what you are getting. The PAS 100 specification also includes the amount of food waste (including meat waste) that is allowed, plus the amount of nutrients that are contained in it. WRAP[2] state that 1 tonne of PAS 100 contains around 8.1 kg of nitrogen (as N), 3.3 kg of phosphate (as P_2O_5), 6.6 kg of potassium (as K_2O), 2 kg of magnesium (as Mg) and 1 kg of sulphur (as S). Each producer and each batch will differ slightly, so be sure to get an analysis for the actual material you are using. Whatever the nutrients in the compost, it needs to be taken into calculation when considering applying other fertilisers, especially if you are in a NVZ. Although the nitrogen is the largest nutrient in the compost, because it is contained in a form which requires time to break down, it is released slowly. The same goes for the other nutrients. Also note that PAS 100 has a slight neutralising effect upon the soil (about 10 per cent as effective as limestone), so pH levels will be raised, especially in the area immediately below the mulch. If you are using woodchip it ideally needs to be made from hardwood, not pine, and the chips need to be fairly coarse, otherwise they also break down too quickly.

As for the cost of using loose mulch, this very much depends on such factors as the cost of delivery of the material, whether you have your own spreader, the width and thickness to which you spread it and the time interval between applications. PAS 100 compost is almost free at the point of production, but the haulage can be expensive, depending on how close you are to your nearest plant (WRAP publish a list of producers). Road-going tipper lorries, especially large, eight-wheelers, will not usually go very far off a metalled road and will only tip on to good firm ground. If you happen to be within tractor and trailer distance of a compost plant then you win twice: you can pick the compost up, transport it cheaply and can probably get it very close to the point of use. To get PAS 100 compost to one Kent vineyard cost around £200 per 20 tonne load using a tipper lorry which was able to get to good hardstanding near to the vineyard site where it could

unload. Twenty loads were delivered (around 400 tonnes) and spread, using a professional contractor who came to the site with spreader, loading machine and drivers. The compost was spread in a strip about 600 mm wide and to a depth of 75–100 mm in the centre of the strip and the 400 tonnes covered 6,900 running metres. The cost of the machinery was £780 (13 hours at £60 per hour), plus £160 haulage each way to and from the site, a grand total of £1,100. The total cost of this (2009 prices) was £5,000 or 72p per metre or 87p per vine planted at 1.20 m apart. Given that these costs are over 12 years old, they probably need increasing by at least 15-20 per cent to today's values. The latest price I had for PAS 100 was £12 +VAT per tonne delivered, but this was to a site quite close to the source. The distance the compost has to be hauled makes a considerable difference to the price.

PAS 100 contains quite a lot of small material (25 per cent is less than 10 mm in size) so it consolidates and breaks down and is absorbed by the ground relatively quickly. In the above example it was really used as a soil conditioner and vine nutrient rather than as a pure weed suppressant and some additional herbicide treatment was required in the second year. However, the spreading rates and costs are probably quite accurate for coarser material which would break down more slowly.

For woodchip, the costs were slightly higher, although the material used was organic. On a vineyard near Guildford, a large articulated lorryload of around 90 m³ of organic hardwood woodchip cost around £1,150 per load delivered and one load covered around 2,000 running metres. It took two people around three days to spread woodchip on 2 ha (5-acres) of vines planted at 2.00 row width (so 5,000 running m-ha) which means that the whole job cost around £4,000-ha (£1,619-acre) which equates to 80p per metre or almost £1 per vine at 1.20 m. This includes the cost of the material, unloading it and stockpiling it, the hire of the spreader and spreading it. In this example, as the woodchip started to weather and break down, particularly in areas where there was a heavy weed infestation, some weeds grew through and had to be pulled out by hand before the situation was bad enough to warrant a second application of material. Once the weed situation came under control, an application of a fresh layer of material was needed around every two years, although the sides of the mulch needed an annual weeding. Mulching in this vineyard was kept going for a few years, but eventually they decided it was costing too much and bought an under-vine weeder. As these are somewhat historic costs, they need to be updated.

In newly planted vineyards the use of loose mulch, such as PAS 100 compost, is probably as good a way as any of making sure that the vines establish well. Not only are weeds suppressed without the use of herbicides, but the vines will get both extra humus and a small amount of slow-release nitrogen to help them develop. Remember that with loose mulches there is the same problem of scion roots as you get with under-vine cultivators, and therefore

2 Waste & Resources Action Programme www.wrap.org.uk

from time to time (every two years perhaps) every vine where the graft is not visible must be checked for scion roots and, if found, they should be removed. Planting the vines slightly higher out of the ground than normal, say, with the graft 75–100 mm above the ground level, would probably prevent this happening. Using high-graft vines certainly would.

Other mulch materials such as spent mushroom compost or hop waste might work just as well, but given that most of the cost is in the transport and spreading, it doesn't really pay to use materials that are second-best. In Germany and France I have seen straw used, but it tends to blow away if it doesn't get well soaked as soon as its spread – probably less of a problem in Britain than warmer regions – and it also takes a long time to settle down and form a weed-suppressing mulch. Unless you are very lucky, there are usually quite a few weed seeds in the straw, which will germinate. Straw will also often attract mice which can damage the shoots of newly planted vines, and will ring the bark on the trunks of more established ones. It is also a fire hazard. However, in low-density, wide-rowed vineyards, straw mulch might work very well and would certainly be cheaper than compost. Side-spreaders which take large round bales are often used on strawberry beds and with some adaptation could probably be made to work in vineyards.

I have also considered the 'mow-and-throw' system much used in the low-cost bulk-wine regions of Australia and California where a grass crop is encouraged to grow in the middle of the rows in the spring and then mown and the cut grass is thrown (by the mowing machine) under the vines as a mulch. Whether this would work completely in Britain's damp summer climate is open to question and there would bound to be some growth of weeds under the vines. However, given that the grass keeps on growing in most summers (unlike in Australia), the constant toppings up of the mulch might help keep the weeds supressed. With systems such as GDC and Blondin, where some weed growth below the vines can be tolerated, this system could be considered. You would need a mowing machine, however, that was suitable and really did lay the cut grass in a tight strip beneath the vines. There are also swing-arm under-vine mowers, but these suffer from the same problem as under-vine cultivators (the propensity of the activator arm to strike the vine on the trunk at the same point each time) and are not to be recommended in most vineyards. In organic and biodynamic vineyards, swing-arm mowers might well be a pragmatic solution to under-vine weed control.

Whatever loose organic mulches are used, they will in time break down and be taken into the soil by worms and other organisms which will raise both the levels of organic matter in the soil, as well as the numbers of worms – both admirable things in themselves. However, loose mulch covering the soil around the vines will tend to keep the soil wetter and therefore cooler, and this is a factor that ought to be considered. As has been pointed out at the start of this chapter, the warmer the soil, the earlier the bud-burst and the greater the production of cytokinins.

Weed control – a summary

Whatever system of weed control you decide to use in your vineyard, there are only two requirements: it has to work and work first time. I have so often seen vineyards in their first season completely swamped by weeds which have got beyond the point of spraying and where

manual weed pulling and hoeing, or mowing and hand strimming, have had to be used in order to get the weeds down to a point where they can be tackled with herbicides. My preferred route is the herbicide one, but this is only my personal preference and with the right equipment and know-how, one of the other methods can be made to work, although none are cheaper or easier alternatives.

Chapter 15

Protected vinegrowing

Protected vinegrowing – a history

Plants have been grown under protection of one sort or another since at least Roman times. Pliny wrote about the use of 'muscovy-glass' (his name for mica) to ripen both fruit and vegetables in the first century AD and in the 1200s the Vatican had what we today would call 'cloches' made from small glass panels mounted in metal frames, similar to stained glass windows. Although fairly primitive, the first true greenhouses appeared in France and the Netherlands in the 1500s and were soon being tried in Britain. The first large heated greenhouses and conservatories appeared in Britain in the early 1600s and became common in the gardens of royalty, the nobility and the well-to-do by 1630 when glass production became industrialised. Exactly when the first vines were grown in British greenhouses is not known for certain, but by the mid-1600s table grapes were being grown and are mentioned in several books. Although John Rose, gardener to King Charles II, who wrote *The English Vineyard Vindicated* in 1666, doesn't mention growing vines other than in the open air, I feel sure that he would have had vines in his greenhouses. He was after all the man who had, in 1671, is alleged to have presented the King with the first pineapple to be grown in Great Britain and therefore had great experience of growing fruit under glass.

Over the next century or so, the growing of vines under glass became something of a craze and dozens of different varieties were being grown and even new ones developed in Britain. In 1769, a Black Hamburg vine was planted at Hampton Court under the direction of Lancelot 'Capability' Brown with a cutting taken from a vine at Valentines Manor, Ilford, Essex which had been planted in 1758. The Hampton Court vine was, until an even older vine was discovered at Maribor in Slovenia, the world's oldest and today it has a girth of 4 m (13-feet) and the longest cordon is 36.5 m (120-feet) long. It often produces over 300-kg of fruit. The Victorian era was undoubtedly

John Rose, who wrote the 'English Vineyard Vindicated' in 1666, presenting King Charles II with the first pineapple to be grown in England

the high-point of protected grapegrowing in Britain and dozens of books were written about how to do it. The best known was written by Archibald Barron, Superintendent of the RHS Gardens at Chiswick, whose *Vines and Vine Culture* first appeared in 1883 and ran to five editions, the last being updated by his widow in 1912. In the final edition there was a chapter on 'Commercial Grape Culture, or the Growing of Grapes for the Market', in which (Mrs) Barron writes about the extraordinary increase in the cultivation of 'grapes for sale or market purposes over the past few years'. The book says that the popularity of grapes is due to the public's sudden demand for tomatoes, which were then becoming very popular. Growers were able to erect glasshouses, plant them with vines and during the first four or five years while the vine was making wood and not producing fruit, crop the glasshouse with tomatoes. Once the vines were established and fruiting, tomato growing would stop.

Wine Growing in Great Britain

Barron says that the centres of table grape production were the Channel Islands, Hertfordshire, Worthing, Finchley and Galashiels in Scotland and that: 'several of the Vineyards or Grapegrowing establishments are of a leviathan character'. One grower, Rochfords at Cheshunt in Hertfordshire, had over 50-acres (20-ha) of glass in total, with half of it devoted to vines and expected to send between 300 and 400-tonnes of grapes annually to Covent Garden. At 2 shillings a pound, this equated to a return of £224 an acre – worth around £63,500-ha (£25,700-acre) at today's values. Grapes were also sent by *fast steamer* to New York and arrived about ten days after cutting in good condition. Sadly, many of these enterprises ended during the First World War when both labour and fuel became short and expensive, although a few growers in Worthing did keep their greenhouses producing grapes until the 1930s. The chapter in Barron ends with a paragraph on *Grape Growing in Belgium and France* and reports on similar vine and tomato establishments to those found in England that sent large quantities to the markets. It ends by saying that the French had put a tax of five francs a kilo on non-French grapes coming onto the Paris markets, the result of which was that the Belgians sent all theirs to the English markets.

Vines in British greenhouses also had their darker side. It was in 1831 or 1832 that Powdery Mildew was first seen in a greenhouse in Margate, Kent and named *Oidium tuckeri* after Mr Tucker, gardener to Mr John Slater, whose greenhouse and vine it was. The first time *Phylloxera* was seen in Europe was in another greenhouse, this time in Hammersmith, London in 1863, when a vine appeared to be suffering a mystery attack. Some damaged leaves were sent to Professor John Westwood, the *Gardener's Chronicle's* 'Insect Referee'. He reported back that he could come to no firm conclusion about the insect 'for want of a knowledge of a male' [sic]. This lack of a male also baffled the French for several years, they too having discovered it in 1864 on some vines in Pujaut, a village a few kilometres north-west of Avignon, belonging to a wine merchant, Monsieur Borty.

In more modern times, glass cloches were used by two pioneers of twentieth century British grape growing: Ray Barrington Brock at his Oxted Viticultural Research Station; and Edward Hyams in his garden at Molash, near Canterbury. In 1952 Hyams wrote *Grapes Under Cloches* which drew heavily ('rather too heavily' Brock told me!) on Brock's two published works: *Outdoor Grapes in Cool Climates* (1949) and *More Outdoor Grapes* (1950). Hyams also wrote *The Grape Vine in England* (1949) and edited *Vineyards in England* (1953), a masterly work which consisted of twenty chapters, many written by different specialists (although Hyams himself wrote six of them and Brock, two) covering every conceivable aspect of grape production in the British Isles. It included chapters on using cloches and the greenhouse cultivation of vines.

In the era of commercial British vineyards, several growers have tried growing vines under cover, mostly under polythene-covered, aluminium-framed tunnels. According to Gillian Pearkes in her 1982 book *Vinegrowing in Britain*, Colonel George Nangle 'pioneered tunnel-grown vines in South Devon', although I can find no other information about the Colonel. One of the earliest commercial growers to use tunnels must have been the Sampsons at Loddiswell Vineyard, near Kingsbridge, Devon. Dairy farmer, Reg Sampson and his wife Betty (who was a well-known amateur winemaker, wine judge and author of *The Art of Winemaking*) had started a trial plot of vines (outdoors) in 1972 and then in 1977 had decided to 'go professional' and they planted 6,000 vines. Following some very wet summers they experimented with polytunnels and found that the vines did well, giving more regular yields and higher sugars. Their wines were well received and they won several medals including the SWVA's 'Best Wine' trophy in 1986 with their 1985 Loddiswell Medium Sweet. According to my 1989 book *The Vineyards of England* they had 'approximately 0.9 acres' (0.36-ha) under plastic. I remember visiting the vineyard one wet summer – we camped near Dartmouth for a hateful week during which it never stopped raining – and remember seeing the tunnels on a steep hill above the farm house. They had closed sides, but open ends, which helped with the ventilation as the Sampsons were concerned lest disease take over. The tunnels were quite low and each contained two rows of vines. I don't believe that there was enough room to drive a tractor inside, so if they were sprayed, it was probably by hand with a lance or knapsack sprayer. The Sampsons always put the quality of their wine, which was very good for the time, down to the grapes coming from their tunnels.

Another Devon grower, who has been growing vines under tunnels for many years, is Mark Sharman at Beenleigh Manor, near Totnes. The Cabernet Sauvignon and Merlot vines were planted in 1979 by Mark's cousin, Marian Ash, who, with her husband and with some help from Gillian Pearkes (who probably supplied the original vines), grew them unprotected until they were established and then covered them. The vines struggled for moisture inside the tunnels, so an irrigation system was installed. The wines produced were, according to Mark 'thin, pale and acidic'. After several years' neglect, Mark bought the vineyard in January 1989 and says he 'managed to produce a small vintage of surprisingly good quality' in that year. Over the years, his wines have won many major awards in the UKVA's competition including the Bernard Theobald trophy for 'Best Red' with the 1990, 1995, 1998 and 1999 wines, the President's Trophy for 'Best Wine' in the small volume section for the 1990, 1995, 1999 and 2001 wines and a Gold Medal for the 2003 vintage – an impressive tally by anyone's standards. Today, the changes in Britain's climate has meant that better red wines are being made out of doors from varieties such as Dornfelder, Pinot noir, Regent and Rondo and red varieties do not need protection. However, Mark told me in 2018 that he still makes between 300 and 500 bottles a year from which sell at around £25 a bottle and he says that the income 'pays for a week's holiday somewhere nice for me and my partner'.

On the Isle of Wight, an island well known for its glasshouse industry, Anthony Goddard grew 0.4-ha (1-acre) of Gewürztraminer in polytunnels on the grounds that he loved the wine from this variety (albeit from Alsace) and as he already had 4.13-ha (10.2-acres) of the standard varieties growing in the open and knew all the pitfalls of growing vines in the open, he thought he would cover his bets. From memory, his tunnels were of a very superior type, around 5.50 m (18-feet) wide and high enough for a tractor to be driven through them. They also had sides that could be lifted up to about 1.00 m off the ground to aid ventilation. The cost, as I recall it, for the 0.4-ha (1-acre) of tunnels was somewhere in the region of £10,000, excluding the cost of the vines and the trelliswork. Although the vines grew well, fruited well and were disease-free, Goddard admitted himself that the grapes were never very special and didn't do much to enhance his wine. My guess is that it was the wrong variety to grow under cover as its acidity is fairly low to start with and the conditions inside a tunnel, even a tunnel in Britain, were just too warm for it.

In the 1980s, Martin Seed at his small vineyard at Worthenbury in Wrexham, North Wales, grew around 0.25-ha (0.60-acres) of Pinot noir, Chardonnay and Sauvignon blanc in small tunnels. He planted the vines between 1991 and 1999 and until around 2005 he grew quite respectable crops of high sugar grapes. 11–14 per cent natural alcohol was regularly achieved and his wines, made at both Tenterden Vineyards (by myself) and at what was then called Valley Vineyards by John Worontschak, won several awards. Because of the northerly position (latitude 53°N) there was always a danger of spring frost, so each row of vines had its own hot water pipe running along it, heated by a gas-fired boiler in one of the tunnels. The vineyard has now been grubbed.

Protected vinegrowing – current practices

Using some form of netting or textile cover over your vines to protect them from frost, poor flowering weather and as an aid to ripening, is something that has been considered often in the past. I recall a New Zealand company marketing something called the 'Kerilea Cloche'[1] which was in effect a mini-greenhouse that covered every row of vines. Whilst it undoubtedly worked, the cost of the frames and supports, the cost of the covering, and the cost of putting it on and taking it off at the appropriate times of the growing season, made it far too uneconomical for British grapegrowers. For other crops – cherries, late-season raspberries and other high-value crops – overhead covers are quite widely used. The Vöhringer company, which make Voen cherry covers, state that they cost around €75,000-ha (£30,000-acre) to install and with cherries give a 15 per cent increased

Potted vines in polytunnels at Haygrove Evolution

crop of much larger cherries which, because cherries are priced according to size, made the crop worth over twice as much and paid for the installation in two years. In vines, the difference might be in having almost no crop to having a good crop and if you say grapes are worth around £1,500-tonne after picking costs, and say you can get an extra 5 tonnes-ha (2 tonnes-acre), an additional income of £7,500-ha (£3,035-acre) per annum might also seem worthwhile. However, this doesn't account for the costs of putting the covers up and down or the cost of replacing the covering. The relatively high cost of these covers is because they require an additional post and wire system to be erected over the plants/trees. In vines however, there is already an existing trelliswork and this is an advantage that could be used.

In around 2012, Richard Smart, in conjunction with a company called Capatex[2] developed a covering made of an 'almost transparent film with special properties for light and UV' which could be attached to the training wires and draped either side of the vines to trap them in an envelope. The film was 400 mm wide and covered the fruiting zone. It was proposed that the vines were first covered 'from around bud break to late May' which 'could help mitigate a late frost'. The covers could then be rolled up and 'stored' on the wires until flowering, when they were put on again and then on again from *véraison* to ripening. The capital cost was around £9,000-ha (£3,642-acre) for vineyards planted with a 2.00 m row width, with annual costs for the three moves of £1,200-ha (£486-acre). The film was designed to last for 'three to four years continual exposure' so by using it as outlined above, the film might well last longer. Therefore, assuming that you got five years use, the annual cost was around £3,000-ha (£1,200-acre) which – if it worked – might be quite cost-effective. An initial report from Richard Smart, based upon covering one 100-metre row of Seyval blanc in 2013, showed a very significant increase in yield (up by a factor of three) with sugar and acid readings about the same as outside uncovered grapes. I am aware of a few growers who trialled these covers,

---

1 The Kerilea Cloche is today sold by Redpath in New Zealand www.redpath.co.nz

2 www.capatex-agro-textiles.com

Wine Growing in Great Britain

Simon Day unearthing the terroir of his potted vines

but they all found it too much additional work to keep raising and lowering the plastic, plus wind damage was an ever-present problem. I know of nobody now who uses them on anything like a full scale.

Although I personally don't think that protected growing of vines to make what one might call 'normal' wines will ever take off in Britain, I can quite see that some British growers might consider using one or other of the above systems of protection, especially if they were in one of the more challenging parts of these islands. Maybe if you were starting a new vineyard, where you could allow the space needed for the supports and could make the structure used to carry the overhead covers part of the overall trellising system, then the cost of the 'cherry cover' type of protection could probably be lessened. Once one grower proved that better and larger crops of grapes suitable for winemaking could be achieved on a regular basis, then this type of protected growing might well be taken up by others.

In a relatively recent development, Simon Day, of Sixteen Ridges Vineyard, is working with Haygrove in Ledbury, Herefordshire, to grow a range of vine varieties in pots under cover. Haygrove are not only one of the world's major manufacturers of polytunnels, but also a substantial user of them for growing soft fruit on a large scale all around the world. Some of the varieties Day is trialling are ones that we don't currently grow out of doors in Britain – Cabernet Franc, Cabernet Sauvignon, Gewürztraminer, Merlot, Riesling and Sauvignon blanc – and he is getting some good results. Time will tell whether this system is financially viable and much will hang on the quality of the wine. It wasn't that long ago in Britain that blueberries, cherries, raspberries, strawberries were all outdoor crops. Today, most large-scale commercial growers of these fruits wouldn't dream of it and in fact, many supermarkets will only take cherries and strawberries if they have been grown under cover. Maybe, if both yields and quality are good enough, vines in pots under polytunnels will become more widespread?

Chapter 16

Pest and disease control

Pest and disease control

Note:
- The term 'pesticide' covers all plant protection products used to control pests, diseases and weeds in the vineyard and immediate surroundings.
- Before buying or using any controlled pesticide, advice should be sought from a BASIS registered adviser.[1]
- All operators of equipment for spraying or otherwise applying pesticides are required to have the necessary qualifications and/or supervision before using any controlled products.
- Before using any pesticides, spray operators are required to read the statutory information contained on the label of the product or, for off-label products, the relevant EAMU[2] approval notice. For vines, these notices are contained in the WineGB 'Green Book' (see below).
- The advice given below relates to pest and disease control on grapes grown in the open air for wine production. Products used on grapes grown under protection (glass or polythene) or table grapes for eating may be subject to different rules on dilution and application rates and Harvest Intervals (HI).
- The storage of pesticides and the disposal of pesticides, pesticide residues and empty containers must be carried out in an approved manner.
- The official British 'Code of practice for using plant protection products' (PB 11090) can be downloaded from the DEFRA and HSE (Health and Safety Executive) websites www.pesticides.gov.uk
- By law, all users of pesticides are required have a written Integrated Pest Management (IPM) plan, although many smaller growers (and even some larger growers) do not. DEFRA is reviewing this requirement in 2020 and is likely to reinforce it rather than weaken it, and growers who do

not have one can be fined. In the event of a spray-drift event, where a neighbour complains about a grower's spraying practices, the lack of a written IPM plan would definitely count against the grower.
- WineGB's *Full & Label Extension Approvals* (known as the 'Green Book') is published annually and lists all approved materials, plus suggested spray programmes for conventional, organic and biodynamic producers. It also contains a considerable amount of detail about pest and disease control and other matters concerning the use of pesticides. It is free to WineGB members, but not available to non-members.

Notifiable pests in vineyards

There are currently two pests and one disease which, if discovered in vineyards, must be notified to DEFRA:
- Grape Phylloxera (*Daktulosphaira vitifoliae*)
- Spotted Lanternfly (*Lycorma deliculata*)
- *Xylella fastidiosa* (Pierce's Disease).

Introduction[3]

All other things being equal, a mastery of pest and disease control (particularly disease control) is probably the most important skill a vinegrower in Britain can acquire. Why? Because Britain has plenty of rainfall, often during the height of the growing season; because our cool climate fosters vigorous shoot and leaf growth at the expense of fruiting; and our canopies are often shaded. These are all ideal conditions for the development of fungal diseases. Pests and diseases not only damage and destroy the current crop, they can also damage the future crop by reducing the fruiting potential of the buds for next year. This is one of the reasons why yields in vineyards in Britain are (a) very variable and (b) relatively low. However, pest and disease control are not only about chemical control. Under what is termed an Integrated Crop

---

1 BASIS is an organisation that manages the Professional Register for qualified pesticide advisers. www.basis-reg.com
2 EAMU stands for 'Extension of Authorisation of Minor Use' and is discussed more fully later in this chapter

3 Pest and disease control in organic and biodynamic vineyards are discussed in Chapter 18 – Organic and biodynamic viticulture in Britain.

Management (ICM) approach to the problem, vineyard hygiene, vineyard layout, canopy management and under-vine and alleyway management are also part of the mix and these will also be discussed in this chapter.

ICM (and its counterpart, Integrated Pest Management – IPM) are terms often used, but which have a myriad of different meanings. DEFRA, on its Integrated Crop Management website, offers several meanings, one of which I believe best suits British vineyard owners. It states that ICM is: 'a management system which employs controlled inputs to achieve sustained profitability with minimum environmental impact, but with sufficient flexibility to meet natural and market challenges economically.' Therefore, whilst a reduction of inputs in the form of pesticides, fertilisers and other carbon-based products (fuel for instance) is a desirable aim, and care of the environment and its inhabitants an obvious goal, one has to bear in mind the 'profitability' of the enterprise and the 'market challenges' that confront it. One could have a vineyard in Britain planted entirely with a proven disease resistant variety (Seyval blanc comes to mind) and use mechanical weed control (supplemented by hand hoeing when required) and carry out most operations using manual labour and maybe horse-power. This would certainly be a very environmentally sustainable vineyard in the sense of having the lowest inputs for the maximum outputs, but whether this would be the best way of creating a profitable, sustainable enterprise is open to conjecture. In 2020, LEAF (Linking Environment and Farming) published a new booklet *Simply Sustainable Integrated Pest Management* which is well worth reading. It covers the 'Eight Simple Steps' to IPM and allows you to score yourself 'Poor', 'Medium' or 'Good' against easily understood and well laid out criteria. The leaflet is available from www.leafuk.org or from www.englishwine.com/wgigb.htm.

WineGB instigated a *Sustainable Wines of Great Britain* (SWGB) scheme in 2020 which issues a Certification Trade Mark to those growers and producers whose vineyards, wineries and wines have been approved. The aim of the scheme is 'to minimise our impact on the environment and promote environmental conservation and biodiversity'. Members of the scheme will be audited on entering the scheme and then every three years. A wine made in a certified winery and containing at least 85 per cent of grapes coming from certified vineyards, will be able to carry the 'Sustainable Wine of Great Britain' logo. To date, some 30 growers and wineries have signed up to the scheme. The objectives of the scheme (see text box) are laudable and one would hope relatively easily achievable in well-managed vineyards and wineries, although there is probably rather too much paperwork and record keeping for some. I counted twelve separate diaries, maps, plans and records which have to be kept. One element which has been missed out under 'Reduce (and optimise) pesticide inputs' is to suggest using disease-resistant vine varieties. Under the LEAF scheme mentioned in

---

Sustainable Wines of Great Britain
Vinegrowing
- Maintain and improve soil health
- Manage vineyard canopies and yields optimally
- Reduce (and optimise) pesticide inputs
- Conserve the vineyard (and surrounding) environment and promote biodiversity
- Reduce vineyard energy input, greenhouse gas emissions and carbon footprint per hectare

Winemaking
- Improve winery design to reduce environmental impact
- Increase energy efficiency to reduce the energy footprint per bottle of wine
- Reduce the volume of water used per bottle of wine
- Deploy an environmentally responsible wastewater management system
- Reduce the environmental impact of wine packaging
- Reduce the level of greenhouse gas emissions and carbon footprint per bottle of wine

---

the preceding paragraph to this one, 'disease resistance is a key consideration when choosing varieties'. Maybe this was a step too far for the SWGB committee?

There is sometimes a reluctance on the part of British vineyard owners to realise that by planting most varieties of vines in what is a relatively inhospitable climate they have signed up to the task of looking after them with some form of chemical control, even when they have chosen to grow their vines organically or biodynamically. And, if they are keen to safeguard the considerable investment of time, energy and money they have ploughed into their enterprises, they must become good at the one task which can really help them do this, that of adequate pest and disease control. I quite understand the reluctance to spray – it is messy, time consuming and expensive – but in my experience it is one of the major factors between success and failure in growing grapes in Britain. Whilst it may be desirable to spray as little as possible and to act on information about the likelihood of disease due to climatic factors, the reality for most vineyards is that spraying is like house insurance. You don't wait until you see the first signs of subsidence or the first flickers of flame before you take out house insurance: you act pre-emptively and insure against loss every year.

Vineyard owners must also be aware of the 'honeymoon period'. Like many marriages, everything in the vineyard can often be hunky-dory for the first year or two (or even three) and when vines are young, root systems are still developing and the amount of wood left at pruning may not be filling the fruiting wire, disease pressures may well be light and crops can be exceptionally clean and ripe and of good quality. However, a few years down the line, problems start and you have to work a bit harder. When the

vines have got a bigger root system with more access to water and nutrients and you have laid down the full complement of fruiting wood, the canopy can suddenly become crowded and diseases will very quickly start to show. Be prepared for this and be prepared to spray preventatively.

Pests and diseases fall into several distinct categories. There are pests such as rabbits, hares, deer, badgers, foxes, wild boar and any other mammals that take a shine to vines and grapes and these have been dealt with in *Chapter 6: Site planning and preparation* in the section on *Fencing*. Luckily for British vineyard owners, many of the wide range of insects and mites which, in their various forms, can damage vines and grapes, are absent from our shores – at least for the present – and the very few that are found in British vineyards are discussed in this chapter. Wasps, snails and slugs are then covered, followed by *Phylloxera*, which is probably the worst pest of all, since attacks are nearly always fatal if vines are grown on their own roots, a practice most commercial growers avoid in regions where *Phylloxera* is present (as it is in Britain). The final pest to be dealt with is birds, not quite a pest in the same sense as the others, since eradication is not a solution, but nevertheless, a pest that can be extremely costly.

Diseases too, fall into several distinct categories. There are the fungal diseases, of which the three major ones – Powdery Mildew, Downy Mildew and botrytis – are discussed in detail, plus a range of others less commonly seen in British vineyards, but nonetheless to be guarded against. Finally, viruses are covered. Trunk diseases (TDs), some of which have been present in British vineyards for many decades and others which have appeared more recently and which have been the subject of considerable debate amongst British vineyard owners in the past decade, are dealt with in *Chapter 17: Trunk diseases*.

Vineyard conditions

The job of good pest and disease control – and more so with diseases than pests – starts with getting the right physical conditions in the vineyard. Selecting a low-vigour rootstock will, for most varieties, be the first important step towards creating an airy and open canopy. Your overall canopy height should be appropriate for your row width so that the fruiting zone gets the maximum amount of light and is not shaded. Your fruiting wire should be at the appropriate height for the row width and kept tensioned so that the vines can be tied down correctly. Your training (catch) wires should be fitted with chains so that moving them up and down is easy and you should consider wire spreaders if you feel that your canopy is too dense and could be improved by being opened up.

Pruning must be correctly carried out so that the canes do not overlap and the spur for the replacement cane is below the fruiting wire (assuming a two-cane flat *Guyot* pruning system). Although there is some debate about its contribution to disease control, some growers feel that burning the prunings, either in the vineyard in a *chariot de feu* or taking them out and burning them on the headland

or composting them, helps keep the vineyard healthier.[4] In addition, if TDs are of a concern, then painting the pruning wounds (if permitted) may also be helpful.[5] When vines get older, and the wood to be cut out gets thicker, the use of powered secateurs helps pruners to make the pruning cuts exactly where they want them and helps keep pruning wounds clean and at the correct angle. This reduces the chances of TDs entering the vine via wounds and makes wound painting more effective The aim should be to keep a weed-free strip of around 600–750 mm under the vines at all times as allowing weeds and grasses to build up directly beneath the vines will only increase the humidity in the fruit zone. Your alleyways, unless you are growing a green manure crop, should be kept mown short during the growing season, right up until harvest, so that moisture retention is at a minimum and the vineyard floor dries out as quickly as possible.

Two to three weeks after bud-burst, if the crowns of your vines are crowded with excess shoots, selective shoot thinning should be carried out to open up this area. This will help get air and light on to next year's replacement canes. Canopy management, in the form of deleafing and shoot trimming, is an essential part of disease control and the aim must be to keep the canopy thin, permitting light and air to get on to both the crop and the canes for next year's fruiting wood and to aid maximum air circulation and spray penetration. An open, airy and light-penetrating canopy is the foundation of good pest and disease control. The advice given in the 2011 Luxembourg study on methods of controlling botrytis, already mentioned in the section *in Chapter 12: Management – cropping years* in the section on *Deleafing*, is worth repeating as I believe it reinforces the rôle of canopy management in disease control:

'Manual leaf removal in the cluster-zone just after bloom [flowering] provided a significant reduction of bunch rot infestation without any input of chemical substances. Subsequently, this treatment can be recommended as an important tool in any bunch rot protection strategy for integrated as well as organic viticulture.'

Nutrition

The nutrition of your vines, whilst really a secondary consideration, can also play a part in the health of your vines if things are not correct. The leaves of your vine are their power-plant and unless they are well looked after, well maintained and running at full speed, the vine will suffer. Apart from grapes that are less ripe, stored reserves of carbohydrates will be lower and consequently, in the spring and at times of stress (which in Britain, with our cool, wet springs can often occur together), your vines, like humans

---

4 Please note in England you need to apply to the Environment Agency for a 'D7 Exemption Certificate' if you want to burn prunings in the place where they were produced. In Wales you also need to apply for one, but for a slightly different permit.

5 There is more on wound painting in Chapter 17 Trunk Diseases on Wound paints.

who are under-nourished and over-stressed, will be more likely to succumb to diseases. In particular, just because the crop is nearing ripeness and even after picking in years when the harvest is not very late, the vine can still continue to photosynthesise and can produce carbohydrates until the leaves senesce and fall off. The leaves must therefore be kept in good condition so that photosynthesis can continue for as long as possible. Obviously making sure your vines are well balanced with the correct nutrients, both major and minor, will help keep the leaves in perfect working order. As already mentioned in *Chapter 7: Vineyard nutrition*, keeping your vines adequately supplied with trace elements and spraying with products containing phosphites (such as Frutogard) will help the vine build a natural resistance to diseases.

Machinery requirements

In *Chapter 10: Machinery and equipment* I said that: 'Your sprayer is the most important tool in the vineyard and will help you win the battle to keep your vines and your crop clean and pest and disease free' and for good measure added: 'Do not try and economise' on your sprayer. Both of those statements are true. Apart from your tractor, which also needs to be reliable and trustworthy, your sprayer is really your insurance against crop loss and must perform on demand. Unless you have a large vineyard, your sprayer will probably only get used for one or two days per fortnight and in dry years, even less than that. In addition, it will sometimes be used to carry and spray corrosive chemicals and therefore must be kept clean and well maintained.

There are a large number of sprayer types and sprayer manufacturers and the type you buy will depend upon several factors. The size of your vineyard and vineyard layout may well determine whether you have a three-point linkage mounted or trailed sprayer, although your budget may also be part of that decision-making process. My advice is that you want to try and cut the non-spraying time of your spraying day down to a minimum and this means keeping the filling time to a minimum. Given that you have a spray tank of water conveniently situated in or near the vineyard – a tank mounted on a stand which can, in a few minutes, deliver under gravity the same amount of water that your sprayer tank holds – then mixing up materials and filling your sprayer can be kept to the smallest time possible. Therefore, when buying a sprayer, try and get the largest size tank that you can afford and which suits the area of vines you are likely to have to spray. Most sprayers work in much the same way, having a pump which pressurises the water (which carries the pesticides) in the system, and blows it out through nozzles which are facing the crop. There is also a fan which propels the spray coming out of the nozzles, and forces it onto the crop and into the centre of the canopy. There are sprayers with other modes of action (described below), but the one just described is by far the most common in vineyards.

Mounted sprayers seldom have tanks larger than 600 litres as a fruit or vineyard tractor's three-point linkage will not carry more weight, whereas trailed sprayers can have tanks up to 3,000 litres. Mounted sprayers are of course cheaper as they do not have a chassis and wheels. They can also turn in much less space so are suitable for the smaller vineyard or in vineyards where the headlands are tight. However, they are more difficult to hitch up and attach to a tractor, whereas a trailed sprayer is easier and quicker. The size of pump and fan that you need will depend on the row width and layout of your vineyard, and the type of sprayer you have. Sprayers with axial fans, with the nozzles placed around the outside of the fan, typically have fans around 600–700 mm in diameter which is fine for 2.0–2.5 m wide rows. There are also sprayers with 'snakes' that can be directed to different parts of the canopy where the air is delivered down the snakes with the spray coming out of fingers placed in the air stream. These tend to have smaller fans – around 450–550 mm. Finally, there are also sprayers with tangential flow (or cross-flow) fans mounted vertically. Despite claiming various benefits – reduced drift and better penetration – sprayers with this type of fan are seldom seen in vineyards (although some of the tunnel sprayers, mentioned later, do have cross-flow fans). For the larger vineyards there are also sprayers of different types that can spray three or four rows at once. Pictures of sprayers

On most sprayers, pump output, spraying pressure and fan speed are adjustable, so it is worth remembering that you can always reduce the air output from a larger fan: you cannot increase it from a smaller one. Many sprayers, designed principally for orchards with wider rows that vineyards, produce too much air and growers might find that using something like the Andrew Landers 'Cornel Doughnut', especially early in the season when the vine's canopy is less dense, can improve spray penetration.[6] Most modern sprayers have dual nozzles which can easily be switched from one output volume to a lower or higher one (or turned off completely) making it very simple to adjust sprayer output to the canopy size and density. The canopy will expand as the season progresses, requiring larger and larger liquid volumes, and selecting the correct nozzles for the right product or occasion is important and switching between air-induction nozzles and hollow-cone nozzles is all part and parcel of effective sprayer use.[7] The best advice is to go to a dealer who is used to dealing with orchard and vineyard sprayers and seek their advice. Under the 2012 Sustainable Use Directive (SUD),[8] your sprayer will have to have had an inspection by 26 November 2016 (unless it is under 5 years old) and pass an MOT, so keeping it serviced on a regular basis is essential and having a helpful mechanic on your side will pay dividends. Sprayers will then require an MOT once within five years up to November 2020 and every three years after that. New sprayers will need an MOT once before they are five years old. Full details can be found on the National Sprayer Testing Scheme's website www.nsts.org.uk

---

6 See Andrew Landers' paper Improving spraying efficiency which is available at www.englishwine.com/wgigb.htm.

7 The French firm Albuz are the leaders in nozzle technology. http://www.albuz-spray.com/en/

8 See www.englishwine.com/wgigb.htm for this document.

In *Chapter 10: Machinery and equipment* I said that in recent years, so-called 'tunnel' (or 'recovery') sprayers have been used in British vineyards with the Lipco[9] brand being the most popular (although several manufacturers make them). Although they are expensive to buy and really only suitable for the larger vineyard, they are undoubtedly cost-effective in the long-term, owing to the amount of spray material they save. Andrew Landers, in his book *Effective Vineyard Spraying*, states that a saving of an average of 35 per cent (much more in the early growth stages) can be made. Given that spray materials will come to at least £1,000-ha (£400-acre), then on 10-ha (25-acres), this is an annual saving in this example of £3,500 a year. Also, because much of the spray is recovered, you can work for much longer between fill-ups, thus saving time and expense. However, they are not universally popular.

Whether tunnel sprayers are as effective at controlling disease is a question often asked. Some say that because the spray is re-circulated, two things happen: one is that the spray gradually loses its potency and effectiveness by this repeated use; the other is that as the sprayer passes around the vineyard, it tends to spread disease about. Landers says in his book (p. 92–6) that these are stories put out by competing sprayer manufacturers, and are not true. Whilst a loss of 4 per cent concentration in the spray has been observed, this is well within the tolerances allowed by spray manufacturers when setting rates (such as in case of over-filling) and if fungal spores are being spread, they are being spread by a droplet comprised of a fungicide, so unlikely to thrive. Some makes and models of tunnel sprayers also do not have fans inside them, relying on hydraulic pressure only to carry the spray into the centre of the canopy. Whether this is sufficient in a crowded, dense canopy such as is often seen in British vineyards is also open to question and I would always advise getting a sprayer with cross-flow fans. It should also be noted that for tunnel sprayers without fans, under the LERAP (Local Environment Risk Assessment for Pesticides) scheme, the default buffer zone (of which, more later) for pesticides for which a buffer zone is required, is 5 m from any watercourse (wet or dry) and cannot be reduced even following a LERAP assessment. Full details can be found in the LERAP *Broadcast Air-Assisted Sprayers* guide (PB 6533) available from the DEFRA website.

Vineyard weather stations

In some vineyards, growers have installed vineyard weather stations which monitor a large amount of weather data and, in theory, can be used to predict the likelihood of certain diseases. Using variables such as rainfall, temperature, humidity, leaf wetness and wind speed, these weather stations claim to be able to predict when Powdery and Downy Mildew and botrytis are likely to occur. This gives growers advance warning of likely disease events, enabling them to time their sprays for maximum effect. However, the

sensors can only sample the situation in one spot in what might be quite a large area of vines, and therefore there are bound to be inaccuracies. I also suspect that the algorithms most vineyard weather stations are using are not based upon either weather data or vineyard conditions in Britain and therefore greatly mis-read the likelihood of disease. Although we are wetter, cooler and windier than many warmer climate regions, this does not mean that we suffer from disease as much as the data would suggest (see the section on botrytis below). The weather stations that I see tend to suggest much more frequent spraying than in reality actually occurs, which is quite the reverse of what is intended. On balance, having a weather station in your vineyard is something of a luxury and there are plenty of good growers who do not have them. Whilst they can record all the weather data you will ever need, how you use it is another matter. For frost sensing, a standard weather station is probably not that suitable and the requirements are discussed in *Chapter 13: Frost protection*. A weather station will be useful if you have problems with spray drift in that it will help confirm what the wind speed was at the time of spraying should you need this as evidence.

Spraying practices and techniques

How you spray, what you spray and when you spray will depend very much on a range of factors. The sprayer you have, the vine varieties you are growing, and the plant densities, row widths and trellising systems will all play their part in determining your spray programme. The weather during the year will also be critical in determining both spray frequency and timing. Dry years will be generally good for growers and disease pressure will probably be lessened, although do not get lured into the trap of thinking that because there has been little rain, there will be no disease. Powdery Mildew is spread by wind and actually favours hot, dry humid conditions. In general, one should aim for disease prevention rather than disease control. Keeping vines clean, at least in the early part of the season, up until the start of flowering ought to be straightforward. Once the canopy fills out and side-shoots start to appear, spray penetration will be more difficult and timelier deleafing and shoot trimming will help open up the canopy to light, air and sprays.

The first task of good pest and disease control is to get help and advice from a BASIS registered adviser for your vineyard, your varieties and your sprayer. He or she can then produce a suitable spray programme which you can, at least to start with, stick to. Your spray programme should be designed to take account of the varieties you grow and if you are growing disease-resistant hybrids your spray requirements will be very different from the one you would use if you were growing disease prone varieties. However, in general terms the spray programme will be much the same for all *vinifera* varieties. Once you get through a cropping season or two, you will hopefully get a feel for how your vines and your site behave in respect of pests and diseases and learn to adjust your spray programme accordingly. In time, and with experience, you will discover which blocks

9 117: www.lipco.com/english/products/TUNNEL_Sprayers/tunnel_sprayers.htm

require less spraying, which require more and which weather conditions favour your vines and which don't. Until then, spray first and ask questions afterwards. Of course, this is not to advocate spraying too often, or using more material than necessary – the cost of time and chemicals is high enough not to want to waste them. But allowing diseases to develop and then attempting to halt them, let alone rid your vineyard of them, is not a recipe for an easy life and certainly not a way of saving time or money. Timing is often the key to good disease control, all other things being equal. The saying 'the difference between a good farmer and a bad farmer is ten days' is nowhere truer than in the area of disease control.

Standard spray programmes generally allow ten to fourteen days between sprays, but this interval needs to be adjusted according to the prevailing weather conditions and disease pressures. If it so happens that the day you intend to spray on is going to be at the start of a five-day wet period, then you might need to bring the date forward and perhaps spray on day nine or day eleven rather than leave it until it stops raining. There may also be occasions when disease pressure is very high and the vines are growing rapidly when a seven day spraying interval might be appropriate. If, on the other hand, the weather is dry, the vines are in balance and the canopy management is up to scratch, then allowing the interval between sprays to lengthen may not cause any problems. Of course, the size of your vineyard and the time it takes for you to hitch up, fill up and get a spray on your whole vineyard will also dictate when you spray. With a small vineyard that can be sprayed in a few hours, finding a rain-free gap to get a spray on with time to let the spray dry and be absorbed in the case of systemic materials, is far less a problem than in a large vineyard which might take a day or two (or more) to spray completely.

Using the right water volume for the growth stage and for the disease is also often a large part of getting good disease control. Using sulphur to control Powdery Mildew requires good coverage with spray, as sulphur is a contact material and has no systemic action. This is especially true in the later growth stages when canopies can become crowded and shaded. In Germany, the following recommendations are made:

- Budburst to appearance of inflorescences: 100-400 litres-ha
- Pre-flowering: 200-800 litres-ha
- Mid-flower to post-flower: 250-800 litres-ha
- Post-flower: 300-800 litres-ha
- Pea-size: 1,400-1,600 litres-ha
- *Véraison* and post-*véraison*: 1,600 litres-ha

The range of volumes is given to accommodate different row widths, different trellising systems and canopy management techniques. If your rows are 1.5 m wide, then per hectare volumes are going to be higher than if your rows are 2.5 m wide. Likewise, with a well-trimmed, deleafed canopy, lower water volumes will be required than with a crowded canopy. In Britain we tend to talk about low volume and high volume. Low volume (LV) spraying is taken to mean 200-600 litres-ha (18-54 Imp. galls-acre) and high volume

(HV) spraying at least 1,000 litres-ha (89 Imp. galls-acre) and often as high as 1,200 litres-ha (107 Imp. galls-acre). In 1977, when I started farming, 100 gallons an acre (1,125 litres-ha) was always considered 'normal' for spraying apples and pears with a full canopy. However, in most vineyards, even when vines have a full canopy, what might be termed a medium volume of around 650-800 litres-ha will be adequate.

In Germany they also express spray amounts not as in Britain, where fixed amounts of product per hectare are given, but in quantities per 100 litres of water. This means that as the water volume per hectare required to spray the ever-growing canopy rises, so does the amount of product, keeping the concentration of product that hits the vines the same. Therefore, with a product such as Switch, just after flowering, a German *winzer* might use 180 gm of product in say 500 litres-ha of water, whereas in the final spray before the start of the 21-day HI, the amount of product would have risen to 960 gm in 1,600 litres-ha of water. In Britain, the recommendation for Switch is 800 gm-ha in at least 500 litres of water, whatever the state of the canopy.

Apart from lengthening spray intervals, another way to save on both sprays and time is, in the early stages of the season, before the first tucking-in, is to spray more frequently, but spray every other row. Assuming your row widths are not too far apart – I would not consider this technique would work well with a 3.50 m GDC vineyard – and you have (a) a sprayer with a suitably sized fan, (b) a sheltered vineyard and (c) the wind is not blowing too hard, then perhaps spraying every other row on a seven day cycle might save a couple of spray rounds over the whole season without jeopardising vine health. Of course, you need to remember which rows you have sprayed (having every row numbered definitely helps with this) so that you don't end up spraying the same row twice.

Weather conditions

When you spray must also take account of the prevailing weather (especially wind) conditions in order to avoid spray drift. The 'Code of practice' (PB 11090) suggests spraying when the wind speed is: 'a steady 3.2–6.5 km-hour' (light breeze) and blowing away from 'any sensitive areas or neighbour's land'. It also suggests avoiding spraying: 'when there is little or no wind', 'when there are low winds on warm sunny afternoons when humidity is low' and 'when temperatures are above 30°C'. And of course, when it's not likely to rain! Vineyards with neighbours, especially ones not involved with agriculture, need to be careful on two counts. Obviously (and legally) it is unwise to allow spray to drift from your land onto neighbouring property if people (especially children) are likely to be in adjacent fields or gardens. The same goes for livestock, domestic animals and horses, donkeys etc. By the same token, it is always advisable to tell your neighbours, domestic and agricultural, that if THEY are using sprays, especially weedkillers, that they also have obligations with regard to spray drift and any damage arising from it. As has already been pointed out in *Chapter 3: Site selection* under the heading *Other factors*, greenkeepers on golf courses, people with pony paddocks and farmers spraying hay fields in the early part of the season, especially against thistle, nettles and docks, love using cheap and cheerful hormone weedkillers which unfortunately can vaporise some hours after application and drift several hundred yards to neighbouring fields. This can cause widespread damage to vines. It is always best to point this out to neighbours and just gently ask them for the name of their insurers 'just in case' you need to make a claim.

Legal requirements

As might be imagined, there are many regulations governing every aspect of pesticides – purchase, handling, storage, use and disposal – and every grower must make sure that they are fully up to speed with the current ones. You should also note that under certain conditions and when using certain products, it may be necessary to give notice to beekeepers, neighbouring property owners and to put up notices to warn visitors and footpath users that spraying is taking place or that spraying has recently taken place. There may also be 're-entry' periods (see below). Pesticide users must also have somewhere to fill and clean the sprayer and this must be so constructed that any spills of products, unused spray or wash-down water does not reach a watercourse. Collecting these and having them removed by a licensed contractor is probably the easiest way for the small producer to deal with these legally.

Operators

All users of pesticides in a professional capacity have, as of 26 November 2015, been required to be trained and hold the relevant 'certificates of competence'. These are issued by the National Proficiency & Training Council (NPTC) and most vineyard owners will require the following certificates: PA1, the foundation module which covers the safe use of pesticides, environmental factors and legislation; PA3A for broadcast sprayers with air assistance, i.e. most fruit sprayers; PA3C for boom sprayers without air assistance, i.e. most weed spraying booms; and probably PA6 (option A or AW) for hand-held applicators, i.e. knapsack sprayers, both motorised and non-motorised. If you were born prior to 1 January 1965, 'grandfather rights' existed which exempted you from these requirements, but these have now been ended, and whatever your age, you must obtain the correct certificates for the equipment you are using. The courses and tests mentioned in this paragraph are offered by several different providers and full details can be found on the NPTC's website. In any event, before taking any NPTC courses, the precise equipment to be used must be discussed with your training provider. Such things as axial flow fans, CDA (Controlled Droplet Application) and ULV (Ultra Low Volume) sprayers may well require particular modules.

Under certain conditions, a person holding the relevant certificate can supervise someone who has not yet obtained their certificate, but you must: 'be able to see the person doing all parts of the job including: preparing and mixing the pesticide; filling equipment and making sure the dose levels are correct (calibrating); applying the pesticide; and cleaning equipment and disposing of washings, leftover pesticides, and the containers'. In other words, you cannot 'supervise' unless you are physically present during the complete process of using the pesticide. As might be imagined, there are large fines and even imprisonment if you fail to comply with any of the above. If you are using a contractor to apply pesticides (which includes weedkillers), you must make sure that the operator (and not the contractor) holds the relevant certificates and always ask for written confirmation that they do hold them. If you do not, and something goes wrong, it might count against you in the event of a prosecution or an insurance claim.

Products controlled by the pesticide regulations may only be bought if the purchaser is sure that they are for use by someone holding the correct certificates. After 26 November 2015 it is an offence to purchase controlled products if the buyer has not ensured that the user holds the correct certificates. Suppliers of pesticides may also refuse to sell you products if you cannot show them the correct certificates.

Products

There are many different products that can be sprayed on to, under or around vines, including pesticides – a term which covers weedkillers, fungicides, insecticides, acaricides, molluscicides and probably a few other –icides as well – plus wetters, spreaders, foliar feeds, fertilisers, physical barrier-forming products, growth regulators and various other materials. Although not all of these will be covered by the pesticide regulations, it is safest to adopt a policy of 'everything is illegal unless specifically approved'. Just because a manufacturer suggest that a product is permitted, doesn't always mean it is, especially at the less-chemical end of the spectrum. Those products that have clearance on vines, and are definitely allowed, are all listed in WineGB's annual 'Green Book'.[10] Here you will find all products that either have a full label approval, i.e. their use on outdoor grapes will be on the product packaging itself and on any data-sheets about that product, or the product will be what used to be called a SOLA product (Specific Off Label Approval) but which is now called an EAMU product. These products will not have outdoor grapes listed on their packaging and any user must, by law, be in possession of the appropriate EAMU documentation (all the current ones are in the 'Green Book') and follow the instructions for such things as dose rate, water volume, tank mix for specific application, maximum number of applications, minimum interval between applications, method of application, re-entry period and HI. Products also have to be both stored and disposed of in the correct manner and the relevant section in the 'Code of practice' (PB 11090)

---

10 The 'Green Book' is officially called: Full Label and Extension Approvals and is published by WineGB.

should be consulted. There is a wide range of products which have 'all edible crops' approval and which might be suitable on vines. Before using these, it is advisable to ask the supplier and/or manufacturer for their specific approval before using them on outdoor wine grapes.

Adjuvants – spreaders, wetters, surfactants, buffers

There is a wide range of additives which can be added to the spray tank in order to do a number of different things: enhance pesticide performance; reduce drift; lower the surface tension to help adhesion; reduce spray droplet size; help create bubbles if using air-induction nozzles; help the spray stick on to the target and make it more rainfast; and change the pH of the solution (buffering). Some products already have adjuvants added and other products work better if the correct adjuvant is added to the tank. Protectant and contact products sometimes work best without adjuvants as they are designed to work by remaining on the surface of the leaf, although latex-based adjuvants can assist some contact materials in sticking to the surface of the leaves. Before using an adjuvant, it is best to obtain advice from your BASIS adviser as not all products are compatible with all adjuvants. The legislation governing adjuvants states that all permitted products of this type can be used up until fruitset, but after that date specific conditions apply for each product. Oregon State University has a good fact sheet on Agricultural Spray Adjuvants which can be found on www.englishwine.com/wgigb.htm.

Re-entry periods

Re-entry periods – the time between product applications and re-entry of people (owners, employees, and visitors) into the vineyard – have started to appear. To date (2020) there is only one product which has one – Unicorn DF has a 7-day re-entry period – during which time nobody may handle the foliage or the crop. This is undoubtedly the first of more re-entry periods to come. Rules regarding spraying when there are footpaths crossing or adjacent to vineyards have also yet to be agreed. It is good practice, however, to put up notices warning people with access to vineyards, that spraying is about to take place, is taking place or has recently taken place. In that way, you have everything covered. You should also warn people that their dogs should be kept on a lead in case the dogs decide to lick, eat or brush against recently sprayed vines or weeds.

Harvest intervals and residue levels

Most controlled pesticides have a harvest interval (HI) which is the number of days which must elapse between the last application of the product and when any crops may be picked. The main reason for having a HI is for the safety of the public so that pesticide residues do not get into the food chain. However, this is not always the whole story. Teldor, for instance, an anti-botrytis fungicide, has a twenty-one-day HI for grapes, but for strawberries, raspberries, blackberries and loganberries its only one day and for currants, black, red and white and for gooseberries, the HI is three days. The reason? On fruits mainly eaten raw and sold very much on the basis of appearance, damage by botrytis is financially painful and Teldor helps prevent it in transit, storage and display until purchase. Currants and gooseberries are usually cooked before eating, so that, plus three days HI is deemed to be enough to get rid of any residues, but on grapes, the problem is that residues would have a damaging effect upon yeast growth and therefore might upset fermentation, especially if a low-vigour, naturally occurring, yeast was being used. HIs are not always completely about public safety.

The HI can also depend upon the rate of product used and/or volume of water applied per hectare (the concentration rate). Mancozeb for instance has an HI which varies between 63 days for HV application and 28 days for LV application. Some products do not have a HI at all, but for sound winemaking reasons it would be unwise to spray them too near the harvest. Sulphur, which until about ten years ago used to have a 56 day HI in Britain, now has none at all. Since the use of coal in the home has diminished and its use in many power stations has now ended, the amount of sulphur in the atmosphere – and hence falling on the land – has dropped massively, meaning that many soils are deficient in this element. It is therefore widely used on a range of agricultural crops as a foliar feed and has ceased to be classed as a pesticide. However, in other countries, the same sulphur approved for use on grapes in Britain (Kumulus DF for instance) has a 21 day HI in Germany and France, but in Britain, the label merely says 'do not apply near to harvest on grapes intended for wine'. In Germany, copper-containing sprays for vines also have a 21 day HI, whilst in Britain they have a zero day HI. The reason why it would be unwise to spray both sulphur and copper too near to harvest is that residues can give rise to problems in the winery: hydrogen sulphide in the case of sulphur, and a copper casse (*casse cuivreuse*) in the case of copper.

The correct HI should also be adhered to – and indeed in some situations, very dry years perhaps, even exceeded – as residue testing by supermarkets and other large wine retailers is now commonplace. Almost all controlled pesticides have a MRL (maximum residue level) which varies according to the product and the crop it is used on; for wine, the MRL is that for wine grapes. Current GC-MS (gas chromatography-mass spectrometry) testing can detect residues as small as 0.01mg-kg which, according to Geoff Taylor, the well-known wine industry analyst, the equivalent of 'one grain of salt in 100,000 litres' of wine and is way below all MRLs. The MRL on Scala for instance (active ingredient of which is *pyrimethanil*) – and being an anti-botrytis product is probably one of the last to be applied and which has a 21-day HI – is 5 mg-kg, some 500 times greater than the current GC-MS detection rate.

However, even though residue levels are below the

permitted MRLs, there are concerns that a cocktail of residues, each of which are below permitted levels, taken together might be shown to be detrimental to human health. The last thing a supermarket wants is a news story that reads 'Pesticides in wine can lead to brain damage' (although, of course, the headline 'Alcohol in wine causes brain damage, liver failure, domestic violence, car crashes, deaths etc. etc.' would be treated as an April Fool's joke!). If it were to be found that grapes and/or wine had residue levels higher than the MRL for the product concerned, undoubtedly your local Trading Standards would prosecute and naturally all wines would have to be withdrawn and, in all probability, destroyed. What this would do to the reputation of English and Welsh wine doesn't bear thinking about.

Buffer zones

Many products have buffer zones – the distance you have to leave between your sprayer and the 'top of the bank of any static or flowing waterbody or ditch which is dry at the time of application' – and these have to be adhered to. Products fall into two classes: LERAP A, products which are not eligible for a buffer zone reduction; and LERAP B, products which are eligible for a buffer zone reduction. If using a horizontal boom sprayer (such as might be used for herbicides) you cannot spray either LERAP A or B products 'within 5 m of a watercourse or within 1 m of the top of a bank of a dry ditch'. However, you can, with certain LERAP B products, reduce the 5 m buffer zone if the assessment has shown the products and your spraying technique to be safe. For 'broadcast air-assisted sprayers' (i.e. most vineyard sprayers), there is no distinction between LERAP A and B products and the buffer zone required will be on the label or the EAMU. Buffer zones will also vary according to whether you are spraying a full or reduced rate of the product which has a buffer zone. Buffer zones can also be reduced where there is a windbreak between your sprayer and the watercourse, but only when the windbreak 'is formed from broad-leaved trees or shrubs, not conifers'. The windbreak must also be managed to protect the crop from the effects of wind or spray drift, extend the full length of the boundary between the crop and the watercourse, be 2.0 m higher than the crop, have no gaps 'including those resulting from systematic stripping of lower branches' and have leaves visible over its entire length. If you are spraying pesticides at less than 75 per cent of the approved rate, the buffer zone can be reduced by 3 m; if using at 50 per cent of the approved rate, the buffer zone can be reduced by 6 m; and if below 25 per cent, the buffer zone can be reduced by 12 m. Note however that reducing the rate of the material might seriously affect its effectiveness and increase the risk of resistance. For non-air-assisted sprayers, such as some tunnel sprayers, and for hydraulic-only sprayers with vertical booms, there is a default 5 m buffer zone. The LERAP *Broadcast Air-Assisted Sprayers* guide, available from www.pesticides.gov.uk, gives full details.

Resistance

If you use the same product repeatedly over several seasons and on the same crop, it is possible that the problem you are trying to control will become resistant to the product. Given that insecticides are little used in British vineyards, this problem is likely to occur with fungicides and most likely to occur with what are called 'single-site' fungicides. Margaret McGrath in *What are Fungicides?* explains:

Single-site fungicides which are active against only one point in one metabolic pathway in a pathogen or against a single critical enzyme or protein needed by the fungus. Since single-site fungicides are highly specific in their toxicity, having little effect on most organisms, they can safely be absorbed into plants, thus these fungicides tend to have systemic properties. As a result of this specific activity, fungi are more likely to become resistant to the fungicide because a single mutation in the pathogen usually allows it to overcome the action of the fungicide, such as by preventing it from binding to the active site in the fungus. Typically, older contact fungicides have multi-site activity and thus usually affect many fungi in different classes. Through the development of in vivo screens, and due to the increase in the stringency and number of regulatory tests required to register a new active ingredient, fungicide manufacturers have found it easier to develop single-site systemics recently. As a result, fungicide resistance has become a more important concern in disease management.

In the past, there have been several instances of almost complete resistance to certain fungicide products. In Britain, growers used to use a product called Benlate (containing benzimidazole) against botrytis but it gradually became less and less effective and scientists even discovered strains of fungi that needed benzimidazole to survive! After thirty-three years of use it was withdrawn from the market.

The factors that cause resistance are very varied, but include frequency of use, dose rate, whether the product is used singly or repeatedly, the target disease, and the strength of the disease to build up resistance. These are, in turn, influenced by other factors such as the method (and quality) of application, the timing of application and adverse weather conditions. In short, the causes of disease resistance are many and varied and must be guarded against. In order to make this easier, all approved pesticides have been given Mode of Action (MoA) codes which are grouped according to their biochemical mode of action. Products with a MoA code that starts with an M – such as sulphur which has the MoA code of M2 – are multi-site acting and generally less likely to cause resistance if used repeatedly, although they do not necessarily give the best level of disease control. Products with a MoA code that consists of a number – such as Teldor (the active ingredient of which is Fenhexamid) which has a MoA code of 17 – are single-site products and are therefore more likely to lead to resistance if used unwisely. Therefore, in designing a spray programme, it is best to alternate products and not repeatedly use those with the same MoA codes and, when tank mixing, try and use products having different MoA codes. It should also be noted that some products contain more than one active ingredient and therefore have more than one MoA code. Switch, for instance which contains the active ingredients Cyprodinil and Fludioxinil – has two MoA codes: 9 and 12. Anti-resistance strategies have now more or less been built into the product recommendations which only allow a certain number of applications per season, thus forcing growers to vary and rotate products. For smaller growers, having to buy 5-litre cans of several different products in order to perform one season's round of anti-botrytis sprays can be expensive. A product such as Fytosave, which works by stimulating the plant's own defences, and which can be used up to eight times a season can be helpful. It will not get rid of disease if its present, but will help defending the vine against getting it. It is also cleared for use by most organic and biodynamic regulatory authorities.

Disposal of unwanted pesticides and empty containers

The disposal of unwanted pesticides, whether diluted for use or undiluted, and of empty pesticide containers, is strictly controlled. The best way to use already mixed pesticides is, of course, to spray them on to the crop and only mix up the correct amount. In theory this is fine: in practice less so. Given that the ends of the rows often get under-sprayed compared to the rest of the vineyard, I always think it legitimate to finish spraying by giving the outside of the vineyard one (extra) round of spray. But if you genuinely have excess spray to dispose of, then you need to follow the rules. For empty containers, the easiest way, after triple-washing them, is to use a licensed disposal contractor to take them away, or take them yourself to a licensed disposal site.

Record keeping

The 'Code of practice' (PB 11090) gives guidance on record keeping:
- Storage records – Not mandatory, but advisable.
- Treatment records – Mandatory.
- COSHH assessment – Depends on the product used and circumstances.
- Environmental Risk Assessments – Depends on the product used and circumstances.
- Exposure monitoring – Might be necessary if staff exposed to certain products.
- Maintenance records of Respiratory Protective Equipment (RPE) – Only if RPE used.
- Health Surveillance Records – Only if using staff that require health assessment.
- Disposal records
- Required if moving or disposing of waste.

Identification of pests and diseases

Before embarking on any pest or disease control, correct identification is essential. With experience and practice, many of the major pests and diseases can be quickly and easily identified without recourse to hand-lenses or books. Others, however, such as a light brown spotting on the leaves, might be caused by a disease, a herbicide spray or insect damage and require proper identification before treatment can be prescribed. As a minimum I would advise vineyard owners to invest in a small hand-lens, say 10x magnification and with a lens of at least 18 mm diameter (23 mm is better), plus a book or two, although there are of course multiple pest and disease identification websites which can be consulted. I use the following resources:
- *Diseases and Pests* – Edited by Phil Nicholas, Peter Margery and Malcolm Wachtel. Winetitles, Ashford, South Australia, 2011.
- *Compendium of Grape Diseases* – Edited by Roger C. Pearson and Austin C. Goheen. APS Press, Minnesota, USA, 1988. There is now a 2nd edition of this available for $139.
- *UC IPM Online* – This is aimed at Californian growers, but is a very good resource for pests, diseases, weeds and many other vineyard problems. www.ipm.ucdavis.edu and follow link via 'Agricultural Pests' to the 'Grape' pages.
- *Ontario Grape IPM* – www.omafra.gov.on.ca/IPM/english/grapes. Look for Diseases & Disorders.

Pests

Until relatively recently, vineyards in Britain were remarkably free of vineyard pests and, compared to vineyards in warmer climates, were almost pest-free. The University of California's UC IPM website for instance, lists twenty-four insects and mites to watch out for. Britain escapes lightly, partly due to the relatively cool climate (as most vineyard pests prefer warmer conditions), but also because vineyards in Britain tend to be isolated from one another; our vineyards have a wide variety of plants in their alleyways providing plenty of habitat for predator species; and because most vineyards are part of a mixed agriculture landscape that also provides habitat for predators. However, this situation – as it is in many other spheres of both agriculture and everyday life – is changing and there have been several instances in the last twenty years of new vineyard pests causing damage at levels where treatment becomes necessary. The European Grapevine Moth (*Lobesia botrana*), called *Eudemis* in France, Vine Moth (*Eupoecilia ambiguella*), called *Cochylis* in France, Grape Berry Moth (*Paralobesia viteana* or *Endopiza viteana*), Light Brown Apple Moth (*Phalaenoides glycinae*) also known as LBAM, Spotted Wing Drosophila (SWD) (*Drosophila suzukii*), and Summer Fruit Tortrix Moth (*Adoxophyes orana*), also known as SFT, have all made appearances in Britain, particularly in warmer years. There have also been warnings about spittlebugs, the vectors for *Xylella fastidiosa* (Pierce's Disease), the Brown Marmorated Stink Bug, the Spotted Lantern Fly and the Asian Hornet, although none of these last three has actually been seen in Britain.

RAK mating disruption capsules

To say that these moths and other insects are regularly found in British vineyards would be an exaggeration (although SWD can now be considered a regular guest in many vineyards), but if growers see physical damage to any part of their vines and grapes, then they must be considered as a possible cause and, if it is felt that the damage is of a level to cause commercial damage, sprayed against. Almost all of the insecticides permitted today are specific to the insects they are trying to control and therefore will not harm other predator insects which may well be useful allies in the vineyard. Hallmark, a pyrethroid (synthetic pyrethrin) will control a wide range of pests, and thus should be used sparingly. In order to monitor moth numbers (and not trap them), pest-specific pheromone traps ought to be employed to decide whether to spray an insecticide or not.

Overseas, several types of pest (mainly the *Eudemis* and *Cochylis* moths) can be controlled with what the French call *confusion sexuelle* and which in Britain we call 'mating disruption'. This is carried out by hanging small tabs around the vineyard, containing the pheromones of the female of the species to be confused, so that the males, instead of mating with real females, wear themselves out hunting down plastic capsules. In order for this technique to work correctly, a large area needs to be covered – BASF say a minimum of 5-ha for their RAK product – and 500 capsules per hectare (200-acre) need to be distributed. Costs are high, with the materials costing around £300-ha (£121-acre), plus the labour to hang the tabs out. Because of the requirement to cover large areas in order for this technique to work, it is being used successfully in regions where co-operatives dominate and where growers are able to work together. In isolated vineyards, such as we typically have in Britain, it is less likely to prove effective. In Britain, some apple and pear growers are using RAK mating disruption capsules against tortrix and codling moths, but their use is not widespread as they

only control these specific pests, whereas insecticides control other pests at the same time.

Mites have been present in British vineyards for many decades, but until recently, did little economic damage. The Grape Erineum Mite (*Colomerus vitis*) also known as an Erinose or Blister Mite, was occasionally seen on leaves, especially in hotter, drier years, but apart from looking a little bit like *Phylloxera* leaf galls, did little damage and always seemed to be easily controlled by sulphur sprays, However, in recent decades, Grape Vine Bud Mites (*Eriophyes vitis*) have started to make an appearance – in my experience especially on Chardonnay – and once established, they seem quite difficult to eradicate. The clues are in the first few weeks of budburst and, if Bud Mites are present, buds can be blind, shoots stunted and leaves deformed and there may well be lower than normal levels of inflorescences. In other countries these mites are often a contributory factor in what is called Restricted Spring Growth (RSG). Unlike Erineum Mite, these Bud Mites spend most of their lives inside the buds and are therefore more difficult to spot as their damage is largely hidden and much more difficult to control. The recommendation to control these is two to three early sprays of sulphur at a high rate – 10 litres-ha – one at woolly bud, one at bud-burst, and maybe another after that. Whether this is a long-term solution remains to be seen. Red spider mites, of which there are two types, European Fruit Tree Red Spider Mite (*Panonychus ulmi*), and Two-Spotted Red Spider Mite (*Tetranychus urticae*), are quite common in some crops in Britain – hops, blackcurrants, cane fruit, protected strawberries – and have been seen on vines in Britain, but only very occasionally. They like warm dry conditions (which is why they can be found on vines in greenhouses), so perhaps a pest to look forward to!

Other insect pests such as capsid bugs (of which there are several types) and thrips (thunderbugs) have always been seen in British vineyards, but their damage is not usually bad enough to be of economic importance and are therefore not worth spraying for. In sites near to woodland or established scrubland, where there are plenty of hosts plants for these insects to live in and multiply, damage can be quite severe and, in these instances, worth spraying against. Capsids, which like living in clumps of nettles and other tall weeds, will sometimes cause damage to leaves, especially on young vines. Thrips can cause scarring and russeting on grapes which could make berry splitting more likely, but again, at non-commercial levels. Capsids however, are also predators of several species of insect, so may be doing some good as well as harm. However, leaf quality comes first if you want your vines to produce and ripen good crops.

The Spotted Wing Drosophila, first seen in Europe in French vineyards in 2010, has been causing considerable damage in Britain in soft and stone fruit crops, and has been seen in many vineyards. It appears to prefer early varieties, both white and red, but it especially likes early ripening reds, Rondo and Blauer Frühburgunder (Pinot noir Précoce) are favourites, and warm years. It was seen quite widely in 2015, 2017 and 2018, but not so much in 2016 and 2019. Reducing

SWD habitat around vineyards, especially blackberry brambles and elderberry trees, helps keep numbers down. Vineyards near plum and cherry orchards, especially abandoned ones, are especially vulnerable to attack. The female SWD prefers laying her eggs in the shade, therefore early, and strict, deleafing of susceptible varieties will help prevent damage. Growers, especially those with early ripening, softer-skin red vine varieties, should use detection traps (such as the Biobest Droso trap) for monitoring numbers and helping them decide when to take action. In 2020, WineGB is funding the hire of four hi-tech camera traps to monitor SWD in four vineyards around the country in order to support the 'emergency' application to use the pesticide Exirel against SWD on vines. There are also several other pesticides which can be used (Tracer and Hallmark), plus Surround, a kaolin-based spray, and Wetcit, a spreader, can also help deter egg-laying. The WineGB 'Green Book' has a whole page on SWD with some good advice on monitoring this pest, plus measures to prevent egg-laying and what needs to be done should it be found.

The multi-coloured Asian Lady Beetle or Harlequin Ladybird (*Harmonia axyridis*) has also been seen in British vineyards, but not in numbers large enough to cause problems. This ladybird, when disturbed, exudes a substance called 'reflex blood' which, if it gets into juice, can possibly taint the wine, depending on numbers. However, given that the numbers of this insect are currently very low and given that harvesting machines are not often used in Britain, no problems have been reported with this pest.

Pollen beetles (of which there are many different sorts) are often seen on inflorescences during flowering, but it is not thought that they do any harm and may even possibly aid pollination. They are generally only seen during warm weather, so perhaps ought to be generally welcomed rather than worried about.

Other pests which might appear include scale insects, of which there are two main types – brown and woolly – which might affect vines, which usually gather on the trunk or under the bark on older vines. They don't appear to cause much harm on outdoor vines (although can be a problem on greenhouse vines) and before an insecticide is used, an assessment of the scale (no pun intended) of the damage should be made. A petroleum oil (Croptex Spraying Oil) has an on-label approval against scale insects and this product, sprayed on to trunks during the dormant season, might help. In the past, both Vine Weevil (*Otiorhyncus sulcatus*) and Dock Sawfly (*Ametastegia glabrata*) have been known to attack young vines in Britain, but on established vines are seldom a nuisance. The literature mentions aphids as being a problem, but I have never known them to be. It is quite probable that the use of relatively high rates of sulphur on vines, which starts early in the season, coupled with our sometimes-challenging weather, keeps most pests down to levels where they are not noticed.

So far, Britain has not seen any of the wide variety of sharpshooter leafhoppers which in other countries are the vectors for pathogens that cause ailments such as *Flavescence*

dorée (Grapevine Yellows) and Pierce's Disease (*Xylella fastidiosa*) as our climate is not yet that warm. There is more on *Flavescence dorée* under the section in this chapter on *Viruses*. Pierce's has been found on one imported plant (which was destroyed) but DEFRA is concerned that as incidences of this disease travel ever further north in mainland Europe (from southern Italy where it is devastating olive trees), and reach southern France where many vine nurseries are located, there is every chance for imported vines to be affected. DEFRA has advised vineyard owners to watch out for spittlebugs (the insects that leave what is generally called 'cuckoo spit') as they could well be vectors of Pierce's.

Wasps will always be a problem when there is ripe fruit about, especially if the fruit is damaged or rotten, and they have an easy entrance to the sweet flesh. In vineyards in Britain, wasps have usually been a problem in years when the vintage is early and sugars are high; on early ripening varieties such as Siegerrebe, Ortega, Optima, and Rondo, especially when individual grapes are damaged and start to rot; when poor flowering conditions have given rise to *millerandage* and there are a significant number of small seedless grapes which ripen fully, well before the seeded grapes on the same bunch; and before the weather starts to turn cold. If your vineyards are also near apple orchards where there is a supply of fallen fruit, this too may attract them. The remedy against wasps is to firstly track and destroy their nests. How you do this exactly is part of country folklore, but looking in likely spots – dry banks and old farm buildings – is a start. Following their flight paths is probably the best way as they tend to fly in a fairly straight line from food source to nest. Capturing one and tying a piece of white cotton to it so that you can follow it home is another, but I have never seen this done, but can believe it might work (although probably illegal). Having found nests, they can be eradicated using approved materials. The second method of controlling wasps is to trap them *in situ* with wasp traps. However, Chris Cooper in WineGB's 'Green Book' points out that this technique is now construed as 'mass trapping' and is not authorised for grapes in Britain. Therefore, wasp traps now have to be renamed as 'wasp monitoring devices' and the technique referred to as 'precision monitoring'. The question of whether you can actually use an insecticide against wasps is problematic. None are specifically cleared for use as a 'wasp eradicant', although there are some pesticides approved for use against other pests, SWD and capsids for example, which will also control wasps. Both Hallmark +ZT, and Markate 50 mention wasps on their EAMU notices.

In young vineyards, where vines have yet to develop a woody trunk, slugs and snails can sometimes cause problems, damaging both canes and leaves and stunting extension growth. In this instance, slug pellets are often effective. Slug and snail damage will often occur inside rabbit guards of the tree guard type and slug pellets may need to be sprinkled inside them as soon as they are put on. I have also known snails to attack mature vines, but this only in a vineyard that had been repeatedly flooded and where incoming silt off neighbouring fields had carried with it huge quantities of snails and snail eggs. Getting rid of these was much more of a problem and despite trying various molluscicides it was only the hard winters of 2009–10 and 2010–11, when temperatures fell to -10°C, that killed them off.

Phylloxera

Although *Phylloxera* (which is more correctly known as *Daktulosphaira vitifoliae*) has to be classed as a 'pest', it is not one in the same sense of any of the above. It cannot be eradicated by spraying or soil drenching and the only practical remedy is to use *Phylloxera* tolerant rootstocks. It has been found in Britain many times since vines were first imported for commercial vineyards in the 1950s, in both open vineyards, as well as in garden centres on potted vines. Given that it is endemic in virtually all European vineyards (as well of course in most vineyards worldwide), eggs can easily be missed when vines are being sorted after lifting and, if they are pot-grown, can be hidden in the soil. Therefore, anyone buying vines from European nurseries (which is all but a very small number of amateur vineyards) the possibility of *Phylloxera* being present in your vineyard must always be considered.

If you are growing pure *vinifera* vines on their own roots (something you would only do in certain circumstances and where you appreciated all the risks involved), the first signs of *Phylloxera* may well be just a general lack of vigour in a few plants, building up over several seasons from a nucleus to a wider circle of vines. If this happens, and there is no other obvious cause (such as poor drainage) and no signs of leaf galls, then get a spade out and expose a few roots near the surface. *Phylloxera* does not like tunnelling deep – indeed it cannot – so finding them is relatively easy if you know what to look for.

If you are not growing vines on their own roots, the appearance of *Phylloxera* is likely to be as leaf galls on the leaves of older hybrid varieties such as Cascade, Seyval blanc, Triomphe etc, or, and much more likely, on the leaves of rootstock shoots which have come from below the graft. This typically happens when earth or mulch has been allowed to build up over the graft giving rise to both roots and shoots coming directly from the rootstock. *Phylloxera* finds the leaves of American rootstocks much more palatable than pure *vinifera* leaves (which it almost never touches). You may also find that the vine has put out roots from above the graft, i.e. scion roots and these are highly susceptible to the root form of *Phylloxera* as they are pure *vinifera* roots. Depending on how many seasons the leaf galls have been there, you may or may not have root infestations. The root form of *Phylloxera* is, in its youth, a fairly bright yellow and can just about be seen with the naked eye. If the *Phylloxera* have been there for a year or more you will probably also find root tuberosities (also known as nodes or knots) which are dark brown and can often be mistaken for just pieces of old root or general soil debris attached to the roots.

If you find (or even suspect) that you have *Phylloxera* in your vineyard, it is a legal requirement that you notify FERA (Food and Environment Research Agency), the agency whose task it is to deal with 'quarantine pests' for DEFRA, and inspectors will be sent to investigate. If *Phylloxera* is confirmed, although FERA has the power to make you remove infected vines (don't worry – it has never done so), it is likely to impose a movement control on your vineyard, restricting the movement of both plant material in and out, plus the movement of visitors to the vineyard, and let the infestation run its course. It will also probably ask that you spray infected vines with an approved insecticide (see below). If you only have leaf galls on a few rogue shoots of rootstock, then the answer is to make sure that every graft union is exposed and inspected and the graft made completely free of extraneous shoots and roots. If you have pure *vinifera* vines growing on their own roots and they have root galls, then they will weaken and die and unless the *Phylloxera* has new hosts to migrate to, they too will die out. If you have Seyval blanc or other older hybrids either grafted on to rootstocks or on their own roots, then it is quite possible for *Phylloxera* to stay on the leaves for several years, although they can be sprayed with insecticide and they do not like high levels of sulphur. Currently, one product (Batavia) is cleared for use in British vineyards for the control of *Phylloxera* (and scale insects). This is a 'unique two-way' product which moves throughout the entire plant system and reduces the breeding habits of adult females and 'limits the survival of their offspring'.

It is also worth noting that not all rootstocks are as tolerant of *Phylloxera* as others. Those with pure Berlandieri, Riparia and Rupestris blood are very tolerant and unlikely to be damaged. However, as was seen in California with the rootstock AxR1 (Aramon x Rupestris Ganzin No.1), there are several rootstocks which are *vinifera* crosses and therefore not as tolerant of the pest. The rootstock 41B is a Chasselas x Berlandieri cross and therefore does not have 100 per cent American species blood and there are reports of it suffering from *Phylloxera* damage in overseas vineyards. Given that in Britain 41B is being used on high-pH soils (as it has a tolerance of 40 per cent of active $CaCO_3$) and is also 'low-medium' for vigour, growers using it should make doubly sure that *Phylloxera* does not establish itself in their vineyards. The only rootstock that is currently actively resistant to *Phylloxera* is Börner which is a Riparia x Cinerea crossing. This means that instead of supporting, but not encouraging, the root-gall form of *Phylloxera*, it actively rejects them. For this reason, it is becoming more popular in Germany.

Birds

Bird damage in vineyards is a worldwide problem and Britain is no exception. In many parts of Europe, where vines have been grown for centuries and where many wine areas and regions consist of hundreds, even thousands, of hectares of vines and no other crops, the native bird populations have been substantially reduced by shooting and trapping. In German winegrowing regions, and very likely in other Old World winegrowing regions as well, pensioners and

Wine Growing in Great Britain

Pheasants will take the lower bunches

teenagers will be allocated their own patch of vines and during the four to six weeks prior to harvest, will spend all their daylight hours patrolling their vines with rattles (what we would call old-fashioned football rattles), lighting bird-banger ropes and generally keeping birds on the move. When flocks of starlings threaten, it is not uncommon to see quite large numbers of vineyard owners appear as if from nowhere, armed with shotguns. In other countries where vineyards are more isolated, birds can be hugely damaging and will, if left to their own devices, take almost all the crop.

The extent of bird damage will depend on many factors and no two vineyards, and indeed no two years, will suffer the same level of damage. Such factors as the proximity of woods, buildings and other safe havens to which birds can escape and roost; the availability of other crops which might either make the area attractive – thus drawing them in – or provide easier to obtain or more satisfying food – thus diverting them from eating grapes; and the presence of overhead high-voltage wires which make ideal perching spots from where to plan their attacks will all affect the level of bird damage. The presence in the area of birds of prey can also deter birds, but in my experience, they cannot provide anything approaching even partial protection from damage. Maybe if a falconer and his or her falcon took up residence in your vineyard, birds of prey might be a solution.

In years when other crops and natural foods (hedgerow fruit) are plentiful, bird damage to grapes tends to be less, as these other foods are generally ripe before grapes, easier to access, and, at least in the case of hedgerow fruits, are not protected. In early years, especially with early varieties, and even more especially with red varieties or varieties that colour up as they ripen (such as Schönburger grapes which turn dark pink to tawny as they ripen), bird damage will always be greater than with those high-acid, white varieties that tend to ripen later. The main attractant to the birds appears to be a combination of the sugar level and the softness of the grape –

which typically go together – with darker grapes being more visually attractive (or perhaps just more noticeable) than green grapes.

In the early decades of the revival of viticulture in Britain, birds were often cited as the main problem facing vineyard owners. The first vintage at Hambledon in 1955, the first vintage of the modern era, was largely eaten by birds. Major-General Sir Guy Salisbury-Jones had engaged a winemaker from Bordeaux, Monsieur Chardon, to come and help with the winemaking. Unfortunately, the 1955 *vendange* in Bordeaux was very late and by the time Chardon arrived, the birds had eaten most of the Hambledon grapes. Salisbury-Jones recounted the story:

Every day the lovely green bunches grew less. Here and there, tucked away under the foliage, a few bunches still defied the invaders. Elsewhere, only the naked hideous stalks remained. It reminded me of moments in the First World War, when some lovely wood, under the ceaseless pounding of the guns, would slowly be transformed into a gaunt array of skeletons that had once been trees.

Salisbury-Jones was not the only vineyard owner to face the same problem. Growers had limited options: pick early before the bird damage got too severe, the result of which was some fairly green, high-acid, thin wines; double (and often re-double) efforts to scare birds off with a wide array of devices, static, audible, and visual; mount early morning and early evening patrols armed with a shotgun; or net. Most tried a combination of all, although picking early was never the right option. Scaring devices work for a time and work better with some species, but audible ones (gas cannons, bird scaring 'banger' ropes and recorded distress calls) tend to upset the neighbours. The National Farmers Union (NFU) Bird Scarers Code advises that they should not be used more than four times per hour, not within 200 m of 'sensitive buildings' and not before sunrise or after sunset. Vineyards with footpaths running through or near them may also find that they receive complaints if audible devices are placed too near to the footpaths.

Devices such as black cotton thread strung along the rows (used widely in fruit orchards to deter bullfinches from taking apple and pear buds) worked against some species but proved

Allsopp Helikite about to be launched

a nightmare to both pickers and machinery. Humming tape, spinning discs, magic mirrors, CDs strung along rows and a range of other devices barely worked and if they did at all, it was very short-lived. Two devices which do appear to work are helium-filled kites (such as the Allsopp Helikite) and bird-scaring 'hawk' kites. These devices, which can 'fly' above the crop, are relatively maintenance-free and are cheap enough – at least in relation to the value of the crop they are protecting – for there to be several per vineyard (depending of course on the size of vineyard to be protected) for them to provide some protection. The kites are made by Scarem (www.scarem.co.uk) and there are three different models: the 'hawk', the 'vineyard & orchard' and the 'monster', each of which is mounted on a 7–10 m telescopic pole, depending on the size of the bird and the area to be protected. The company makes various claims for the area that one 'bird' will protect, but for grapes I would suggest at least one per 2-ha (5-acres) to start with and to move them to different locations on a weekly basis. The 'Vineyard & Orchard Complete' model, including a 10' pole, costs £127.50 + VAT (2020)

In the early '80s, a product called Mesurol, which was widely used at the time in New Zealand vineyards, proved very effective at discouraging birds, and several British growers used it for a few years until it was withdrawn from use for this purpose. It was known as 'electric fencing for birds' as once they had tasted grapes sprayed with Mesurol, they never touched them again. According to the literature at the time, it was said that it constricted the bird's throats and made them gasp a bit (this is probably not an accurate technical description of the effect). Unfortunately, it wasn't good news for wildlife if it got into watercourses, so it was banned. Mesurol (the active ingredient of which is Methiocarb) until quite recently continued to be sold in Britain and used as a seed dressing – to stop birds eating seeds after they are sown – and as a slug and snail bait (i.e. killer). If you have ever wondered why the birds don't eat slug pellets after you have sprinkled them on your garden, this is the reason! I also note from Bayer's Australian website that it is permitted in vineyards against snails and slugs in most regions. The Australian pellets also contain Bitrex which makes them unpalatable to mammals. Other sprayable products based upon such things as chilli pepper and garlic

were (and still are) sold as 'natural' sprayable bird deterrents, but the risk of the product persisting on the grapes and possibly passing into the wine and tainting it was always too great. In the USA, a product called Avian Control, which contains methyl anthranilate, is approved for a wide range of bird species and different crops including grapes. It claims that 'one treatment can last up to three weeks' and it can be applied 'up to the day of harvest'. They recommend a rate of around 23.5 to 47 litres-ha (9.5 to 19 litres-acre) at a product cost of $361 to $722-ha, so by no means cheap. Whether it works or not is another matter and the advice leaflet for the product ends, somewhat ominously: 'Be realistic. Nothing is 100 per cent. There are no silver bullets.'

Netting, on the other hand, erected correctly, does work 100 per cent. Lamberhurst Vineyards were probably the largest users of whole-vineyard overhead netting and at one time had over 16-ha (40-acres) worth of Lobrene netting (made by Low Brothers of Dundee – now known as Don and Low) which was laboriously erected over the top of their vineyard rows, stitched together where the 6.0 m wide strips met and weighted and pegged down along the sides and ends. Whilst it undoubtedly worked, the cost, not only of buying it and erecting it, but also of removing it, mending it, and storing it in vermin-free storage, was prodigious.

An alternative to multi-use netting such as Lobrene was a Swiss product called XiroNet which was sold for one-off use (although some growers claimed to get two seasons out of it) and was cheap enough to tear off and burn after the harvest. It was a very lightweight yellow product, which came on a 1 m roll which, when held aloft on a purpose-made dispenser on the back of the tractor, expanded to 10 m wide. Each roll cost around £150 and covered 0.2-ha (0.5-acres), although you lost some of that area where it draped down to the ground at the sides and ends. For several seasons I used it to great effect at Tenterden Vineyards, wearing my fingers to the quick with stitching the edges of the net together where they met in the middle of the rows, and making twice-daily patrols to disentangle hedgehogs and mend holes where dogs, foxes and badgers had found their way through it. However, XiroNet does not appear to be made any more.

The main problem with all types of overhead netting, but XiroNet in particular (probably because of its colour it was difficult to see), is that birds, especially songbirds and starlings, are easily trapped in the net. They often hang there for hours emitting loud distress calls which, although this frightens off other birds, causes visitors and neighbours to complain (quite rightly, of course) and also attracts other birds such as owls and kestrels which are after the trapped birds and then themselves get caught. Some growers have tried side-netting, hung on to the existing training wires and clipped together at the top and bottom, but birds easily find that they can hang on to the outside of the net and peck the grapes through the netting, so this technique is not completely successful. However, with the right rolling and unrolling equipment and by using cheap netting (big bale wrap works well), this is a cost-effective way of making a vineyard semi-bird proof, although it does tend to hamper air movement within the canopy and make late-season disease problems, especially botrytis, more of an issue.

In New Zealand, where vineyards are often isolated, bird netting is very widely used and (according to Mike Trought) its use is as high as 90 per cent of vineyards in certain regions. Given that there are almost 39,000-ha (96,000-acres) of vines in New Zealand that is a lot of bird netting. The cost of netting, including all labour to put it on and remove it, and writing the netting off over (an optimistic) 10 years, is said to be around £700-ha (£280-acre). This sounds like an under-estimate to me and given that £700 represents maybe 500 kg of grapes, it would soon save you money. Growers in New Zealand mostly use overhead netting which covers several rows and drapes down to the ground or netting covering whole vineyards (or at least much larger areas) which is supported by an additional wirework frame erected above the vines. If your bird problem is severe, then this latter solution is probably the easiest and most effective. The overhead wires act as 'rails' along which the netting can be dragged to put it on, and the wirework is sufficiently high enough above the vines for it not to get tangled up with the growing shoots, a real plus when it comes to taking the nets off. Damage to the nets is also kept to a minimum. Overhead netting also allows easy access for tractors for spraying, and eventually picking, after the nets have been put on, by raising up the ends for access. There are several suppliers of bird netting in New Zealand, and plenty of information available on various websites.

In Britain, the widespread use of nets to protect vineyards against birds started to decline in the late 1980s and today it is seen far less often. The reasons for this decline are various. In the days when many of the grapes being grown were relatively low-value German crosses such as Müller-Thurgau, Reichensteiner, Bacchus, and Schönburger – Seyval blanc never seemed to suffer so much, as it was both later ripening and less sweet – the cost of netting versus the costs of other systems of bird control made it prohibitive. The increase in the area down to the later ripening varieties Pinot noir and Chardonnay also probably helped as birds appear to be less interested in these. Some more isolated vineyards use it – Camel Valley still nets almost all of its vines – and it is often used on vulnerable, early ripening varieties, in situations where vineyards are remote from habitation, and in relatively small vineyards where the owner wants 100 per cent peace of mind and the task of putting it on can perhaps be achieved in a weekend with a few friends.

A major factor in the reduction in the use of bird netting in Britain has been the decline in bird numbers over the last few decades. The early 1980s saw the start of the decline in certain bird species including starlings, blackbirds and thrushes, the birds that caused the most damage in British vineyards. The British Trust for Ornithology's (BTO) Common Birds Census and Breeding Bird Survey shows that Britain's population of starlings, by far the most damaging species in vineyards, because of their habit of attacking vineyards in large numbers, has declined by over 80 per cent since the mid-'60s, and that the song thrush

and blackbird populations have more than halved in the same period. Also, the rise in bird species such as magpies and jays, which typically do not eat grapes, but which do prey upon birds which do, has further reduced pressure on grapes from these garden birds. In addition, there have been significant rises in the population of birds of prey in Britain. The BTO records that two commonly found British birds of prey – buzzards and red kites – are all showing 'rapid increase' which in some vineyards has definitely contributed towards the lack of damage from other bird species. Taken together, these factors have resulted in far less bird damage in British vineyards. However, with natural sugar levels rising seemingly with each vintage and as some species of birds start to recover their numbers, one cannot rule out the re-appearance of netted vineyards in the future.

In some vineyards, pheasants (and occasionally partridges) can be a problem, but these are almost always vineyards situated near to a farm or estate with a shoot, and especially so if the shoot's release pen is within 1.6 km (1 mile) of the vineyard. Pheasants do not generally perch on the wires and get amongst the foliage like most other birds, but take the grapes by pecking them from the ground or leaping up and taking them. It stands to reason, therefore, that, if your vineyard is liable to be damaged by pheasants, the higher the fruiting wire the better. However, an adult pheasant can easily jump to 800 mm or more and you may still lose fruit at this height. Pheasants, especially juvenile ones, don't really like flying and tend to enter vineyards by walking along the ground from the ends and sides. Netting the ends and sides therefore, especially the perimeters facing where the birds come from (adjacent woodland or towards where the release pens are situated), may well deter some of them.

For many vineyards, the capital costs, plus the costs of putting on and removing nets, is just too much to consider, especially those with fluctuating yields where the value of the crops to be protected is none too certain. Therefore, a combination of static, silent methods such as balloons and kites, and, in areas where neighbours will not be annoyed, a limited number of audible devices such as gas guns and banger ropes, plus patrols at feeding times, especially with varieties susceptible to bird damage, provides adequate, though not total, protection. It may be in the future, if vineyards start to investigate protected growing systems in order to overcome such hazards as spring frosts and poor flowering conditions, that the framework used to carry the protection might also be used as support for bird nets. After all, if you are going to the considerable expense of covering your crop, you certainly want to make sure you pick every grape you grow.

Airport operators – including military airports – have developed a wide array of techniques to scare birds as a 'bird strike' can easily damage an engine and a quick look on Google under 'airport bird scaring' will bring up almost 10 million results. When you start reading them – the Wikipedia page is a good place to start – you will see that they all agree on one thing: most sensible methods work for a while, but only a while. Birds get used to everything short of being killed, so if birds are a problem, try several different methods and keep rotating them on a regular basis.

Under certain conditions, certain species of birds damaging fruit crops may be killed (or trapped and destroyed) under what is called a 'general licence'. This is allowed under the provisions of the 1981 Wildlife and Countryside Act and managed (in England) by Natural England. The list of species that may be killed under licence is relatively short and includes wild pigeons and doves, rooks, crows, jackdaws, magpies and jays, plus certain species of geese and gulls. It also includes parakeets (Monk and Ring-Necked) which, if your vineyard is near London (or other large conurbations), you will know all about the damage they can do. There is currently (2020) considerable confusion with the 'general licence' and there is a lot on uncertainty about exactly what one may shoot, and when. If in doubt, contact Natural England, the RSPB or the NFU. Blackbirds and thrushes, which are partial to grapes, are protected by law and may not be harmed. Starlings, also very partial to grapes, are, somewhat surprisingly, protected by law in England (but not yet in Northern Ireland, Scotland or Wales) and may also not be disturbed or harmed. Natural England has full details on its website.

A recent article in the *Australian Wine and Viticulture Journal* showed how drones, programmed to fly in formation up and down vineyards in between the rows of vines, emitting distress calls and any other noise that might scare birds, will soon be available. A USA company called Bird-x also has anti-bird drones, plus a very wide range of other anti-bird solutions.

Diseases

Successful disease control in vines really requires a two or three stage approach and it is not just about spraying the right chemicals at the right time (although that helps). Most diseases are made worse when canopies are crowded, shaded and slow to dry out. Light and air, those most valuable of (free) natural resources, must be encouraged, through good and timely canopy management, to penetrate into the centre of the leaf-wall. Leaves should look clean, with a good glossy finish (depending on variety) and be uniformly bright green in colour. It is only the lower, older leaves that should be starting to fade. The yellowing of leaves inside the canopy is a sure sign of a shaded canopy and indicative of poor light penetration. Good canopy management also requires that winter pruning is carried out correctly, excess shoots are removed after budburst, and deleafing and trimming are attended to. Excessive weed growth below the vines and long grass in the alleyways can also add to the level of moisture in the atmosphere. Any mechanical damage, such as that from hail, animals, insects, or damage by wires or machinery, will allow disease to get a hold on the vine, especially from botrytis.

Powdery Mildew – Oidium

Powdery Mildew (PM) or *Oidium*, which used to be known as *Uncinula necator* but is now more correctly known as *Erysiphe necator,* is a fungal disease native to eastern North America where it did little damage to the native vine varieties and caused little excitement. Its first European appearance was in Margate, on the north Kent coast, in 1831 or 1832. Mr Tucker, the gardener to a Mr John Slater, noticed a white powdery discolouration on a few leaves of a vine in a greenhouse under his care. These he sent to the Reverend M. J. Berkeley (a famous fungologist who wrote the *Outlines of British Fungology*) who sent an account of the 'new' disease to the *Gardener's Chronicle and Agricultural Gazette* and named it (incorrectly as it happened, as it is not a true *Oidium*) *Oidium tuckeri*, after gardener Tucker.

PM then took a few years to reach France, but by 1846 the keeper of the Versailles grape-forcing houses complained of a new disease and it soon spread throughout the vineyard regions causing widespread damage. Tucker had already noticed the physical and visual similarity between his mildew and that of peach mildew which was then (as it still can be today) controlled by sprays of sulphur and it was found that this remedy worked on vines. The vintage of 1854 was the smallest in France since 1788 (a year when the French *paysannerie* no doubt had other things on their minds), caused by nation-wide attacks of PM. Only then did growers take to treating their vines with sulphur – sprayed in cooler regions and dusted[11] in warmer ones – which proved to be a very cheap and effective remedy. It was found that sulphur had the effect of increasing yields and bringing the harvest forward by 7–10 days, two factors which made the uptake of sulphur as an effective remedy quite rapid. Sulphur also works against other diseases and mites, so its protection against PM was but one of a number of benefits.

PM is one of the few fungal diseases that are worse in dry years than they are in wet ones as, unlike most fungal diseases, it does not require water to convey it from one part of the vine to another. Air with sufficient humidity is enough. It prefers warm weather (18–30°C) and still, humid conditions and thrives inside shaded canopies, as the spores are inactivated by bright sunlight. It overwinters on old wood and, given the right conditions, will often start producing spores early in the season, which are spread by the wind and will attack any part of the vine, especially tender tissue such as young leaves and inflorescences. If untreated or if treatments are ineffective, the first signs of PM will be a slight discoloration on the leaves and if allowed to progress the leaves will be dusted with a white powdery covering. In Britain, leaves often remain in quite good condition and the first signs of PM will be a dusty mauve-white covering to the berries, especially those inside the most shaded part of the canopy. Quite why this should

---

11 Dusting of sulphur is preferred in warmer regions as it is less likely to cause scalding or russeting of the grape's skin. Dusting is often carried out very early in the morning when there is still dew on the grapes so that the sulphur sticks to the skin, before being absorbed as the skin dries out.

187

be is an interesting question and it has been suggested that a combination of our relatively low temperatures, coupled with leaves that are often wet, means that the disease is not able to attack the leaves as easily as it does in warmer, drier and more humid climates. Depending on the state of development of the bunch, the grapes will either just wither away or, if large enough, they will split open and the seeds will pop out like two eyes – what I call the 'Mekon' look.[12] Usually, as PM attacks early in the season, damaged berries never reach the point where they ripen. In machine harvested vineyards and where there are a large number of PM infected berries remaining on the vine, there is always the danger of the juice and/or wine becoming tainted.

Unless you are growing a variety such as Seyval blanc, which is fairly resistant to PM, it is safe to assume that your vines will get the disease. Some varieties are very susceptible, Chardonnay being one of them. When to start spraying will depend on whether you have bud mites to control. If you have, then at the pre-bud-burst 'woolly bud' stage, you should get two applications of sulphur on at a high rate and at around ten days apart. Otherwise, aim to get the first application on when the first leaves are just open and keep on spraying on a 10–14-day cycle until just before flowering. By this time, you should have gained the upper hectarend. Delaying spraying until you see the first signs of the disease is unwise as, by the time you see the disease, it is too late to prevent it and you will have to use more expensive curative products. It is often said that sulphur's effectiveness is temperature related and that below around 15°C is far less active that in the 20–25°C range typically seen in British vineyards. However, Wayne Wilcox, in his *Cornell grape disease control notes* states that they found no difference in the effectiveness of sulphur when spraying it at 15°C or 28°C, so this looks like another old wives' tale. The *Cornell grape disease control notes* are available at: www.englishwine.com/wgigb.htm

There are many different products available to prevent, control and eradicate PM and the current (2020) 'Green Book' lists around fourteen of them. These include some 'biological control agents' (BCAs) with sprays based upon strobilurins,[13] other fungi and yeasts. Which product you choose to use against PM will depend upon several factors. However, given that sulphur is (a) cheap, (b) effective, (c) helps control mites and other insects, and (d) is often deficient in soils, it is the first choice of most vinegrowers. The other thing to remember when using sulphur is not to skimp on the volume of water used. Sulphur is a contact material, has no systemic action and can only work if the product, in suspension, gets to all the parts of the vine it is trying to protect. It is therefore very important that spray rates are high enough, especially when the canopy reaches its full size, so that all parts of the vine are fully wetted. The most important sprays are those that

go on to the open inflorescence during flowering and up until fruitset. In countries warmer than Britain, it is usual to avoid using sulphur directly in the open flower as it is claimed that it can burn foliage and soft tissues, i.e. inflorescences. Given that temperatures during flowering in Britain seldom rise above 30°C (86°F), let alone the 38 °C (100°F) at which sulphur is said to damage vines, it is probably fairly safe to continue using sulphur throughout the flowering period. However, on the basis of better safe than sorry, if temperatures are high during the flowering and immediate post-flowering period, it is much safer to switch to another material.

As has already been pointed out in the section on *Harvest intervals* earlier in this chapter, sulphur applied to grapes in Britain used to have a HI of between 21 and 56 days, depending on the exact product used, but now it has been re-classified as a fertiliser and there is no HI. But, please note that in other European winegrowing countries, sulphur still has a 56-day HI on wine grapes and it is advisable that applications of sulphur should finish in good time before the harvest. In any event, if you haven't got good control of PM by 56 days before harvest, it is probably too late. The maximum rate of sulphur suggested for use in British vineyards is far lower than in other European countries and you cannot use higher rates than allowed by British legislation except under certain conditions. The rate for Kumulus DF, one of the most widely used sulphurs, is 250 g per 100 litres (which is only around 3 kg-ha at high volume), although other brands allow 4 kg-ha. As a contrast, in Germany, 4.8 kg-ha of Kumulus is permitted at certain growth stages. However, if you are in a sulphur deficient area or can show, by reference to soil samples, that sulphur is deficient in your soil, and can persuade a FACTS adviser that your vines need more sulphur, then 10 kg-ha can be applied. Potassium bicarbonate, which has EAMU approval for control of PM (and a zero-day HI) can also be used. New for 2020 is Thiopron, a sulphur which has been very finely ground and comes in liquid form, which is said to have increased activity and, at the full rate of 5 litres-ha, is claimed to have activity against *Phomopsis*. There is more on *Phomopsis* in *Chapter 17: Trunk diseases*.

The main point to remember about PM is that it is relatively easy to prevent the disease from starting, but relatively difficult to eradicate it once it has got going. If you leave spraying until you see the first signs of the disease, you are probably two sprays behind the curve and you will have to use more expensive, curative, rather than cheaper, protective materials. An open canopy will greatly aid fungicide applications, and deleafing immediately following flowering will allow pre-bunch close sprays to reach all parts of the bunch.

Downy Mildew

Downy Mildew (DM), also known as *Plasmopara viticola*, Peronospora and (in France) *le Mildiou*, originated, like PM, in North America. It was first seen in Europe in 1878 and

12 The Mekon was the evil leader of the Treens (from northern Venus) in the Dan Dare story in The Eagle comic of the 1950s and 1960s.
13 Strobilurins are sprays developed from a toadstool called Strobilurus tenacellus.

quickly spread, and is now endemic to almost all the world's vineyards. It overwinters on fallen leaves and, once present in a vineyard, is impossible to eradicate, although with regular sprays it can be kept under control. It likes warm, damp conditions, between 20–25°C is considered optimal, and will attack any green part of the vine once the weather warms up. The first signs of DM are either light discolorations on the surface of the leaves, which in time will turn darker – the so-called 'oil spots' – or small patches of mould on the underside of the leaf which appear in between the veins of the leaf. Inflorescences can also be infected, shrivelling and turning brown with white mould on the stem of the inflorescence. On the berries, DM will show itself as a white fungal covering or, later on in the season, will shrivel the grapes and turn them leathery (hence the German name for the disease *lederbeeren* – leather-berry). If present in sufficient numbers at harvest time and if machine harvested, infected berries can impart a characteristic mouldy taint to the wine. A common pattern to the spread of the disease is that chemical control is gained by the grower in the early and middle parts of the growing season, but, as the leaf-wall increases in thickness and density and spray penetration into, and air movement within, the canopy becomes more difficult, late-season outbreaks of the disease occur. In these instances the leaves, especially the younger ones at the top of the canopy, which by this time are contributing significantly to the overall level of photosynthesis, become damaged and sugar production is reduced. This leads to a slowing down of the ripening process and it is therefore important to keep up protection against DM until as near to the end of the season as HIs will permit.

Control of DM is helped by the usual physical ones – opening up the leaf-wall to light, air and drying winds – but these alone will not eradicate the disease and spraying will be necessary. In 1885 it was discovered that sprays containing copper were effective against DM and Bordeaux Mixture (a blend of slaked or hydrated lime and copper sulphate), was developed.[14] Bordeaux Mixture is still a potent protective spray against the disease and its characteristic blue/mauve dusty traces can be seen in vineyards throughout the world, although it is not permitted in commercial vineyards in Britain (although it can still be bought in garden centres for domestic use). The disadvantage of Bordeaux Mixture is that it needs preparing freshly each time it is used and can, if over-used, lead to increased levels of copper in the soil, which is toxic to worms and other soil organisms Copper in the form of copper oxychloride is currently (2020) permitted in vineyards in Britain, although most growers will only use it for a post-harvest winter wash, or for early pre- and post-budburst 'clean up' sprays. Cuprokylt, the most widely used copper-containing product in Britain against DM, has an HI of 21 days, but best practice is not to use it anywhere near harvest. This is because of the danger of traces of copper

getting into the fermentation is something to be avoided and, as has been already mentioned, can lead to a copper casse (*casse cuivreuse*), which causes a haze in wine. Copper oxychloride is about to be phased out and will be replaced by copper hydroxide which, although it contains less actual copper, is more effective at controlling DM. There are a number of other products currently available for DM, both protective and curative, and the choice of which material to use will depend upon your spray regime and disease status of the vineyard. Valbon (which contains mancozeb and is likely to lose its approval relatively soon) and Percos are the most widely used protectants in Britain. SL567A has been used as a 'fire brigade' treatment when DM becomes a problem, but its over-use in some vineyards has led to resistant strains of DM developing and it is no longer effective on these sites.

Botrytis

Botrytis (*botrytis cinerea*), also known as Grey Rot or Sour Rot, is a very common fungus that occurs widely in nature and attacks all manner of vegetable, salad and fruit crops. Lettuces that turn brown and start to liquefy and strawberries and raspberries that subside into a mass of grey/mauve mould are all being attacked by botrytis. It particularly likes wet and damp conditions and thrives where sugar is present, hence its liking for grapes, especially when grown in climates with summer and autumn rain. It is also a problem for table grape growers, if they have to store their grapes, and for nurserymen who store wood for grafting. botrytis is also one of the most damaging of diseases in vineyards as, given its preference for sugar-rich and decaying fruit, it often only appears late in the season when grapes are in their final stages of ripening. This makes chemical control difficult, as almost all anti-botrytis chemicals have a HI of at least 21 days and some have a 28-day one. Having said the above, I actually believe that Britain suffers fewer problems with botrytis than in vineyards in warmer climates. Why? Our temperatures, especially night-time temperatures during the month before harvesting, are lower; grape sugars are lower, giving the fungus less to live on; acid levels are higher; and pH levels are lower. All of these tend to make the growth of botrytis less than it would otherwise be in warmer climates with sweeter, lower-acid grapes. In many vineyards, with good canopy management and good chemical control, the occurrence of rampant, grape-devouring botrytis, such as we used to see in crops of Huxelrebe, Müller-Thurgau and Reichensteiner in the 1980s and 1990s, is a thing of the past. Having said that, some varieties, especially early ripening varieties such as Optima, Ortega and Siegerrebe (which have high sugar levels when both daytime and night-time temperatures are higher than they are in say mid-October), may well be attacked.

In the winery, if white grapes are heavily infected, some sulphur dioxide (SO_2) added to the picking bins (in the form of potassium metabilsulphate), better pre-fermentation settling and a prompt fermentation using a high dose of active yeast, may well produce a wine without any noticeable botrytis taint. However, some fruit aromas and

---

14 It is said that growers with vineyards by the roadside would spray their vines and grapes with copper sulphate (which is both visible and bitter to the taste) in order to stop passers-by helping themselves. Once Downy Mildew appeared, it was noticed that these vines did not get the disease.

flavours will have been lost, the colour may well be a shade or two darker and more SO_2 will be required in the finished wine in order to achieve a good level of stability. In red wines, where fermentations take place in the presence of skins and pulp, more serious problems may arise including off-flavours and mouldy taints which will need charcoal fining to remove them. Wines made from botrytis-infected grapes will also be more difficult to filter and clarify because of the presence of glucans.

The damage botrytis can do if left to develop is not to be underestimated, not only to the current year's grapes, but also to the buds holding next year's crop. It is my belief that one of the reasons why Seyval blanc crops so much better than botrytis-prone *vinifera* varieties is that it rarely suffers bud loss from this disease. In Britain we also used to see a lot of late-season stem-rot (Bunch Stem Necrosis), especially in years with high yields. Botrytis will attack the stem of the bunch so that the bunch will eventually drop to the ground. It is not untypical to see the ground beneath the vines carpeted with healthy-looking bunches that have had had their stems destroyed. Mineral deficiencies (especially of calcium and magnesium) can help weaken stems, so vineyard nutrition also plays a part in the control of botrytis.

The control of botrytis is helped by creating an open canopy, by shoot removal and deleafing in the fruiting zone, and by good air movement beneath the vines. Spores are transmitted from one part of the vine to the other by dripping water, so the quicker the leaves and shoots dry out, the less the disease will spread. An open canopy will also allow chemical sprays to reach their target more easily.

Botrytis is a disease that has a reputation for becoming resistant to certain chemicals. In the late 1960s and well into the 1970s, as has already been mentioned, the main products used against botrytis contained an active ingredient called benzimidazole (marketed as Benlate or Benomyl) which for some years were very effective. However, growers found that after using them for several years on the same fields, not only did their worm populations start to dwindle, but also the effectiveness of the products fell away and it was realised that their particular strain of botrytis (it has a habit of becoming site specific) had become resistant.

The next generation of anti-botrytis products were based upon active ingredients called dicarboximides – Ronilan, Rovral WG and Sumisclex were the best known – and these proved reliable for a decade or more until, once again, their effectiveness tailed off and resistant strains of botrytis surfaced. In 1995, a completely new product called Scala, based on an active ingredient called pyrimethanil, was introduced. This is a systemic product (all the previous products are contact only) and its use is restricted to a maximum of only two applications per year. It is, however, one of the most effective materials against botrytis, but it has a 21-day HI (although in Australia and New Zealand it only has a 7-day HI). Other products based upon different active ingredients, such as Teldor, Switch and Prolectus (all of which are limited to one or two applications per season), are also permitted for use in Britain. Prolectus has a 14-day

HI, and some winemakers have suggested that its use so near to harvest has resulted in problems in the winery, so caution is advised. In Germany, it has a 21-day HI and it is suggested that this be adopted in Britain as well.

In addition to the pesticides listed above, there are some BCAs permitted for use against botrytis on grapes in Britain. Serenade ASO is a bio-fungicide based upon *Bacillus subtilis* and has a zero-day HI, making it useful for the final anti-botrytis spray, although it is one of the most expensive products on the market. Prestop is another bio-fungicide, based upon *Gliocladium catenulatum* which also has a zero-day HI; others are Amylo-X WG, also fungus based; and Romeo (a yeast-based product). Some growers will also use Karma, which contains potassium bicarbonate, to help control botrytis. How effective these non-chemical BCAs are is open to question and not everyone who has tried them has been totally happy. Serenade and Amylo have not proved that effective, and the jury is out on Botector, said to be very effective, but which is new for 2020.

The third anti-botrytis recommendation is for a growth regulator called Regalis Plus which, if sprayed at 30–50 per cent cap fall, will open up the bunch by reducing the number of berries on the bunch. In years with good flowering conditions, where bunches are expected to be large, and especially with tight-bunched varieties (Chardonnay and some clones of Pinot noir for instance), this could be considered. Growth regulators are nothing new in grape growing, and all table grapes are sprayed with them to produce the large, perfectly shaped bunches of disease-free grapes that we expect to see in supermarkets. Whether many growers in Britain are using Regalis Plus is a good question. It is certainly not widely used (or even widely talked about). Many believe that bunch decompaction, whether induced chemically or triggered by de-leafing – which interrupts the supply of carbohydrates during flowering – is a major help in botrytis control.

Botrytis is usually present in the vineyard throughout the year. It overwinters in the shape of *sclerotia* (hard, dark brown encrustations of the dormant fungus) which are present on canes and on grapes that have not been picked or which have fallen to the vineyard floor. As soon as spring conditions are suitable, the *sclerotia* will provide the nucleus for the fungus to infect the young green tissue in the vineyard. Given that Britain has relatively damp growing conditions and most of our varieties are susceptible, in most years spraying against botrytis will start with a good pre-flowering spray, followed by at least one during flowering and two to three after. On the other hand, there may be years, when growing conditions are dry and when the variety in question has a degree of in-built resistance, when one spray prior to flowering and two after will suffice. As most anti-botrytis products are relatively expensive, their use needs to be carefully targeted in terms of timing.

As with many diseases, the damage botrytis can do early in the season, if uncontrolled, will make life much harder later on. Botrytis will infect the flowers and in a year with wet flowering conditions, especially when the flower caps struggle to detach themselves from the flowers, the disease can remain trapped in the centre of the bunch. As the grapes expand and the bunch closes, the disease will be deep inside and no amount of spraying will have any effect. In varieties with tight bunches – the Pinots and Chardonnay for instance – berries can actually burst inside the bunch due to the pressure from expanding grapes and this is often where internal botrytis starts and is impossible to control. This will lead to classic botrytis bunch rot and may result in substantial, even total, crop loss. The most important single spray to help control botrytis is the one applied as flowering finishes. This will hopefully do two things: blow out and disperse any fragments of flower caps and stamens left within the bunch and coat the remainder with a protective spray. Botrytis is also a secondary invader and will take advantage of grapes that have been damaged by such things as machinery, hail, wasps, other insects or birds. In these situations, where the skin of the grape has been split and the pulp exposed, botrytis will develop and gradually infect the whole bunch.

The use of botrytis to produce sweet wines (Sauternes, Tokay, German sweet wines etc.) through the action of what the wine trade calls noble rot (and known as *Pourriture Noble* in France and *Edelfäule* in Germany) is certainly rare in vineyards in Britain, although not unknown. Most of the very good, even excellent, sweet wines made in Britain have been made using high-sugar varieties such as Ortega, Optima and Huxelrebe, and many have been outright accidents where growers have seen rampant botrytis devour their crop, but realised that what they were witnessing was classic noble rot. The best of them have a luscious fruit character with a backing of superb acidity (Denbies' Noble Harvest is generally excellent).

Once grapes reach a potential alcohol level of around 7 per cent (56°OE), botrytis will affect the grape in a different way. Feeding on the skin of the berry, the fungus will puncture the skin without splitting it and will allow water to leach out. This happens in a slow controlled manner and over time – perhaps up to six weeks – individual grapes on the bunch will shrivel considerably. Such juice as is left in the grape will be much lower in volume, but, in terms of sugar content, will be much higher. Acids will also be preserved, something really sweet wines need plenty of if they are to be balanced on the palate. For noble rot to happen, several things are necessary: the crop must be clean up until the point where the sugar level rises to the correct level; the weather should be warm, even hot; and humidity is required for the botrytis spores to develop. In Britain, we quite often have two of these – the rising sugar level and the humidity – it is just the hot weather that sometimes eludes us.

Other diseases

If control of the mildews and botrytis has been achieved, then there are few other diseases that will make life difficult for British vineyard owners. *Phomopsis* (dealt with in *Chapter 17: Trunk diseases*) and what the Germans call *Roterbrenner* (caused by *Pseudopezicula tracheiphila*), plus Black Rot (caused by *Guignardia bidwellii*) and Anthracnose (also called Black Spot) are all occasional visitors, but cannot be classed as 'important' in British vineyards and will all be controlled by the same fungicides already being used.

Viruses

Jack Ward, in his 1984 book *Vine Growing in the British Isles*, saw fit to devote a whole chapter on 'virus diseases' listing *Court-noué*, Yellow Mosaic, *Flavescence Dorée*, Grape Leaf Roll Virus, Pierce's Disease, Yellow Vein, Asteroid Mosaic and Vein Banding as being of danger and there were also another seven which had been 'identified and reported' (although not, I assume, in British vineyards!) How times have changed. Today, you would have to look quite hard to find many virus-infected vines in British vineyards, such has been the improvement over the last thirty years in virus control throughout the whole production process of vines. Both rootstock and scion varieties have been subjected to 'heat treatment' (which is not the same as hot water treatment – HWT – of which more later) to remove viruses from a single vine of each rootstock variety and each scion variety and then using that single vine as the basis for all future mother blocks of that variety. This technique, together with the enforcement of the pant passport system and the ELISA[15] tests for viruses, has reduced the incidence of vine viruses to a point where, in many European regions, they are almost unknown. Insects such as mealy bug and various types of sharpshooter, plus nematodes (of which there are several types), all of which can act as vectors for both viruses and diseases, are luckily absent from Britain, mainly on account of our relatively low temperatures and our isolation.

In other parts of the world where temperatures are higher and controls are not so strict as in Europe, and in *Phylloxera*-free areas where the practice of taking wood from vineyards in order to produce rooted cuttings is common, viruses, especially ones like leaf roll and corky bark, can cause problems. I have in the past seen instances of crown gall and occasional leaf and shoot deformities (flat, double bud canes) which might suggest virus problems, but these were very

---

15 ELISA stands for Enzyme-Linked ImmunoSorbent Assay and is a test for viruses in many different areas: plants, food and human tissue.

Chapter 17

Trunk diseases

Note 1: The discussion of trunk diseases in the confines of this book can only hope to provide an overview of the subject and to offer some practical advice to vineyard owners. Links to further reading on the subject can be found throughout this chapter.

Note 2: This chapter was originally written in 2014 (although it has been updated and edited for this version of the book) when the spectre of trunk diseases seemed much more real than it does today, in 2020. Now that we have had a few years to see how things worked out, in general terms I would say that today, trunk diseases are not the major issue we were led to believe they might become. Most nurseries have taken measures to make the situation better and I like to think that most vineyard owners (but by no means all) in Britain now pay more attention to the basics: site selection, site preparation, planting and post-planting care. However, six years in the life of a vine is very little and as vines age, their vigour declines and the amount of older wood on the vine builds up, both conditions which favour trunk diseases. Therefore, with regard to trunk diseases, growers should never be complacent and must always be on the watch for the first signs of damage.

Trunk diseases

In the summer of 2011, winegrowers in Britain were alerted to a new threat to their vines, a collection of conditions generically known as TDs. Dr Richard Smart, as a WineSkills mentor, had visited 'around 30 vineyards' and had found 'symptoms of *Botryosphaeria* [one of the major TDs] in all but one'. Smart also reported that in most vineyards the level of infected vines was 'less than 5 per cent', but in a few 'the disease had spread rapidly with over 50 per cent of vines infected'. Whilst there were some who thought he was being unnecessarily alarmist and many whose vineyards appeared to be unaffected by TDs, the fact is that they are a worldwide problem, and becoming more so, especially in young vines

(Young Vine Decline), and therefore need to be discussed in some detail.

Many perennial species suffer from diseases which affect their permanent woody parts and in recent decades in Britain we have been introduced to Dutch elm disease, sudden oak death, red band needle blight, horse chestnut bleeding canker, beech tree phytophthora and ash dieback, all of which affect woody plants, namely trees. Exactly why British trees have been invaded by these new threats is, in many cases, still uncertain, although most authorities believe that climate change plays a part. We therefore need to be aware that vines have not been singled out for special treatment and view TDs in the light of other changes in our environment, learn their causes and learn how to manage them.

TDs of vines of course, are far from new and one of the major diseases, Esca (also known as Black Measles in the USA), was noted by Pliny the Elder in his book on natural history – *Naturalis Historia* published in AD 77–79 – and has been with us ever since. In slightly more recent times, the late 1800s and early 1900s, the famous French viticultural researchers Pierre Viala and Louis Ravaz carried out studies on Esca, then causing widespread problems in the Midi, France's great bulk-wine producing region. Eutypa dieback (*Eutypa lata*) is another similar TD which has been present in vineyards for many centuries. Historically, TDs tended to affect older vines for a variety of reasons. The build-up of older wood and the presence of large pruning wounds probably allowed pathogens to enter the plant and slowly destroy it. Plus, older vines are typically less vigorous than younger ones and are therefore more likely to succumb to problems which in their youth they might have survived. Before vines were grafted, the practice of layering[1] vines to

---

1 Layering (or provignage in French) is where a suitable cane from one vine is laid down on the soil and buried at the point where a new vine is required. Quite often the cane is slightly cracked and a rock placed on top of it to keep it in position. The cane then sends out a shoot which can be staked and trained and in time, the cane from the 'mother' vine can be removed.

Rootstocks bundled up prior to grafting

replace a dead vine or even to re-invigorate a fading one, was commonplace (and still is in regions without *Phylloxera*) and this tended to obscure the problem of TDs. In more recent times however, TDs seem to have been making something of a comeback in vineyards all around the world and are today a source of concern, especially as they appear to be responsible for the already mentioned Young Vine Decline. The major TDs, as well as Esca and Eutypa, are *Botryosphaeria* canker, the most widespread of the many TDs, also known as Bot Canker or Black Dead Arm; Black Foot, caused by *Cylindrocarpon* fungi; Petri Disease, also known as Black Goo, Young Esca or black xylem decline, and caused by *Phaeoacremonium* and other fungi; and *Phomopsis viticola* also known as Excoriose and Dead Arm. Associated with Esca and Eutypa is also 'apoplexy' (sudden death of vines) which the French refer to as *folletage* or *tylosis*.

Exactly why TDs have become a modern problem in vineyards worldwide is a question to which there is no simple answer. One contributory factor may be viruses – or to be more precise – the lack of them, plus a general improvement in the pest and disease status in most vineyards. In the decades immediately after the Second World War, one of the most widespread problems in grapevines, which could not be tackled anywhere else but at source, i.e. in mother blocks and in nurseries where vines were produced, were viruses. Since the introduction (in the EU at least) of Plant Passports and mandatory ELISA testing for viruses, plus mandatory HWT for vines produced in regions where *Flavescence dorée* (Grapevine Yellows) is a problem, the overall health of vines has improved hugely. Higher quality grapevines, plus other viticultural improvements such as better trellising, better canopy management, better weed control and – probably most importantly – much better chemical pest and disease control in almost all vineyards, has, somewhat paradoxically, allowed TDs some space in which to work. This is somewhat

like the increase in cancers affecting humans. Now that we no longer die of smallpox, polio and several other once-common ailments, and life expectancy has risen (at least in most First World countries), cancer rates have also risen. Another often cited reason why Esca in particular has become a major problem in some regions is the banning, in 2001, of sodium arsenite, a fungicide which, when applied as a winter wash, reduced the incidence of Esca (as well as *Phomopsis*), and was widely used in mother blocks.

To understand and to try and get to grips with the problem of TDs, I have divided the topic into three areas:
- The source of TDs and what can be done to reduce the incidence of infected vines being brought into a new vineyard.
- The measures that should be taken when establishing a new vineyard that will allow newly planted vines to survive TDs and thrive.
- What practical steps can be taken in an established vineyard to reduce the likelihood of TDs becoming a problem, and dealing with vines suffering from TDs?

The source of trunk diseases

There can be no doubt that infected mother blocks of both scion and rootstock wood are two of the major sources of infected material. Whilst it is a simple matter to state, the solution is not so easy to find. Having spoken to several French, German and Italian nurseries on the subject, some of whom deny knowing anything about the subject at all – 'what, my vines, trunk diseases?' – the facts are that the organisms causing TDs are, in many instances and to a greater or lesser extent, present in all mother blocks.

What nurseries will tell you is that NO vine is sterile and is 'infected' by many different organisms and even if it was sterile, it would soon be colonised by organisms, beneficial and non-beneficial, once planted in its final resting place. What nurseries will also tell you is that they have been selling vines for many years, many decades even, to growers whose vineyards thrive and who continue to come back for more vines – both of which are true – and whether or not their mother blocks are infected, nurseries live or die by the quality of their vines and their reputations and they do care about this.

My experience, having been buying vines from the same two nurseries since 1977, is that if correctly handled, planted and cared for, most vines survive, and losses after three years are typically below 2 per cent and often below 1 per cent. However, there is also no doubt that the TD problem is a real one and nurseries remain in the firing line as one of the main sources of infected material. To put the problem into perspective, one needs to know the process and timeline of the production of a grafted vine, a living plant that comes together from several sources and usually takes more than fifteen months to produce.

Production process for grafted vines

The worldwide vine nursery business produces many hundreds of millions of grafted vines per year and is a multi-faceted industry. Smaller nurseries may well be growing very little of their own scion or rootstock wood, buying in what they need from suppliers, but grafting and raising all their own vines which they then sell on to vineyards. Others will grow much of their own scion and rootstock wood, buying in what they are short of, and producing grafted vines for sale. There are also specialist rootstock producers who may sell the bulk of their production to nurseries, but who also produce grafted vines. Finally, there are specialist growers who produce grafted vines under contract and return them to nurseries who sell them as their own production. In short, it is a complex industry. One of the largest vine nurseries in the world, VCR in Italy, in 2020 produced: '60 million grafted vines, in 4,000 different combinations of variety, clone and rootstock, for distribution to 28 vine growing countries' and therefore one can imagine the complexities of keeping tabs on everything from start to finish. In Europe, most of the production of rootstocks is confined to the frost-free southern regions of France and north Italy and in regions where hail, which can seriously damage wood, is less of a problem. The production of scion wood is more widespread as varieties and clones are often only grown in certain regions.

The first stage in the production process is the obtaining of (a) wood for the vine variety to be grafted (the scion wood) and (b) the wood for the rootstock. The vineyards that provide this wood may be owned by, or under the direct control of, the vine nursery, or may belong to suppliers who make a living by trading in scion wood and rootstock wood. In the case of scion wood, the vineyards will typically be planted with 'propagation' grade plants which are certified to be true to the variety and clone specified and free of viruses. However, these are also producing vineyards, from which grapes will be harvested, and are managed and kept pest and disease free in the normal way. These vineyards are inspected at least twice during the growing season for pests (especially the leafhoppers which are the vectors for *Flavescence dorée*), signs of diseases or abnormalities – *Phylloxera,* viruses, TDs – and those vines from which cuttings may not be taken are tagged. During the dormant season, scion wood is then taken from the non-tagged vines, bundled up and by degrees, makes its way to the nursery. Whilst scion wood is typically taken before the vines are pruned, I have known it to be taken from the wood left over after pruning, which may have been lying on the ground for some while. Once back at the nursery, the scion wood is sorted, cut into one-bud sections, disinfected, bagged and placed into cold storage.

Vineyards for rootstock wood are also planted with propagation grade plants, guaranteeing the variety and clone of the rootstock. Traditionally, vines in rootstock vineyards are not trellised, and are allowed to sprawl across the earth, forming, in the summer, a green mat across the surface of the vineyard. Rootstock vineyards are usually in warm to hot regions, where summer rainfall is low and therefore most are irrigated. These vineyards are, like their scion wood counterparts, inspected and vines from which no wood may be taken, are tagged. Pest and disease control is fairly basic, given that rootstock varieties are typically quite disease resistant. In more recent years, and partially in response to problems such as TDs, some nurseries are now growing their rootstock vines lifted off the ground on 'tables' with the canes supported by wires. As the canes are not in contact with the soil, this makes air circulation better, and, after leaf-fall, they are not lying on top of a mulch of decaying leaf matter which improves their health status. However, this is a more expensive way of growing rootstocks. The amount of wood produced is lower and the cutting and collection costs are higher and nurseries are naturally reluctant to increase their costs when their competitors may not be. Vines for rootstock production can be grown on the same ground for as long as the vines are producing viable amounts of wood – twenty-five years is not uncommon – but once grubbed, the land has to remain free of vines for eleven years (at least, that is the rule in France) and for this reason, many rootstock producers rent land rather than own it. I have asked several rootstock producers whether their mother vines are ever tested for TDs and have yet to find one that does, even though some researchers believe that TDs such as Black Foot are endemic in many rootstock vineyards. Like scion wood, wood for rootstocks is then harvested after leaf-fall, taken back to the nursery, sorted, put through a de-budding machine to remove all the buds, and then disinfected, bundled up and put into a cold store to await the grafting procedure or for sale to other nurseries.

Grafting starts in the early part of the year. Almost all vines are machine grafted, the Wagner Company's 'Omega' machine being the most widely used, but other types of machine exist and there is even some hand-grafting still carried out. The grafting process is fairly straightforward, although the sizing of rootstock and scion wood is important so that the cambium layers (in effect the 'bark') of the two parts touch each other as much as they can. This makes the callusing better (the physical joining of the two different species of wood), the graft stronger and, most importantly, the percentage of grafted vines that will survive the whole process, and thus be available for sale, higher. What is also important is hygiene at all stages of the production process from the time that both scion and rootstock wood is harvested, to when vines are delivered out to producers. Apart from diseases picked up in the vineyards, storage conditions are important. Wood is often soaked prior to storage and/or grafting in order to make it more pliable and it has been shown that the re-use of water for soaking is probably one of the main points in the production process where disease pathogens can be transferred from one batch of wood to another.[2] During the grafting process, if the

---

2 A recent paper: Soaking grapevine cuttings in water: a potential source of cross contamination by micro-organisms by Waite, Gramaje, Whitelaw-Weck-

Newly grafted vines in the nursery

wood is too wet, it is also thought that the blades on the grafting machines, as they slice through and join the scion to the rootstock, may well be forcing TD organisms into the graft. There appears to be no way (at least, not at present) of spraying a disinfectant or fungicide on to the cut surfaces of the graft as they are joined together.

Immediately after grafting, the vines have their top 50–75 mm waxed. This helps keep the grafts together and keeps them moist and clean. After this they are placed, tightly packed together with damp sawdust and peat around their lower halves to keep them moist, into callusing boxes which are put into a heated greenhouse. The ground that the boxes are standing on will usually have soil warming cables in it which will help the roots start to grow. After a few weeks, the scion shoots will have started to grow, roots from the bottom of the rootstock will have appeared, and the graft will have callused over. At this stage, the vines can be taken out of the callusing boxes, re-waxed and placed out in the nursery fields for the remainder of the summer. The young vines will often be planted through black polythene mulch to help weed control and aid root development. During the summer, the vines will often be irrigated and will be kept free of pests and diseases with a normal spray programme. They will also be inspected for viruses and other diseases and abnormalities and will be rogued accordingly. Land used for growing grafted vines can only be used for one season and then has to be rotated and remain free of vines for six years (again, the French rules) and using GPS, inspectors make sure that plots do not overlap from one year to the next. Again, because of this requirement, land used for this purpose is often rented rather than owned by producers.

At the end of the growing season and when they are dormant, the grafted vines are lifted, and taken back to the nurseries. Here, they will have their roots washed and trimmed, and the growth they have made from the scion bud during the summer cut back to one bud, after which

ert and Hardie (Phytopathologia Mediterranea (2013) 52, 2, 359–368) has shown that soaking wood is a major source of infection and that the: 'development of nursery management strategies that avoid exposing cuttings to water during propagation is an essential first step in reducing the transmission of trunk diseases in propagation.'

they will be re-waxed, bundled into twenty-fives, have a plant passport attached and placed back into cold store until required by producers. How they are stored (wrapped in polythene, in ventilated bags, in sealed bags), the temperatures they are stored at (and what the temperature fluctuations are), and the level of humidity, can all play their part in the development or otherwise of pathogens during storage. Then, when they are needed for planting, they are taken out of cold store, loaded on to transport (for the most part not temperature controlled) and taken to the customer, which, in the case of Britain, might take several days. On arrival at the customer's premises, vines might be placed back into cold store (my preferred option) and only taken out a day or two prior to planting. Some will store them in a cool, dark place awaiting planting; others will heel them into damp soil. Eventually, after what is often an eighteen-month production process, the vine will be planted in its final resting place and face its greatest challenge – a new vineyard!

In some cases, vines will not be sold in the year after production and can be put back into the field for a further year. Whether this is a positive or negative I have no idea, but it certainly happens on occasions. One nurseryman I know thinks he ought to charge more for two-year-old vines as they have more developed grafts and better roots! Nurseries will also trade with other nurseries, buying and selling scion wood, rootstock wood and finished vines. Towards the end of the ordering season, when a nursery may have run out of home-produced vines, I am quite sure that rather than turn down an order, some nurseries will buy in vines from other nurseries, adding another layer of complexity to the business.

As you can see from the above, the production of grafted vines is a multi-layered business and with respect to TDs, there are several occasions when perfectly healthy plant material can be infected by contact with less healthy material: in the scion vineyard, the rootstock vineyard, during the pre-grafting process, in the grafting process and during storage and transport. At all stages, cross-contamination could occur. One also has to remember that vines are composed of wood grown out of doors and therefore contamination from both vine and non-vine species on neighbouring land is always a possibility. The wood is also grown and harvested under conditions where natural rainfall is a common occurrence and where there are several opportunities for the wood to come into contact with soil which may well contain harmful pathogens.

In the last few years, some of the major vine nurseries have accepted that *maladies du bois de la vigne* are indeed a problem that needs addressing and it is good to see companies such as *Pépinières Guillaume* with their 'VigoRhize' process which uses both *Trichoderma* and mycorrhizal fungi in the production process and *Mercier Pépiniériste* with their 'Clean PROCESS' system which results in what they call a 'Clean PLANT'. This patented system is claimed to cover better traceability of the materials used in the production of grafted vines, the sterilisation and disinfection of plants and premises, where appropriate, through all stages of the production process, growing in the presence of mycorrhizal fungi, plus storage under ideal conditions. It also includes 'continuous health monitoring' of the plants. Mercier are also using a strain of *Trichoderma* (*Trichoderma atroviride I-1237*) 'to restore the microbial balance of the plant' and they market grafted bare-rooted vines as 'FORCE-T' vines. Whether the use of *Trichoderma* actually does help control TDs is open to question and some researchers suggest its powers have been exaggerated.

The International Council on Grapevine Trunk Diseases[3] (ICGTD), which was formed in 1998, has been at the forefront of research into the problem and to date has held eleven mainly biennial workshops on the subject (International Workshop on Grapevine Trunk Diseases – IGWTD). It has published a large number of documents, including one titled 'Grapevine propagation; principles and methods for the production and handling of high-quality grapevine planting material' (This document can be downloaded at www. englishwine.com/wgigb.htm). Written by the three foremost researchers on the subject, Helen Waite, David Gramaje and Lucie Morton, it is a very thorough, step-by-step, blueprint for producing, storing, transporting, and planting both rooted cuttings and grafted vines (which includes 'potted' vines) and 'is presented as a common starting point for propagators, grape growers and scientists to support the production of planting material that is of the highest quality' and should be mandatory reading for anyone contemplating buying grapevines and establishing a vineyard. However, whether all its recommendations – at least in the short term – are entirely practical is another matter. One of the most telling paragraphs in the report is the following. Note that the bold emphasis and underlining are in the original:

'**It should be noted that regardless of how well they are managed, the manner of cutting production predisposes mother vines to infection by trunk disease organisms. Infected** <u>cuttings</u> **have no visible internal or external symptoms. Visible symptoms are not apparent for several months after the infective agents (spores or mycelium) move into the tissue of the current season's canes. Therefore, it should be assumed that ALL cuttings have some level of infection.** <u>**Hot water treatment is the only control currently known to be effective.**</u> **The application of fungicides does not effectively control internal pathogens'.**

Much of the problem with getting nurseries to change their production methods is that many of the improvements suggested in this document are (so the nurseries will say) impractical, likely to result in a reduction in the percentage of vines that make it from grafting to sale, cause damage to otherwise viable vines and in almost all cases, raise costs. Subjecting mother blocks to annual laboratory analysis, growing rootstocks on trellises, not soaking cuttings prior to grafting, not processing finished vines in the same area where vines are being grafted, sampling of vines after lifting ('5–10 vines per 1,000 to be dissected'), and the complexities

of effective HWT are all very desirable measures and I am sure, in time, that some will become standard industry practice. For the present though, I suspect they will not.

On the question of HWT in particular, which the report covers in much detail, the authors do not come out and say that it should be mandatory, or even advise that it is best practice, even though in the paragraph above, its effectiveness is underlined. In an article in *Wine & Viticulture Journal* of Nov-Dec 2013, Helen Waite points out that: 'although hot water treatment is currently the most effective control for known trunk diseases, it is not necessarily a completely effective eradication treatment. Very low levels of infection may persist in a small percentage of treated cuttings and act as a subsequent source of cross-contamination if sanitation is not maintained throughout the propagation period.' Nurseries are, except for vines going to regions where *Flavescence dorée* is a problem (as has already been mentioned), usually unwilling to carry out post-grafting HWT as the levels of losses and claims from their customers are too great. By doing it before grafting and planting out, then at least they are going some way towards lowering the risk of infection. It should also be noted that ungrafted vines, i.e. rooted cuttings, which are widely planted in countries and regions where *Phylloxera* is not present, are more successful when subjected to HWT, as they do not then have to go through the whole grafting procedure. It is interesting to note that on the question of HWT, *Mercier*, who do not use it as part of their patented vine production system, state: 'Hot Water Treatment is extremely traumatic for the young vines. It's [sic] consuming most of the vine reserves, that causes bud break delays (15–30 days) and in difficult conditions (poorly prepared soils, hydric stress, strong heat etc), the take rate can be seriously impacted.' How true this last statement is open to question. One researcher I asked about it said that this was an excuse for not being able to carry out HWT correctly, with the necessary pre-HWT acclimatisation times, post-HWT cooling times and subsequent storage conditions being adhered to.

The final chapter of the Waite, Gramaje and Morton document, under the heading 'Pre-planting care of vines' suggests that growers should:

'*Dissect a random sample of vines from each batch. Cut the rootstock, the scion and the graft union and look for dark staining and/or spotting in the tissue. Any tissue discolouration in the wood indicates that the vines have probably been infected with one or more of the fungi that cause Young Vine Decline and should not be accepted'.*

Whilst this may be 100 per cent sound advice, I suspect that there are very few nurseries and vine suppliers (probably none) which would accept an order for vines on this basis. Whilst it is possible to test for the presence of a very wide range of organisms that might cause TDs, there are no numerical limits or levels that can be set to say if one vine is 'clean' and another vine 'infected' which is why a visual inspection is the best that the experts can recommend. The final section of the document headed 'Planting' contains some valuable advice including:

'Vines, particularly the roots, should not be allowed to dry out during planting. However, standing vines with freshly trimmed roots in a bucket of water is detrimental, spreads disease and should be avoided. It is better to cover bundles with a clean damp cloth'.

Given that this is the practice I have always adopted and recommended, I am pleased to see it in writing from such a group of experts.

Growers should also be aware that many of the organisms that cause TDs may well already be present in their vineyards and in the surrounding environment. Existing vines, especially if they are young (less than ten years old) may well have TD organisms in them which are not currently doing any harm. Apart from old pruning wood and general vine debris (including mummified grapes), several weed species can host TD organisms, as well as many species of hardwood trees which might be found in adjacent hedges and woodland. Given that many vineyards in Britain are in close proximity to trees, hedges, copses, woods and even full forests, this source cannot be discounted. TDs can be transmitted from any host (vine or other species) to another vine by both wind and rain splash and any wound or cut can be an entry point. Once present therefore, they are extremely hard to control and almost impossible to eradicate.

Measures to be taken when establishing a new vineyard

In November 2011, following a WineSkills workshop on TDs, the UKVA issued Bulletin No. 50 'Grapevine trunk diseases' authored by Dr Richard Smart, Chris Cooper (WineGB Technical Adviser), Jim Newsome, then a British based viticultural researcher and Duncan McNeill, the viticultural consultant. The report highlighted the fact that 'the problem appears worse in young vineyards, less than 10 years old. Older vines appear more tolerant'. It also pointed out the part played by environmental factors saying that: 'Other stresses on the plant exacerbate the diseases. Poor drainage, poor nutrition, soil compaction, drought, frost, weeds, other diseases all make vines prone to trunk disease.' Bulletin No. 50 can be downloaded at www.englishwine.com/wgigb.htm.

Stress, therefore, is something to be avoided in the vine's establishment phase. Smart has also accepted, in an e-mail written in August 2013 and sent out via WineGB's 'vine list' e-mail circulation system: 'that Britain is not the most favourable environment for vines, being generally too cool. I think that this stress, plus high rainfall which predisposes spread, is also a factor in the expression of poor health due to TDs in Britain.' Smart also stated: 'there is no guarantee that any lot of grafted vines is always the same, nor that any nursery is better than another, nor that any [vine sales] agent is better than another. One lot of vines may be good, another not'.

Given that the organisms responsible for TDs prefer humidity and are spread when conditions are wet and windy, British vineyards would seem an ideal place for them to survive and prosper. In addition, despite an improvement in our climate over the past thirty years, Britain must still be considered marginal for viticulture, especially when one factors in the change in vine varieties from predominantly German crosses plus Seyval blanc to Chardonnay and Pinot noir, both of which are more problematic with regard to fruit setting and ripening. Stress, therefore, in vineyards in Britain is an ever-present threat, especially that resulting from the poor accumulation of carbohydrates brought about by difficult conditions for photosynthesis, especially after harvest. At the time the threat of TDs were first raised in Britain (2011) there has been some quite adverse weather conditions: the very low temperatures in December 2009 and January, February and December 2010; the damaging spring frosts of April and May 2009 and 2010; the very dry winter of 2010–11; the hot and dry spring of 2011; and the strange year that was 2013. The weather during that period had placed a lot of stress upon young vines, especially those poorly sited, poorly planted and poorly managed.

Newsome, who attended the 2012 IGWTD, as well as being one of the authors of the UKVA's Bulletin No. 50 already referred to, made presentations at 2011 WineSkills workshops, at the 2013 UKVA Annual General Meeting and has written a 'Grapevine trunk disease review' dated July 2012 and an 'Update on Trunk Disease Research' dated July 2013 (both available at www.englishwine.com/wgigb.htm). What appears in the next paragraph is taken from these sources:

'We know that it is possible to have two vines with the same fungal profile but a differing expression of symptoms and also that trunk disease fungi can exist latently within the vine, i.e. present but not actively damaging the vine. Therefore, it is likely that there is a rôle for external factors in the expression of the symptoms that cause the vine harm. There has been some discussion of the potentially catalytic rôle of other micro-organisms such as bacteria and this is an interesting area for future research. Stress is also identified as a cause of the expression of TD symptoms in a paper by Brown et al. 2012, who showed that carbohydrate stress caused by partial defoliation of young vines increased the severity of Black Foot infections – a potential concern for Britain where a poor growing year can reduce a vine's carbohydrate stores. There is little doubt that stress can reduce a vine's ability to combat infection so especially with regard to new plantings, every effort should be made to reduce vine stress: to quote from Billones-Baaijens et al. 2013[4] *"...it seems likely that young nursery plants with latent infections can decline and die when planted into the more stressful conditions of vineyards.'* [Jim Newsome's italics and emphasis]

Newsome also wrote:

'It is also hypothesised by several authors that environmental conditions act as triggers to infections that

---

4 Billones-Baaijens R et al. M (2012) Pathogenicity of Botryosphaeriaceae species from New Zealand nurseries. ICGTD 8th International Workshop on Grapevine Trunk Disease 2012.

would not otherwise have meaningful impact on vine yield and health (Bonfiglioli & McGregor 2006; van Niekerk et al. 2006) and the extent of this impact is of particular relevance in Britain where environmental stress is common. Bonfiglioli & McGregor (2006) refer explicitly to vineyards in "*sub-optimal*" areas in New Zealand with heavy, poorly drained soil that are showing the most damage from *Botryosphaeria* infections. Therefore, a logical further control method in Britain is to follow best practice in site selection to maximise drainage, exposure and other positive variables as well as utilise viticultural techniques such as canopy management to minimise negative environmental stresses such as shading.'

Factors to consider in establishing a vineyard

Note 3: This section was largely written in 2014 when the threat of TDs seemed very much more real than it does today in 2020. The advice below should therefore be read in the light of what we now know about the problem.

Much of what it takes to prepare a site for planting has been discussed in detail in earlier chapters in this book, but here is a summary of the most important factors to consider in establishing a vineyard in Britain, in relation to avoiding damage by TDs. It is largely taken from an article I wrote for the UKVA's Grape Press Volume 156 February 2012 (available at www.englishwine.com/wgigb.htm):

- Site selection: Britain is still marginal for viticulture and there is no doubt that the problems faced in many vineyards, not only of TDs, are exacerbated by planting vines on difficult sites and in the more challenging areas of Britain.
- Windbreaks: Vines planted on exposed sites without the benefit of wind protection will suffer in several different ways. Many sites could be improved with effective windbreaks.
- Vine variety: Not all varieties will grow on all sites and attempting to establish hard-to-ripen varieties on lower quality sites will always be stressful for them.
- Rootstocks: Selection of the correct rootstock for the pH and active calcium carbonate ($CaCO_3$) level in the soil is most important.
- Transport and storage of vines: Vines will be in cold store at the nursery and should be transported to Britain, preferably in temperature-controlled transport, and kept in cold store until just prior to planting.
- Preparation of vines: Vines should be trimmed for the planting method being used. Normally, the supplying nursery will trim the vines prior to dispatch.
- Vines should not be soaked prior to planting, but kept damp (and in the dark and cool).
- Site preparation: This should start at least six months in advance of planting and twelve and even eighteen months is better.

- Bio-fumigation of sites using mustard and other brassicas has been shown to lessen the incidence of TDs in vines subsequently planted.
- Nutrition levels (including pH) must be sampled prior to planting and if necessary, adjusted with fertilisers being ploughed down to the root zone.
- Additions of humus, whether via green manures or by compost spreading – or both – will always help vine establishment and lessen stress.
- Poor drainage is without doubt detrimental to the long-term health of vines and must be addressed pre-planting.
- Subsoiling, using the correct implement and to the correct depth, must be done pre-planting, especially if planting by machine.
- Sites must be correctly cultivated, especially if being machine planted, so that roots are fully in contact with the soil after planting.
- Time of planting: Many parts of Britain suitable for vinegrowing can experience spring frosts until the end of May and there is no doubt that frost damage and the incidence of TDs are closely correlated.
- Planting: Whether planting by hand, using a water lance or planting by machine, the vine should have its roots pointing downwards and firmly in contact with the soil. Mycorrhizal root treatments at planting may help.
- Weed control: Effective weed control is one of the major factors in getting vines established and preventing stress. Poor weed control is probably one of the major causes of TDs.
- Cultivations in young vines: The practice of grassing down, whether by natural regeneration or by sowing seed, may stress young vines more than is necessary and therefore keeping a vineyard cultivated for the first year or two may be beneficial. However, not many growers do this and most well cared for vines survive.
- Grow-tubes: The tall, narrow 'grow-tube' type of individual guard may not provide the best environment for the young vine and its graft, especially if vines are not correctly shoot selected and side-shooted. The situation is probably made worse if weeds are allowed to grow up inside the guard. Larger diameter, partially ventilated guards provide a better environment.
- Pest and disease control: Apart from pests such as rabbits, hares and deer, plus slugs and snails, which will make establishment more difficult, if not impossible, vines getting Powdery and Downy Mildew in their first few years will be stressed. Except for varieties known to be resistant to mildews, preventative sprays should be applied as soon as leaves reach a targetable size.
- Early cropping: When I wrote this paragraph in 2014, there was a general view amongst TD experts that early cropping (i.e. in year two or three) might have a detrimental effect upon the long-term health of the vine. However, as I personally thought at the time, if the vine has established well in year one and grown a decent cane which can be tied down to fruit in year

two, then this poses no dangers to the vine. On sites where vines have not established well and/or where good establishment is irregular, then it is probably best to prune all vines back to one bud and not allow them to crop until year three.

Dealing with trunk diseases in an established vineyard

Note: Some of the advice below is slightly contradictory, some probably impractical (or impossible) and some will involve additional expense and/or changes to working habits and practices. Which, and how many, of the measures listed you need to adopt will of course depend on the level of TDs in your vineyard.

The following is an amalgamation of the advice given in the UKVA and Newsome documents already mentioned, plus advice and recommendations I have gleaned from attending two WineSkills workshops on the subject; and the UKVA's 2013 AGM presentation mentioned above.

- Early identification. Get good at recognising the symptoms during the growing season and tag suspect vines.
- Map the distribution of infected vines from year to year so that the progress of infection can be monitored.
- Remove infected vines and burn them downwind of the vineyard.
- Reduce stress on vines by attending to drought, waterlogging, poor drainage, nutrient stress, and weeds. Weeds can potentially carry TDs.
- Do not prune during rainfall or when vines are still wet.
- Double pruning, i.e. basic pruning early in the winter with cleaning up of canes in drier spring weather.
- Prune early in the season when spore levels maybe lower.
- Prune late when wounds are less susceptible and heal more quickly at higher temperatures. Sap flow may hinder entry of spores into wounds.
- Minimise the number and size of pruning wounds.
- Always make cuts at an angle that allows water to drain from cut surfaces.
- When making a large cut on older wood, leave a 25 mm stub protruding from the trunk.
- Paint pruning wounds with a wound paint to seal cut surfaces after pruning in areas with infected vines and on larger wounds on all older vines (see below for more on wound paints).
- Tag suspect vines and prune them in a separate operation to pruning apparently healthy vines.
- Treat vines adjacent to vines known to be infected, as infected.
- Cut back infected vines to well below where disease is present and train up water shoots to replace trunks of infected vines.
- The use of powered secateurs will enable cuts to be

made cleanly and in the most appropriate place.
- Disinfecting secateurs at the end of each day with a suitable product (Milton or propanol suggested).
- Remove and burn prunings.
- Mummified berries and infected wood, if left in the vineyard, can be a source for the TD inoculum. Good vineyard hygiene in general will be beneficial.
- Fungicidal sprays during dormancy are potentially helpful.
- Start protective fungicide spraying early in the season.
- Remove dead wood from trees in surrounding woodland.

Wound paints

Wound paints have been a topic of interest to fruit growers for centuries. My 1771 copy of Miller's 'Gardener's Dictionary' advises that large cuts should be smoothed and then 'to put on a plaister [sic] of grafting clay to prevent the wet from soaking into the tree at the wounded part'. Coming a bit closer to the present day, my copy of the 'Fruit Grower's Guide' by Horace Wright, published in 1924, says pruning cuts 'must be made smooth and brushed over with gas [coal] tar'. There have always been two schools of thought about pruning wounds. One is that if a wound is cleanly made at an angle and left exposed to the fresh air, it will soon seal over with the natural sap from the plant and no other treatment is necessary. These 'do nothing' fruit growers maintain that painting a wound just seals in the nastiness (and of course it saves quite a bit of work and expense). The other school of thought is that painting a wound with something that (a) seals the wound from anything that might land on it and (b) uses a wound paint that has some active ingredient in it to prevent further damage, is the right way of going about things. This school of thought is of course, the one supported by those that make and sell wound paints.

In vineyards in other parts of the world, I would say that whilst one does see wounds that have been painted, it is very much in the minority and nowhere is it considered a 'must do' technique. In Britain, there was some discussion about wound painting in vineyards in 2012 at the height of the trunk diseases scare, but since then it has died down and very few growers do this as routine. In 2019 Vine Works tested three products: Solufeed's 'Garlic Barrier Wound Paint'; Podex from a Spanish company Daymsa; and BlocCade from Hortipro. They used all three and reported their findings in the Vineyard Magazine. Their concluding remarks were that 'it's preferable to use something rather than nothing' although there was no scientific basis for this comment. To carry out a scientific trial on wound paints would be exceedingly difficult and given the nature of trunk diseases would have to extend over years, if not decades. And there are of course multiple reasons why a vine might live or die and deciding whether the culprit was a trunk disease or something else would be well-night impossible. At present (2020), there are no wound paints officially approved for use on vines in Britain. However, there is nothing stopping you from using ordinary emulsion paint and growers could accidentally choose a 'kitchens and bathroom' type of emulsion paint, most of which contain a little mould-repellent fungicide. As with many other things, if painting your pruning wounds makes you feel better, then please feel free to do so.

Ungrafted 'own-root' vines

In July 2013, Richard Smart suggested that as Britain occupied a relatively isolated position in the viticultural world, *Phylloxera* was not in reality a major threat and that the planting of ungrafted 'own-root' vines might be a way of avoiding TDs, given that many of the problems are associated with the rootstocks and the grafting procedure. He suggested that a British nursery should be found that would import mother stock from Europe, maintain a 'mother garden' and produce ungrafted rooted cuttings for British growers. Leaving aside the commercial reality of persuading a nursery to start the production of vines for a relatively small market, and a market which required many different variety, clone and rootstock combinations, I think that when faced with the risk of getting *Phylloxera* versus the risk of TDs, growers with ungrafted vines would prefer the latter to the former. As I have pointed out in *Chapter 16: Pest and disease control* in the section on *Phylloxera*, once discovered, the pest must be notified to FERA and they will place a movement control on your vineyard. How then you would deal with the problem in the long-term is a question to which the only answer I can see is to grub the affected vines and replace with grafted stock.

In the early days of modern viticulture in New Zealand, vines were almost always planted on their own roots and a slight infestation with *Phylloxera* was tolerated and it was even said to help control the vigour of Müller-Thurgau which was then the dominant variety. After a decade or so of cropping, as the *Phylloxera* numbers started to rise and yields started to decline, the vineyard could be grubbed and replanted. Whether this would work in Britain today is another matter. The cost of grubbing and replanting in already established vineyards, i.e. with the trellising and wires already in position, is huge and needs to be taken into account. For small vineyards maybe, but for the larger, more commercial growers, I cannot see own-root vines as a viable alternative. With the control of TDs in the mother gardens and nurseries getting better all the time and with a better understanding of the problems from the grower's end, my hope is that the TD situation will improve to a point where it is no more of a problem than the many others that vinegrowers in Britain have to face.

Replanting vines in a vineyard where trunk diseases have been identified

If you have TDs in your vineyard, have removed vines and have gaps to replant, then you must make sure that the vines are planted into as benign an environment as possible. An assessment must be made of whether the TDs in your vineyard have been exacerbated by something as basic as poor drainage, poor nutrition or some other rectifiable problem and the problem put right before replanting takes place. It goes without saying that good quality vines must be obtained and planted into well prepared soil. Some additional humus in the planting hole, maybe a mycorrhizal fungi dip and a good mulch of fine compost will all help the young vine to thrive. Avoid shading from adjacent vines by not allowing excessively long canes from neighbouring vines to extend towards the newly planted vines. The use of high-graft vines could be considered which will help avoid this problem.

It has been suggested that layering of adjacent vines to fill gaps might be practiced instead of replanting with grafted vines. Whilst I can see this technique being used in small vineyards where perhaps only a few isolated vines have succumbed to TDs and been removed, on a larger scale this practice has its problems. Ignoring the threat of *Phylloxera* – which of course would be able to colonise the roots of a layered vine as there would be no rootstock involved – there is a possibility that a vine adjacent to one that has died because of a TD will also be suffering from the same problem and therefore will, in time, also die.

Trunk diseases in Great Britain – a summary

There is no doubt that in some vineyards in Britain, TDs have been a problem in the past and caused multiple deaths. Whether that is true in 2020 is a debatable point. In vineyards where TDs are affecting a majority of the vines, although there is a reluctance to accept that it is the best route, complete grubbing and replanting is probably the only real solution. However, in many vineyards (in fact I would say in most vineyards) vines affected with TDs account for a very small percentage of total vines and the replacement of a small percentage of vines each year, or every other year, has always been part and parcel of managing a vineyard. If anything, the threat of TDs has brought to the fore one major factor in successfully establishing a new vineyard: the importance of selecting and developing a site so that as little stress is placed upon young vines as possible.

Chapter 18

Organic and biodynamic viticulture in Great Britain

Note 1: Reference is made at the end of this chapter to a New Zealand report on *The Organic Focus Vineyard Project* which sets out in huge detail the process of converting 6 blocks of vines from conventional to organic over three consecutive years. The blocks are in Hawke's Bay, Marlborough and Central Otago, with five different grape varieties and covering over 25-ha of vines. The report goes into every aspect of running a vineyard with detailed costings of running both the conventional and organic blocks (which are side by side), and takes it right through to winemaking and a comparison of the wines. For anyone contemplating growing vines organically, it is THE most useful document I have read. It is available at: www.englishwine.com/wgigb.htm

Note 2: In my book *Viticulture* there is a chapter covering organic and biodynamic viticulture which readers will also find helpful in understanding this subject.

Introduction

The area of vineyards farmed organically and biodynamically across the globe has been rising over recent decades, as it has with most other farmed crops. The world area of vines for all purposes (wine, juice and table) in 2019 stood at 7.4 million hectares (2019 OIV Report) and various sources suggest that around 500,000-ha or 7 per cent is farmed organically or biodynamically. One assumes this is hectares of vines actually registered with one of the many organisations that confer organic or biodynamic status and not just growers who 'follow organic principles'. IWSR, the London-based drinks markets analysts, estimate that global sales of organic wine are set to rise by 9 per cent a year and that by 2022 worldwide sales will have risen to 87.5 million (9-litre) cases or 3.6 per cent of world wine sales. For Great Britain, they say that despite overall wine sales declining, sales of organic wine are on the increase, with consumers (one assumes happily) paying a 38 per cent premium over non-organic wines. Consumer's interest in the provenance, the quality and what they perceive to be the 'sustainability' of their food and drink has never been higher and whilst the Covid-19 pandemic will have a negative effect upon most economies, which is not good for products at the higher end of the price spectrum, it has made consumers think about where the things they buy come from. This can only be good news for organic and biodynamic wines and their producers and these trends suggest a buoyant outlook.

In Britain, the number of organic and biodynamic vineyards is small, but growing. The first vineyard to gain organic registration (with the Soil Association) was Chevelswarde in Leicestershire in 1973, followed by Penberth Valley in Cornwall in 1978, Sedlescombe a year later in 1979, and Avalon in 1981. Since those early days, the number of organic and biodynamic growers in Britain has risen and today there is no doubt that they are now very much part of the English and Welsh wine scene. My UK Vineyards Guide database currently (2020) lists nineteen vineyards registered or in conversion as organic and covering 52-ha (128.5-acres) and 10 registered or in conversion as biodynamic and covering a total of 36.3-ha (89.7-acres). Together these represent just 2.5 per cent of the planted area. Some of them are very small, in reality only hobby vineyards, but of the remainder, most are over 1-ha, several over 2-ha and a few over 5-ha. The best known organic and/or biodynamic wine producers in Britain in 2020 are (in alphabetical order): Albury Organic in Surrey, Ancre Hill Estates in Monmouthshire, Avonleigh in Wiltshire, Davenport in Kent and East Sussex, Forty Hall in Enfield, West London, Laverstoke in Hampshire, Marden Organic in Kent, Oxney Organic and Sedlescombe, both in East Sussex and Quoins in Wiltshire.

And then there is Mark Dixon and his vineyards. Dixon has invested heavily in vineyards in Britain and appears to be following an organic route. He made his fortune with Regus, the global workspace and serviced offices company, and also owns the 121-ha Château de Berne, the Provencal

Rosé producer. In around 2016 he bought Sedlescombe Organic Vineyards, followed shortly afterwards by buying Kingscote Vineyards, East Grinstead. Following the purchase of these, new vines were planted at both places, plus at Bodiam Vineyard, which Sedlescombe had leased since 1994. Additional land in Sandhurst, Kent was also leased (from the owners of the Bodiam site) and 40-ha (100-acres) of vines planted. Not content with this number and area of vineyards, the already-organic Court Lodge and Brookers farms in the hamlet of Luddesdown near Gravesend were purchased and in 2019 some of the land was planted with 177-ha of Champagne varieties. This was followed in 2020 by another new vineyard, this time a mere 38-ha in Essex in the village of Althorne facing onto the River Crouch in the by now much-bevined CM3 postcode. In total, Dixon now has control over 283.5-ha (700-acres) of vineyards, most of which is either organic, biodynamic or in conversion. Quite what the long-term plan is for all these vineyards is not clear. Both Sedlescombe and Kingscote were put on the market in early 2020 as they 'didn't fit in with the business plan going forward'. I heard it on good authority that Dixon's aim was to produce an entry-level (£12.99 was mentioned) organic, Charmat method sparkling wine. It will be interesting to see how this empire of wine develops. See www.mdcvuk.com for more details.

In Britain, there are three organisations which can convey organic status: the Soil Association which has been registering land as organic since 1973; Organic Farmers & Growers who were established in 1973 as a marketing organisation, but have been registering land as organic since 1992; and the BDA (Biodynamic Association) which was founded in 1929 and can certify land as either organic or biodynamic, the latter with Demeter, which is the only organisation that can confer biodynamic status.[1] The Soil Association has the reputation for being slightly less flexible when it comes to applying for derogations for the use of plant protection materials and Organic Farmers & Growers are said to be more down to earth in their approach to practical farming issues.

This chapter in the 1st edition of this book started by saying: 'If the growing of vines in a challenging climate such as Britain's could be described as 'brave', then the growing of vines organically or biodynamically might be considered by some as tempting fate'. Looking back on it, I was perhaps a little bit too negative and some seven years later, with the benefit of hindsight, and with the positive influences of climate change even more evident in our vineyards – organic, biodynamic and conventional – than ever before, its maybe a time for a more positive approach. Having worked with two of Britain's largest organic/biodynamic producers, Albury Organic, first planted in 2009 and now with almost 5-ha under vine, and Oxney Organic, first planted in 2012 and now with almost 14-ha under vine, I can see more clearly what the pitfalls

are and how an organic and biodynamic vineyard can be established and managed.

Until about ten years ago, the wines coming from organic and biodynamic vineyards had a mixed reputation. Sure, there were a few good ones, but on the whole, they didn't really shine. The first organic wine to win a medal in the UKVA (now WineGB) national wine competition was in fact way back in 1985, when the 1984 Avalon Seyval blanc won a surprising gold medal, but this was very much a rarity. However, since around 2010 as some more serious players have come on the scene, the quality of wine coming from organic and biodynamic vineyards has greatly improved, at least judging by the number of gold and silver medals awarded in the WineGB national wine competition. Between 2009 and 2019, a total of eight gold medals and twenty-seven silver medals were awarded with Davenport Vineyards leading the parade (with four gold and five silver) with Albury, Oxney and Trevibban Mill featuring in the gold section, and Avonleigh and Quoins joining them in the silver section. This section on wine quality can only end with the praise given to one organic wine by the well-known (and very well experienced) wine writer Mathew Jukes in June 2020. About the 2018 Oxney [still] Chardonnay he wrote: 'This is the most resonant and beautiful English wine I have ever tasted and I have waited 34 years to say this'. He concluded 'I cannot think of a way in which this wine could be improved, and this sets it apart from every other English wine I have tasted' and awarded it 20 out of 20 – his first every 100-point English wine. However, whilst great wine quality is one thing, and of course is a necessary part of the formula for creating a viable business, it is not the only part. Costs and yields also have to be considered.

Vine varieties for organic and biodynamic vineyards

When organic vineyards were first established in Britain, the climate was cooler than it is today. Many vineyards grew Müller-Thurgau and Seyval blanc, then the standard varieties, plus some of the older disease-resistant hybrids such as Cascade, Léon Millot, Maréchal Foch, and Triomphe. As times changed, and more modern interspecific crosses came along – PIWI varieties as we now call them – these too became part of the organic grower's variety selection with Orion, Phoenix, Regent and Rondo being the most widely grown. However, as the climate has changed, so too has the selection of varieties in both non-organic and organic vineyards, and many of today's organic growers have opted to grow what are now the standard varieties, viniferas such as Chardonnay, Pinot noir, Meunier and Bacchus.

The selection of varieties for an organic vineyard is governed by many different factors. Much will depend on the size of the operation, whether you are smaller and proposing to sell the majority of your output direct to the

---

1 In researching this chapter I was surprised to find that Demeter own the certification mark 'Biodynamic'.

consumer, or whether you are a larger producer, aiming to create a brand that will sell via wholesalers and retailers. Much also will depend on finance and cash-flow. Do you have the appetite for producing bottle-aged sparkling wines where you might have to finance four to five vintages before you start seeing a return?

If more disease resistance and less spraying is your goal, then the modern interspecific crosses have some advantages. Orion, Phoenix, Regent, Rondo, and Solaris are by far the most popular and are used by several organic vineyards in Britain. Some of the newer PIWI varieties, such as Bolero, Divico, and Villaris, are also starting to be planted. Although some of these are not as disease resistant in Britain's climate as they appear to be in say Germany and Switzerland (from where these eight varieties originate), they are in most instances some way ahead of *vinifera* varieties in this respect. As for other non-*vinifera* varieties, as is shown in *Chapter 4: Vine varieties, clones and rootstocks* in the section on *Newer vine varieties*, there is a wide range of new interspecific crosses being developed in Europe and some of these are being trialled in both organic and non-organic vineyards.

If you want to grow today's standard vinifera varieties, then from a marketing point of view, that's fine, but you need to be aware of some of the problems associated with them. Even in conventional vineyards, who will be spraying ten to twelve times a season with effective, systemic plant protection materials, it is not always easy to keep these varieties clean, especially in a difficult year. For organic and biodynamic growers, the problems are magnified and the major issue as I see it, is the difficulty of keeping them disease free right to the end of the growing season, given the relatively limited arsenal of plant protection products allowed to be used. As the season progresses and growers have to stop using copper as their principal defence against Downy Mildew, leaf quality starts deteriorating and with it, the vine's photosynthetic ability. This can lead to ripening issues, especially in a late, cool autumn such as we saw in 2019 (when it was also very wet). In good years such as 2014 and 2018, when the vintage was early and sugars were high, there were of course far fewer ripening issues. And poor leaf quality doesn't just affect ripening. The vine needs reserves to survive the winter and secure the crop for next year, and poor wood (and therefore bud) ripening is one of the reasons why yields in organic and biodynamic vineyards are on average lower than in conventional ones.

Reichensteiner, given its open growth habit, large well separated bunches and its above average sugar levels and yields, is also a variety I would recommend. Again, whilst it may not be as high in the quality stakes as some of the other more classic varieties, its good ripeness levels and neutral flavour make in ideal blending partner. By the number of different organic vineyards growing it, Blauer Frühburgunder (Pinot noir Précoce) is actually the most widely grown variety in organic and biodynamic vineyards, not doubt on account of it ripening around two weeks ahead of Pinot noir.

Yields in organic and biodynamic vineyards

Yields in all vineyards very widely and they are influenced by many different factors: variety, vine density, canopy management, and the year in question all play their part. As has been demonstrated by the data already shown in *Chapter 2: Why plant a vineyard in the Great Britain?* in the section on *Yields*, the variation between the top twenty-five per cent of (mainly conventional) growers and the next fifty per cent is huge with the top group averaging 9.64 t-ha (3.90 t-acre) and the middle group averaging 5.01 t-ha (2.03 t-acre) i.e. the best growers harvest almost twice as many grapes as the next group. Hybrids and interspecific crosses, whilst some wine commentators and wine critics can be negative about them, from a grower's point of view and from an economic point of view, have some definite advantages. Looking at the yield results from 2016–19 (which are in Table 8 in *Chapter 2* in the *Yields* section) Seyval blanc, Regent and Rondo are all in the top six varieties by yield.[2] In my 2017 yields survey I was able to gather enough data to separate the results from organic and non-organic vineyards as shown in Table 31. 2017 was admittedly a year when frost hit many vineyards, so the data is not that reliable, but at least it's an indication. Rondo, which came from a small sample, so statistically not so robust, was able to produced good yields even though in some cases it was frosted, as being an early variety, it had time to set secondary flowers. For many organic growers, despite it having been in vineyards in Britain since the 1950s, Seyval blanc still remains an important variety. Apart from its disease resistance which is usually pretty good, it is an adaptable, high-yielding variety and one that, because of its longevity in British vineyards, has become accepted as part and parcel of Britain's wine offerings. At 6.18 tonnes-ha (2.50 tonnes-acre) its average yield (ignoring Rondo for reasons already explained) were well over twice any other listed variety. Seyval blanc has always been, and remains, one of most widely grown varieties in British vineyards and of the white, still wine varieties grown in Britain, at 145-ha (358-acres) it remains at number two and occupies just over 4 per cent of the total planted area (Bacchus is the most widely planted still wine variety with 295-ha (729-acres)). The fact that some of Britain's best known non-organic producers – Breaky Bottom, Camel Valley, Denbies, Monnow Valley and Stanlake Park – use it to make medal-winning still and sparkling wines is a bonus. As for the other older (all red) hybrids already mentioned, only Triomphe remains popular, although it is very rarely planted today, mainly because of the much better red varieties now available.

Reliable yield data for organic and biodynamic vineyards is certainly hard to find, but in my book *The Wines of Great Britain*, Kristin Syltevik, owner of Oxney Organic, told me that for the years 2015–18, which of course, included the

---

2 This is for all vineyards surveyed, not just organic.

Yields in Organic & Biodynamic vineyards 2017		
Grape Variety Performance	Tonnes per hectare	Tonnes per acre
All varieties	3.21	1.30
All varieties - Not frosted	4.08	1.65
All varieties - Frosted	1.58	0.64
All Champagne varieties	1.98	0.80
All non-Champagne varieties	4.01	1.62
Top 25% - all varieties	7.40	2.99
Top 50% - all varieties	5.17	2.09
Bottom 50% - all varieties	1.33	0.54
Bottom 25% - all varieties	0.81	0.33
Individual Varieties		
Bacchus	2.28	0.92
Chardonnay	1.62	0.66
Madeleine x Angevine 7672*	2.03	0.82
Meunier	2.86	1.16
Pinot noir	1.59	0.64
Reichensteiner	2.80	1.13
Rondo*	12.09	4.89
Seyval blanc	6.18	2.50
Siegerrebe	2.59	1.05
Average of individual varieties	3.78	1.53
Note: Data with * is from a small sample		

Table 31

Davenport Vineyards, with good mixture of varieties both old and new (but no Seyval blanc) admit to a long-term yield of around 3.00 tonnes-ha (1.25 tonnes-acre). Their rows are relatively wide, 2.74-m, and if you planted with a 2.0 m row width and achieved the same yield per vine, this would equate to around 4.1 tonnes-ha (1.67 tonnes-acre). Taking all these yields into account, I would say that for most well-sited, well-planted and well-managed organic and biodynamic vineyards, a yield of 4.0 t-ha (1.60 t-acre), depending on varieties and vine density, would be a good average to aim for and hopefully exceed.

Costs of running organic and biodynamic viticulture

As will be seen at the end of this chapter in the section on the New Zealand report *The Organic Focus Vineyard Project*, whilst the costs of running organic and biodynamic vineyards will vary in certain respects, savings in one area are met by increased costs in another. In organic vineyards, the costs of annual dressings of fertiliser and lime, additional labour and weed control were all higher, but the costs of canopy management and pest and disease control were lower. The lower canopy management costs were due to the lower level of vigour in the organic and biodynamic vineyards and the lower pest and disease control because of using cheaper plant protection materials. Overall, the organic and biodynamic vineyards cost a relatively modest 8.5 per cent more to look after. However, the major costs difference per tonne of grapes harvested is due to the lower yields in organic and biodynamic vineyards. The cost per tonne of grapes was £632 in conventional vineyards and £817 in organic and biodynamic vineyards, an increase of just over 29 per cent which is a fairly significant difference. However, the surplus of income over costs (what we might call the profit) was even more pronounced. The conventional blocks recorded a surplus per hectare of £6,140 per hectare (£2,485-acre) whereas the organic blocks only £3,858 per hectare (£1,561-

high-yielding 2018 (and her vineyards also include some high-yielding Seyval blanc), the average yield was between 4 and 5 tonnes-ha (1.6–2.0 t-acre). Given the spread of varieties and quality of the vintages, I would say these were at the higher end of most organic and biodynamic grower's expectations. Nick Wenman, who owns Albury Organic Vineyard (which is actually farmed biodynamically), between 2013 and 2019 had an average yield of 5.04 tonnes-ha (2.04 t-acre) with the highest yield being in 2018 with 8.28 t-ha (3.35 t-acre) and the lowest in 2016 with 2.82 t-ha (1.14 t-acre). Albury also have some Seyval blanc.

acre), a difference of almost 60 per cent. It is the yield difference that makes organic and biodynamic viticulture financially more challenging. Of course, if one grew varieties with a proven track record of higher yields and managed them as economically as possible, and sold all the wine at a premium price direct to the consumer, the story might be different and the enterprise more sustainable. But that's for those wishing to follow an organic and biodynamic route to decide.

Weed control

After the selection of the grape varieties for an organic vineyard, the next problem which has to be tackled is that of weed control. Given that herbicides cannot be used, the options available are restricted to physical methods, such as mulching and cultivations. Other options, such as grazing and cover crops, whilst they might be made to work in mature vineyards, are certainly not available to owners of newly planted and young vineyards. The options available for controlling weeds in vineyards have been fully discussed in *Chapter 14: Weed control*, so there is no need to go through them here in too much detail, but a short summary of the advantages and disadvantages of each method suitable for organic vineyards would be appropriate.

Cultivations

Since I first wrote this section, the use of under-vine cultivation in vineyards in general in Britain has grown hugely. I would estimate that many, maybe most, of the largest producers, for whom investment in specialist machinery is not a problem, are using under-vine cultivation as part of their weed-control strategy. Most of what I want to say about the subject is in *Chapter 14: Weed control* and there is very little to add here, save to say that of course, with herbicides not an option, organic and biodynamic growers will have to rely entirely on their cultivators, something that can be challenging in a difficult season or on difficult soils. The problems of using under-vine cultivators in young vineyards do not go away because you are organic or biodynamic and care needs to be taken in the first two to three years of a vine's life. With cultivators costing anything from £5,000 to £20,000 and most needing to be fitted to a dedicated tractor, this equipment may not suit smaller growers. However, despite these cost issues, under-vine cultivation is almost certainly the most effective method of weed control for organic and biodynamic growers.

Mulching

Mulches, made from plastic or fabric and laid on the ground with the edges buried, or loose mulches of various natural materials spread around the vines, have been described in *Chapter 14: Weed control* in some detail and there is no need to discuss them in detail in this chapter. From an organic grower's perspective, mulches would seem to be an ideal

Biodynamic vineyards are weeded by horse-plough in Alsace

solution since they are static and do not involve the use of herbicides, and vines may well benefit from their use. However, the main problem with using mulches – a problem which affects both organic and non-organic growers – is the 'edge' problem. Keeping the edges of any mulch weed-free without the use of herbicides involves mowing or cultivating, sometimes both, and this gets problematical if the edges are not to be disturbed or damaged. Inevitably a degree of hand weeding has to take place which can be an expensive process. For small vineyards, where mowing and strimming by hand are practical solutions to keeping the edges of laid mulches clean, these are probably an ideal solution.

Pest and disease control

The first approach to pest and disease control in an organic or biodynamic vineyard must be at the planning stage. Apart from their choice of varieties, growers must plan, plant and trellis their vineyard so that it is as light and airy as possible, although that should not mean selecting a site on the top of a windswept hill so that 'diseases will be blown away' (as I once heard an biodynamic grower explain his poor choice of site). Once established, the canopy management in the vines has to be exemplary, with as much attention as possible paid to shoot thinning, tucking-in and deleafing so that the grape zone is kept as open, light and airy as possible. The vineyard should also be kept as clean as possible, removing as much plant debris as possible including prunings and grapes discarded during thinning or not picked for whatever reason. Inevitably these measures all mean additional expense in running the vineyard.

Unless they have selected really good disease-resistant hybrids and interspecific crosses, organic and biodynamic growers will be faced with the same range of pests and diseases that non-organic growers face, but without their full armoury of chemicals. However, that is not to say that they have no materials to use, as organic and biodynamic does not mean 'not sprayed'. In general, organic and biodynamic growers rely on wettable sulphur and various copper-containing products as their main weapons, supplemented by various other substances. For the 2020 edition of WineGB's 'Green Book', Alex Valsecchi, Vineyard Manager at Albury Organic Vineyard (which is in fact farmed biodynamically), wrote a section titled 'Organic and Biodynamic Spraying Programmes' which sets out two annual spray programmes, one for each discipline. Both spray programmes rely on wettable sulphur to control Powdery Mildew and copper to control Downy Mildew, remembering that copper is restricted to 4 kg a year (of pure copper) for organic growers and 3 kg a year for biodynamic growers, both averaged over five years. These limits on copper are the current (2020) ones and there has been talk of lowering them further, or of even phasing out the use of copper entirely. This last option has been widely criticised as, in many regions, this would make the growing of vines organically or biodynamically almost impossible. The last application of sulphur is suggested to be in mid-September, i.e. around 28 days before an average harvest date, which is the very minimum HI to be recommended and 42 days would be safer. If you are growing early varieties, or it is an early year, you would need to bring that 'last spray' date for sulphur forwards to allow for an HI of at least 28 days. For botrytis control, both programmes use Serenade or BioLife ProS, which are bio-pesticides based upon *Bacillus subtilis*, and up to five times is suggested, depending on disease pressure.

Both programmes also suggest regular inclusions of a seaweed extract (such as Maxicrop) plus Epsom Salts (magnesium sulphate), both as plant tonics. The biodynamic spray programme also (or course) includes the various 'preparations' which are at the heart of this method of agriculture. The preparations are numbered from BD500 to BD508 and are, in numerical order, cow horn manure and cow horn silica, via yarrow, chamomile, stinging nettle and oak bark to valerian and casuarina tea. The first two of these, BD500 and BD501, and the last, BD508, are diluted and sprayed directly on the vines, whilst the others are used in the preparation of biodynamic compost which is an integral part of the process of persuading the vine to fight off disease naturally. In the suggested spray programme, the 'preps' used are equisetum (horsetail), horn manure, comfrey and nettle tea, horn silica, and yarrow. These are sprayed at certain times of the season, and in the case of some preparations, during certain phases of the moon and/or at certain times of the day or night. Organic and biodynamic growers might also be spraying materials such as potassium sulphite, potassium bicarbonate, monopotassium phosphate, dilute solutions of hydrogen peroxide and certain oils (Codacide and JMS Stylet Oil), none of which are classed as pesticides.

Because all of the materials used by organic and biodynamic growers are contact materials and none are systemic, they will have to use higher water volumes than non-organic growers in order to get full coverage. In addition, spray intervals will in most years have to be shorter than in non-organic vineyards as contact materials are more easily washed off by rain. A weekly spray programme is typical in organic vineyards. The number of sprays suggested in the case of both of the above spray programmes is around nineteen to twenty (which includes one post-harvest spray) although I know of some organic growers who would be spraying something more or less on a weekly basis between woolly bud and harvest, thus clocking up nearer to twenty-five sprays in the season. For conventional growers, twelve sprays a year would be considered a maximum, with many growers getting away with eight to ten. Much of course will depend on the varieties being grown, the amount of detailed canopy management taking place and the weather during the growing season.

The New Zealand experience of organic and biodynamic vineyards

Finding the true cost of running a conventional vineyard is hard enough; finding the cost of running an organic or biodynamic vineyard even more so. However, I was lucky to be introduced to a New Zealand study called *The Organic Focus Vineyard Project*. This project was set up to in 2010 to follow three conventional vineyards, one each in Hawkes Bay, Marlborough and Central Otago, as they transitioned some of their vineyards from conventional to certified organic production. Blocks of already cropping vines were selected, and in each location there were two different blocks and two different varieties. In total, over the twelve blocks, there were almost 11-ha of conventional vineyards and 14-ha of organic vineyards. Over three successive years, 2012, 2013 and 2014, the inputs and outputs were monitored and recorded and independent technicians oversaw the project. This was very much an industry initiative, with growers interested in the results and regularly visiting all the sites and having access to the data. Whilst there were anomalies – for instance, not all the vineyards were the same age and not all the vineyards were planted at the same density – they were all managed in similar ways with regard to pruning, canopy management, pest and disease control and weed control. Taking the size of the blocks, the fact that they were followed over three years, and the professional nature of the New Zealand wine industry into account, I believe that the results, both conventional and organic, are a very good guide to the differences in performance in both terms of cost and yields. For the purposes of this chapter, I have averaged all the blocks across all the years to try and arrive at some straightforward conclusions and the results are shown in Table 32. All costs and income have been adjusted for exchange rates, but not inflation.

Growing costs, excluding picking, were £4,334 per

hectare (£1,754-acre) in the conventional blocks, and £4,706 per hectare (£1,904-acre) in the organic ones, an increase of a fairly modest 8.6 per cent. The most significant differences were with the pest and disease control which were more expensive in the conventional blocks, and the weed control which was well over twice as expensive in the organic blocks. The cost of bird netting, where it was used, is included in pest and disease control. Canopy management was slightly cheaper in the organic blocks, owing to the less vigorous growth, but annual fertiliser and lime applications were higher in the organic blocks. However, the major difference in the cost per tonne of grapes was the result of the yield difference.

Yields in conventional blocks averaged 7.97 tonnes-ha (3.23 tonnes-acre) compared to 6.58 tonnes-ha (2.66 tonnes-acre) in the organic blocks, a drop of 17.4 per cent. Picking costs, some by hand and some by machine, were broadly similar per hectare at £707 (conventional) and £667 (organic), but per tonne of grapes harvested, were 16 per cent higher in the organic blocks. The final cost per tonne of grapes worked out at £632 for the conventional and £817 for the organic, a significant difference of just over 29 per cent. Grape values were calculated at an industry standard rate for the variety and year for both conventional and organic grapes, (although one might have thought that organic grapes would command a premium) and the average price per tonne was £1,403. The end result was that conventional growers had a surplus of income over costs of £6,140 per hectare (£2,485-acre) and organic growers £3,858 per hectare (£1,561-acre), a difference of almost 60 per cent. For the sums to be balanced up, organic and biodynamic grapes would have to be worth almost £1,750 a tonne.

Cost comparison between conventional and organic vineyards			
Costs in £ per hectare	Conventional	Organic	Difference
Fertiliser and lime	£ 92	£ 223	£ 131
R&M	£ 38	£ 66	£ 28
Pruning and tying down	£ 1,030	£ 1,025	-£ 5
Canopy Management	£ 1,533	£ 1,452	-£ 81
Other wages	£ 138	£ 300	£ 162
Contract machinery	£ 9	£ 12	£ 3
Pest and disease control	£ 1,305	£ 1,176	-£ 129
Weed control	£ 190	£ 453	£ 263
Cost per hectare	£ 4,334	£ 4,706	£ 372
Tonnes per hectare	7.97	6.58	1.39
Harevsting per hectare	£ 707	£ 667	-£ 39
Harvesting cost per tonne	£ 89	£ 101	£ 13
Cost per tonne of grapes	£ 632	£ 817	£ 183
Value per tonne	£ 1,403	£ 1,403	
Profit per tonne	£ 770	£ 586	-£ 183
Profit per hectare	£ 6,140	£ 3,858	-£ 2,273
Source: Organic Focus Vineyard Project. NZ Organic Winegrowers 2014			

Table 32

Conclusion to organic and biodynamic viticulture

Establishing and running an organic or biodynamic vineyard in a climate such as Britain's is going to be a challenge when compared to running a vineyard conventionally. Whilst running costs might not be significantly more in an organic or biodynamic vineyard, the lower yields, brought about by a combination of factors which have already been discussed in this chapter, make a tonne of British-grown organic or biodynamic grapes cost at least 50 per cent more to produce than a tonne of non-organic grapes and quite possibly even more than that. Whether organic and biodynamic wines can attract a sufficient premium over non-organic wines or whether organic and biodynamic wines have access to markets that non-organic wines do not, to help pay the additional costs of production, is the question that needs to be answered.

Chapter 19

Getting started

Note: There are two useful documents to help guide new vineyard owners and wine producers: *Information for new vineyard owners* published by the Wine Standards (WS) branch of the Food Standards Agency in 2020; and a *Guide for new entrants to the UK wine industry* published by WineGB in 2020. Both are available to download from the appropriate website or can be downloaded from http://www.englishwine.com/wgigb.htm

Starting a vineyard in Great Britain

Starting a vineyard in Britain is remarkably easy when compared to many other countries, certainly easier than starting one in the European Union. No *appellation* to apply to for permission, no planting rights to obtain, and in the main, no restrictions on where to plant, what to plant nor how much to plant.[1] Planting a vineyard also does not require planning permission, even in National Parks or Areas of Outstanding Natural Beauty (ANOB). However, that is not to say that growing and making wine in Britain is entirely unhampered by officialdom.

Vineyard Register

Having planted your vineyard and assuming it is for the production of a 'wine sector' product, i.e. not vine leaves or some other non-potable product, within six months of planting you have to register the vineyard with the WS who will want to know which varieties and clones of both scion and rootstock varieties you have planted, in what quantities, covering what area and where. This is required so that the Vineyard Register, a list of vineyards that every EU Member State must keep, can be kept up to date. Quite what the post-Brexit arrangements will be is, at the time

of writing, not known, but it is assumed that the current regulations will remain in force after 31 December 2020. Forms for new vineyard registration can be downloaded from the FSA's website whose details are in *Appendix 1* or from www.englishwine.com/wgigb.htm.

HMRC Producer's Licence

In order to produce wine in Britain, you should apply to HMRC for a licence so that you can make and store wine in a duty-free environment, export wine without payment of duty and obtain your producer's annual duty-free allowance. To register you must complete form WMW1 'Apply for a licence to produce wine or made-wine' and lodge this with HMRC. Full details of this licence and all aspects of producing wine in Britain are to be found in HMRC Notice 163 Wine Production (updated April 2020) and both available at:
www.englishwine.com/wgigb.htm

Licences to sell wine retail or wholesale

When you get to the point where you wish to sell wine, you will need a Personal Licence to sell alcohol, as well as a Premises Licence, and these are available from the trading standards department of your local authority (LA). Full details can also be found on the Home Office's website. If you sell wine to traders only – wholesalers, retailers, restaurateurs, publicans and the like – who will be selling it on to the public, then you do not normally need a licence, but must register under Excise Notice 2002 the 'Alcohol Wholesaler Registration Scheme'.

---

1 Planting on land controlled by National Parks UK (NPUK), or on land that has not been 'farmed' for 15 years, might need permission from the relevant authority: NPUK or Natural England.

Planning permission

The erection of agricultural buildings solely for machinery and equipment for running the vineyard and for any purely agricultural activity is, subject to siting, access and certain size and design requirements, fairly straightforward and allowed under 'permitted development rights'. Annex E of Planning Policy Guidance 7 (PPG7) sets out the details. The question of planning permission for buildings on a vineyard solely for the processing of grapes grown on the land, i.e. a winery, is less clear cut, although several vineyards have persuaded their LAs that, under the precedent set by the Wroxeter Roman Vineyards case, wineries can be built under permitted development rights. The late David Millington, owner of Wroxeter Roman Vineyards, challenged the Secretary of State and won, confirming that winemaking was ancillary to growing grapes and, therefore, like cider-making, an 'agricultural activity'. The Millington case report and Annex E of PPG7 are available at *www.englishwine.com/wgigb.*

In situations where a building is used for both agricultural and other uses is never clear cut and the planning department of your LA will, in all probability, need to be consulted. Such activities as contract winemaking; making wine from grapes not produced on the land adjacent to the winery; selling non-wine products not produced on the site; and selling tours and tastings may be allowed without full planning permission, but it will depend on the scale of such activities and the attitude of the LA's planning officer. If you bring in grapes from land other than your own, a figure of 15 per cent brought-in grapes is taken as a limit above which you need planning permission. However, this is not a hard-and-fast rule and each LA will be different. It may be that you are leasing or renting land, or buying grapes from another grower under a long-term grape purchase contract, in which case your LA might consider these to be 'your' grapes and not require you to obtain planning permission.

Applications for agricultural accommodation will always require planning permission and Annex 1 of PPG7 sets out the basis for such applications. In general, applications will only be allowed on 'well-established agricultural units', which are 'financially sound', have been 'established for three years', and have 'been profitable for at least one of them'. Applications also have to show an existing functional need, i.e. the accommodation is required for the proper functioning of the enterprise and that there is no other suitable accommodation 'on the unit' or 'in the area' which is suitable. Despite these requirements, several vineyards have managed to get planning permission for houses on site and an application is always worth making if backed up with a professionally prepared financial and a functional report.

Appendix I – Useful addresses

Vineyard and Wine Trade Associations and Organisations

WineGB	National vineyard association	www.winegb.co.uk
WineGB-East Anglia	Regional vineyard association	www.eastanglianwines.co.uk
WineGB-Midlands & North	Regional vineyard association	www.winegb-mn.co.uk
WineGB-South East	Regional vineyard association	www.winegb.co.uk/join/region-se
WineGB-West	Regional vineyard association	www.swva.co.uk
Thames and Chiltern Vineyards Association	Regional vineyard association	www.thameschilternsvineyards.org.uk
Welsh Vineyards Association	Regional vineyard association	www.winetrailwales.co.uk
WineGB-Wessex	Regional vineyard association	www.winegb.co.uk/join/region-wessex
Wine & Spirits Trade Association – WSTA	Wine trade association	www.wsta.co.uk

Official Agencies

DEFRA	Department for the Environment, Food & Rural Affairs	www.gov.uk/government/organisations/department-for-environment-food-rural-affairs
Food Standards Agency - Wine Standards	EU Wine regulations, Vineyard Register, Labelling	www.food.gov.uk/business-guidance/industry-specific-advice/wine
Health and Safety Executive	Pesticides	www.pesticides.gov.uk
HM Government	Wine trade documents and regulations	www.gov.uk/wine-trade-regulations
HM Revenue & Customs	Excise Duty, Producer's Guide and Licence	www.hmrc.gov.uk
Home Office	Licences to sell alcohol	www.gov.uk/alcohol-licensing
National Proficiency & Training Council (NPTC)	Pesticide Training Courses	www.nptc.org.uk
Rural Payments Agency	Single Farm Payment Scheme (SPS)	www.rpa.defra.gov.uk/rpa/index.nsf/home

Wine Education

Association of Wine Educators	Wine education	www.wineeducators.com
Institute of Masters of Wine	Wine education	www.mastersofwine.org
Plumpton College	Wine Education	www.plumpton.ac.uk
Wine & Spirit Education Trust	Wine Education	www.wsetglobal.com

Commercial Wine Websites

Drink Britain	News website	www.drinkbritain.com
English Sparkling Wine	News website	www.englishsparklingwine.co.uk
English Wine	Commercial website	www.englishwine.com
Wine Cellar Door	Visit-a-vineyard website	www.winecellardoor.co.uk
UK Vines	Guide to vineyards in Britain	www.ukvines.co.uk
Great British Wine	Commercial website	www.greatbritishwine.com

Appendix II – Vineyard pre-planting check list

❏ Buy copy of *Wine Growing in Great Britain* by Stephen Skelton MW – read it!

Measure up site accurately, allowing space for:
❏ Windbreaks, internal roadways, shading from boundary trees
❏ Sufficient space for trimming hedges and windbreaks
❏ Fencing
❏ Correct headland distance for tractor and any likely (towed) equipment
❏ End-posts and anchors, 1.5 m - 2.0 m
❏ Sufficient turning space for rows at angles
❏ Removal of internal trees, hedges etc not required
❏ Area for stacking and loading of crop
❏ Parking for worker's and picker's vehicles
❏ Rest and/or changing room and WC facilities

Calculate number of vines required
❏ Decide on row width based upon tractor width +700 mm depending on pruning system
❏ Decide upon inter-vine distance depending on pruning system
❏ Divide area to be planted by row width and inter-vine distance to arrive at vine numbers

Vines
❏ Order required number of vines
❏ Decide where vines are to be stored after they arrive in the country and before planting
❏ Decide whether vines are to be treated with dip (mycorrhizal or *Trichoderma*) at planting and get materials on order
❏ Book planting contractor for early May
❏ If hand planting, make sure team available for early May and assemble equipment: tape measure, measuring strings and stakes, individual vine position markers, spades, vine carrying boxes etc

Windbreaks
❏ Decide whether you need windbreaks
❏ Measure up for number of trees allowing for appropriate spacings
❏ Order windbreak trees. Bare rooted whips (600-900 mm) are best
❏ If windbreaks are not to be inside rabbit fencing, order rabbit guards
❏ Plant windbreaks in advance of vines if possible
❏ Make sure windbreaks are looked after, kept weed-free, fertilised and if possible mulched with compost
❏ Stake windbreaks until they grow a trunk strong enough to support themselves

Nutrition
❏ Take soil samples and send to laboratory
❏ Ask for BS (Broad Spectrum) analysis plus Active Calcium Carbonate if pH over 7.5
❏ Alternatively (or as well as) have site EC mapped (SoilQuest from Agrii)
❏ Depending on soil analysis, decide upon fertliser and lime requirements
❏ Lime best applied before any work done to the site
❏ Other fertliser applied at suitable times, typically half before sub-soiling and half before final cultivations
❏ Would the site benefit from green manure?
❏ Would the site benefit from manure or compost applications?

Machinery requirements for establishment phase
❏ Tractor with sufficient power for sprayer, mower etc
❏ Pesticide sprayer
❏ Mower or cultivator for between rows
❏ Weed control equipment

Other equipment
❏ Secateurs – Felco No.2
❏ Max Tapener tying machine, plus appropriate tape, staples and spare blades

Site preparation

Drainage
❏ Does the site require draining?
❏ Contact drainage contractors for drainage survey and quotes
❏ Book drainage contractor and get date confirmed allowing enough time between drainage and remainder of site preparation

Levelling
❏ Are there humps and depressions on the site that would be best ironed out?
❏ If yes, find contractor with suitable tracked bulldozer to carry out work
❏ Major earthworks best carried out before drainage work

Trees and hedges
❏ Trim up or remove trees around site if these are likely to impede planting machine and/or cast shadows on vines
❏ Remove any internal trees and hedges
❏ Make sure trees and undergrowth are open up at the bottom of a slope if this is relevant to frost problems

Wine Growing in Great Britain

Fencing

❏ Decide upon fencing requirements

❏ Get quotes based upon correct specification. See Forestry Commission and British Deer Society data sheets on rabbit, hare and deer fencing

❏ Make sure rabbit netting is correct height and gauge

❏ Make sure posts are of correct quality for designed life of fencing

❏ Allow for sufficient gates for machinery and entrances for personnel

❏ Place order with contractor

❏ Work to be completed within 2 weeks of vines being planted

❏ If site is not to be fenced, decide how vines are to be protected from rabbits and hares and order vine guards

Water

❏ Is there a requirement for water on the site? It will be needed for both pesticide and herbicide applications

❏ If not already present or nearby, make provisions for a connection to the mains

❏ Install spray tank

Cultivations

❏ Spray area to be planted, including headlands, with suitable weedkiller

❏ Apply lime and fertilisers

❏ Apply composts or manures

❏ Subsoil to a depth of at least 600 mm. Ideally this requires a tracked machine

❏ Plough, disc, rotary harrow to seedbed quality. Before rotary harrow, apply balance of fertliser

❏ If it has been decided to grow green manure crop, then plant it for overwintering. Site will require probably need spraying off again, re-ploughing and cultivating prior to planting.

❏ Note: All cultivations want to be up and down the slope in the direction of the vine planting

Trellising

❏ Based upon planting density, decide upon number of end-posts, intermediate posts and planting stakes required and place on order

❏ If using contractor, get quotes based upon trellising design, number of vines and number of posts, end-posts etc

❏ Also order required number of end-post anchors, anchor wires, trellising wire, chains, spreaders, clips for planting stakes

❏ After planting, order end-post tags with details of row number, variety, clone and rootstock (Daltons ID Systems)

Planting

❏ Make sure vines are on-site or nearby a day or two before planting

❏ Have available 100-200 bamboo canes (1.20 m) for marking out ends of rows

❏ Welcome planting machine and get planting

❏ Put rabbit guards on if using these

❏ Spray pre-emergent weedkiller once vines settled and before bud-burst

❏ Consider using strip-spread compost on sites with poor humus content

Appendix III – Vineyard running costs

The annual running costs of a vineyard will depend on many variables and it is impossible to arrive at a standard costing. The major variables are:

- Row width and vine density. A vineyard planted at 1.75 m row width with 7,000 vines-ha (2,833 vines-acre) will take a lot longer to prune, tie down, tuck-in, deleaf etc than one planted at 2.50 m row width with 3,000 vines-ha (1,214 vines-acre).
- In small vineyards, much of the work will be carried out by hand and/or with small-scale machinery. In larger vineyards, much of the work can be partially and even fully mechanised, with huge savings in time and cost.
- Both methods of, and efficiency of, the weed control will have a considerable impact upon costs of keeping the vines weed-free.
- Costs of pest and disease control will depend upon many factors: varieties being grown, condition of vineyard, quality of sprayer, spraying techniques and the weather during the growing season.
- Whether the vineyard is run using family labour, directly employed, contract labour or by a contractor using his or her own machinery will greatly affect costs.

The costings overleaf have been arrived at based upon a vineyard planted at 2.00 m row width, with vines at 1.20 m apart, a density of 4,167 vines-ha (1,686 vines-acre). The cost per hour of labour (£11.91) and of tractor and driver (£38.12) have been taken from the John Nix Pocket Book for Farm Management 51st Edition (2021).

Several assumptions have been made with regard to machinery and equipment available and the costings have assumed that lower shoot removal, tucking-in and de-leafing would be done by hand.

They do not include the costs of bunch thinning, although have allowed for two rounds of tucking-in, one round of deleafing (by hand), and ten each of mowing and spraying. This number of operations could be considered excessive by some producers, but realistic by others. I have also not included anything for fertilisers or manures.

If anything, these costings are on the high-side and in a vineyard of 5-ha (12.5-acres) or more and with suitable equipment, I would expect savings to be made. In really large, fully mechanised vineyards, I would expect the annual cost to be nearer to £6,200-ha (£2,509-acre).

These costings do not include picking costs, which are covered in *Chapter 2: Why plant a vineyard in the UK?*

Wine Growing in Great Britain 2020 Annual growing costs per ha	Hours per hectare	Cost per hour	Cost per hectare	Cost per acre
Hand work				
Pruning and pulling out prunings	93	£ 11.91	£ 1,108	£ 448
Tying down	26	£ 11.91	£ 310	£ 125
Repairs to trelliswork. Gapping up vines.	10	£ 11.91	£ 119	£ 48
Shoot removal - trunks	26	£ 11.91	£ 310	£ 125
Shoot thinning - vines	40	£ 11.91	£ 476	£ 193
Tucking in x 2	20	£ 11.91	£ 238	£ 96
Deleafing by hand	80	£ 11.91	£ 953	£ 386
Sub-total	**295**		£ **3,513**	£ **1,422**
Tractor work				
Pulverising prunings	5	£ 38.12	£ 191	£ 77
Herbicide applications x 2	10	£ 38.12	£ 381	£ 154
Trimming by machine x 3	9	£ 38.12	£ 343	£ 139
Mowing. 1.5 hrs-ha x 10	15	£ 38.12	£ 572	£ 231
Spraying. 3 hrs-ha x 10	30	£ 38.12	£ 1,144	£ 463
Sub-total	**69**		£ **2,630**	£ **1,064**
Materials				£ -
Ties etc	£ 30		£ 30	£ 12
Herbicides	£ 300		£ 300	£ 121
Pestcicdes excluding herbicides	£ 1,150		£ 1,150	£ 465
Miscellaneous: posts, vines, secateurs	£ 150		£ 150	£ 61
Sub-total	£ **1,630**		£ **1,630**	£ **660**
Total cost: Labour, machinery, materials			£ **7,774**	£ **3,146**

Appendix IV – Vineyard machinery costs

Equipment	Cost minimum	Cost maximum
Tractor – four-wheel drive 70-80 hp	£35,000	£70,000
Orchard mower – rear mounted, rotary blades	£4,500	£4,500
Orchard mower – front mounted, rotary blades	£5,500	£5,500
Flail mower – prunings pulveriser, rear mounted	£4,500	£4,500
De-leafer – single-sided, suck and slice	£15,000	£15,000
De-leafer – double-sided, suck and slice	£23,000	£24,000
De-leafer – double-sided, air pulse	£40,000	£40,000
Leaf trimmer – single-sided inc. fitting frame	£8,500	£8,500
Leaf trimmer - double-sided inc. fitting frame	£15,000	£15,000
Leaf trimmer – four-sided inc. fitting frame	£23,000	£24,000
Stem-cleaner – flail type	£3,000	£3,000
Tucking-in machine	£16,000	£17,000
Pre-pruning machine inc. fitting frame	£18,000	£18,000
Mechanical weeder – Rollhacke, single sided	£5,000	£5,000
Mechanical weeder – Rollhacke, double sided	£9,000	£9,000
Mechanical weeder – undervine hoe, single-sided	£8,000	£8,000
Mechanical weeder – undervine hoe, double-sided	£14,000	£14,000
Mechanical weeder – powered tiller, double-sided	£18,000	£18,000
Herbicide sprayer – 400 litre tank, double-sided	£5,400	£5,400
Pesticide sprayer – 600 litres, mounted, axial fan	£6,000	£6,000
Pesticide sprayer – 1,000-1,500 litres, towed, axial fan	£12,000	£13,000
Pesticide sprayer – 3 rows, axial fan and over-row snakes	£24,000	£25,000
Pesticide sprayer – recovery type (Lipco), two rows	£25,000	£25,000
Static frost fan	£9,000	£12,000
Trailed frost fan	£20,000	£20,000
Seed drill for narrow rows	£8,500	£8,500
Rotary harrow for narrow rows	£5,000	£5,000
Spring-tine cultivator – lightweight	£3,000	£3,000
Spring-tine cultivator – heavyweight	£5,000	£6,000
Sub-soiler – two tines	£4,500	£4,500
Sub-soiler – Shakerator	£6,500	£6,500
Fertiliser spinner – 600 kg, broadcast or strip spread	£4,500	£4,500
Fertiliser spinner – variable rate, GPS enabled	£7,000	£8,000
Dung spreader – narrow, side or rear discharge	£14,000	£14,000
Lifting forks – rear mounted on tractor 3-point	£600	£600
Fork-lift – rear mounted on tractor 3-point	£4,000	£4,000

Appendix V – The Agricultural Flat Rate Scheme

The Agricultural Flat Rate Scheme (AFRS) is an alternative to being registered for Value Added Tax (VAT) and can be joined by most farmers and growers. Wine producers are specifically included in the scheme and HMRC state that the scheme: 'includes the growing of grape vines and the production by vineyards of their own wine, up to and including the sale of the wine'. Full details of the scheme can be found in Notice 700/46 (available at www.englishwine.com/wgigb.htm). Farmers wishing to join the AFRS should apply using application form VAT 98. It should be noted that this is specifically not a VAT scheme and is not the same as the Flat Rate Scheme for VAT.

The essence of the scheme is that instead of being VAT registered and therefore being able to reclaim all VAT paid out on goods and services bought, but having to charge VAT on all eligible sales and pay that VAT over to HMRC, the AFRS does not allow you to reclaim any VAT, but, on sales to non-VAT registered customers, i.e. members of the public and small traders who are below the VAT threshold for registration, you keep 100 per cent of the sales price and on sales to VAT registered customers, i.e. most wine traders, retailers, restaurants, hotels etc you can charge a 'Flat Rate Farming' (FRF) charge of 4 per cent of the invoice value which can include delivery charges. The VAT registered customer to whom you are making the sale is allowed to treat the FRF charge as if it was VAT and reclaim it in the normal way, i.e. it has no effect upon their business or the profit they can make from selling your wine. As I show below in an example where a producer sells 15,000 bottles of both still and sparkling wine a year, two-thirds to the public at full price and one-third to traders at usual discounts, this equates to additional income for the AFRS registered producer of around £30,000 per year, against which has to be set VAT paid out on purchases which cannot be reclaimed.

AFRS registered traders are also allowed to make supplies of goods and services not produced on the farm, which would normally be subject to VAT, as long as the turnover of these supplies alone is below the VAT registration threshold (2020 £85,000). For a vineyard and winemaking enterprise these might include visits, tastings, meals, sales of non-wine items in a farm shop, room-hire, weddings and accommodation such as bed and breakfast. It also includes sales of vines to other vineyards and the public which are not normally liable to VAT, but you can add the 4 per cent on to . This means that you can charge the same prices for these as would be charged by VAT registered providers, but retain 100 per cent of the price instead of paying 20 per cent to HMRC. Therefore, if you were selling say £40,000 worth of non-agricultural goods

and services, you would be £6,667 better off than a VAT registered trader.

AFRS registered traders are also allowed to supply quite a wide range of services as long as they are 'linked' to one of the qualifying activities. Therefore, a vineyard owner can supply 'technical assistance', 'storage of agricultural products' and 'hiring out of equipment' in relation to the farming activity that entitles them to be AFRS registered. As these relate to your farming activities, they do not count towards your non-farming supplies total and again, if you supply these to VAT registered customers, you can add the 4 per cent FRF charge which you retain.

One of the additional benefits of the AFRS is that for vineyard owners having wine made by another winery, no VAT is chargeable upon the winemaking charges, which can include labelling, including the cost of the labels, all storage charges and all delivery charges. This is because a business that is carried on in duty-free (i.e. tax free) premises, such as a winery, may not add a tax to its charges. If, for instance, you decided to sell all your wine abroad, or to a duty-free operator such as a cruise line or sold it in-bond to a British supermarket, then no VAT would be payable. Therefore, AFRS registered wine producers, who would not be able to reclaim the VAT were it to be charged, can get hold of their wine free of VAT.

Most vineyard owners will register for VAT as soon as they start their enterprises and will be able to reclaim the VAT on all the equipment and supplies that they buy in order to establish the vineyard, even though they may not be making taxable supplies for some years to come. Once in production however, and making taxable supplies, i.e. selling wine, there is nothing to stop them de-registering from VAT registration and becoming AFRS registered.

The difference between being VAT registered and AFRS registered are shown in the examples below. They are based upon a producer selling 15,000 bottles a year, 50 per cent still and 50 per cent sparkling, two-thirds being sold to the (non-VAT registered) public and one-third sold to VAT registered traders, i.e. wholesalers, retailers, restaurateurs, hoteliers etc. The retail prices have been taken for this example as £10 per bottle for still wine and £24 per bottle for sparkling wine. Trade margins of 35 per cent for still wines and 30 per cent for sparkling wines, as set out in *Chapter 2: Why plant a vineyard* in the section on *Return from wine sales*, have been used. Both examples use a VAT rate of 20 per cent and Excise Duty rates of £266.72 per hectolitre for still wine and £341.63 per hectolitre for sparkling wine. As can be seen from these examples, the additional income based upon the data above, amounts to

£30,275 per annum or just over £2 per bottle.

However, deducted from this would have to be the VAT paid on all goods, supplies and services used in the vineyard. These would include VAT on repairs, maintenance and fuel on tractors and other farm equipment, plus VAT on pesticides, fertilisers, farming sundries and any other farming expenses that carried VAT. Items such as rent (if the land was leased for instances), plus all wages and salaries are not subject to VAT. Taking the above example of selling 15,000 bottles a year, this might come from say 2 ha (5-acres) and the running costs attracting VAT might come to say £5,000, so a VAT charge of £1,000 to be offset against the additional income. Even allowing for some quite hefty costs of repairs to the vineyard and replacements to machinery over time, the additional income will more than compensate for this and in addition, the additional income from non-farming sales would add further income.

If the wine producer also operated a winery, then this might complicate matters as, unless you were VAT registered, you would be unable to reclaim the VAT for any expenses incurred in the production process such as materials, including bottles, packaging and labels. However, there is nothing to stop you running the farming enterprise and the winemaking enterprise as two legally separate entities. You could for instance have two separate limited companies, or run one as a sole trader and the other as a partnership. In this way, the farming enterprise could be AFRS registered and the winemaking enterprise could be VAT registered and able to reclaim any VAT.

As a slight refinement to the above, you could have one enterprise (A Ltd) that owned the vineyards and a winery and was VAT registered. It owned all the equipment necessary to operate the vineyard and winery and could reclaim VAT on all its outgoings. You could then have a separate company (B Ltd) that rented the vineyards from A Ltd and farmed them and would therefore be eligible to be AFRS registered. Its rent would include use of the farm equipment. This might be challenged by HMC&E as being somehow separate from the rent of the land (which is not subject to VAT), so there might be some VAT to pay at this stage. B Ltd would produce the grapes which would then be processed into wine by A Ltd under bond (and therefore free of VAT). All costs of production, including the cost of all labels, packaging and transport to the customer, would be part of the charge B Ltd paid to A Ltd and would be free of VAT. The figures need checking.

Appendix VI – The story of Wrotham Pinot

The vine that Edward Hyams discovered in Wrotham in the early 1950s was in all probability the variety known as 'Miller's Burgundy' which had been widely grown on walls and in gardens in Great Britain for at least two hundred years. Philip Miller's *Gardener's Dictionary*, first published in 1735, has an entry for 'the black Cluster or Munier [sic] grape, as it is called by the French, from the hoary down of the leaves in summer —- which some call the Burgundy grape'. Archibald Barron writing in his book, *Vines and Vine Culture,* the standard Victorian work on grape growing and first published in 1883, states that the variety was: *found by* [the famous horticulturalist] *Sir Joseph Banks in the remains of an ancient vineyard at Tortworth, Gloucestershire* – a county well known for its medieval vineyards. Hyams took cuttings to Brock at Oxted who said that when compared to supplies of Meunier from France, the variety now known as Wrotham Pinot: *had a higher natural sugar content and ripened two weeks earlier*. Hyams, ever the journalist in search of a good story, claimed that this vine had been left behind by the Romans although provided no evidence for this.

Brock, in his *Report No. 3: Progress With Vines And Wines* of 1961, states that Wrotham Pinot: 'appears to be better adapted to this climate than the standard Pinot grapes' and 'is a consistent cropper, and ripens here in good years in the open vineyard', i.e. not under cloches. He continues that: 'it would probably be suitable for a really highly flavoured Burgundy type of red wine in favoured sites in the south of England, but it is not suitable where the conditions are not very good'. Brock sold cuttings and the variety became quite popular in early vineyards, although it is unlikely that any vines from the cuttings supplied by Brock survive in any of today's vineyards. Jack Ward, writing in his 1984 book *Vine Growing in the British Isles*, says that Wrotham Pinot: 'was given a fair trial in the Beaulieu vineyard and contributed to the production of a rosé wine' but that: 'its performance was adequate but hardly good enough, for it was later grubbed in favour of other varieties'. Gillian Pearkes, writing in the 1989 version of her 1982 book *Vinegrowing in Britain* says that Meunier: 'suffers from the same inadequacies' as Pinot noir which she states: 'is seldom if ever even a mediocre wine; it is thin, pale and lacking in intensity of bouquet and depth of flavour'. Despite the fact that today all plantings of Meunier in Britain stem from French and German nurseries, the name Wrotham Pinot is still a legally acceptable synonym for this variety, although the name is not used at all by any British producers that I know of.

In 2004, a well-known grower from California, Richard Grant Peterson, (who was winemaker at Beaulieu Vineyards in Napa after the famous André Tchelistcheff) announced that he had a vineyard in Yountville in the Napa Valley, planted with Wrotham Pinot vines propagated from cuttings taken from the original Wrotham Pinot vine in 1980. At the time, Peterson was a judge at the International Wine and Spirits Competition (IWSC) and had been told about a vine growing wild against a stone wall in the village of Wrotham. He then claims to have inspected it and found that it had tiny white hairs on the upper surface of the leaves and was unlike any vine he had ever seen. In addition, this vine: 'had developed a considerable immunity to Powdery Mildew'. On his website, Peterson states that the original vine in Wrotham died in the mid-1980s, but that: *there are now at least two new wild seedlings growing a few feet from where the mother vine had stood*. This statement would appear to show that Peterson knows where the original vine was. Despite several attempts to contact Peterson, I have been unable to verify whether he personally visited the cottage in Wrotham or who the 'local winemakers' were that had made the wine he tasted. Ignoring the fact that a 'seedling' cannot possibly be the same variety as the vine variety that produced the bunch of grapes from which it came, I find the whole story slightly unbelievable!

On his website, Peterson states that: 'English viticulture scholars eventually came to the conclusion the Wrotham vine was a natural seedling of Pinot noir vines that the early Romans brought to England 2000 years ago'. Ignoring the 'seedling' question again, quite who these 'English viticulture scholars' were is not known, but possibly a reference to Ray Brock at Oxted, although he never mentioned this to me when I interviewed him (several times) for my UK Vineyards Guide before he died in 1999.

Peterson also states that he tasted a sparkling *Blanc de Noirs* wine from the variety made by 'local winemakers' and immediately recognised its potential. He imported cuttings into California and, 'after a quarantine of many years as required by law', vines were propagated and a 0.8-ha (2-acre) vineyard established which he grows without the need to spray them with sulphur against Powdery Mildew. Peterson had a vine analysed by the University of California's Davis wine department who pronounced the vine's DNA to be 'identical to that of Pinot noir'. Peterson made a bottle-fermented rosé sparkling, which sold for $60 a bottle and its quality is obviously excellent as the 2001 vintage won a Double Gold at the US National Women's Wine Competition. On the website www.princeofpinot. com is says that he now produces a sparkling wine and a still wine in alternate years 'because of the small number of vines available'. WineSearcher list a $60 *Richard Grant North Coast Blanc de Noir* which doesn't mention Wrotham Pinot, although the label shown does, and says it's a '2014 Reserve' from the Oak Knoll District which is at the southern end of the Napa Valley. WineSearcher also says that the still Pinot noir was 'last available in November 2017'.

The Wrotham Historical Society published a booklet called *Farming in Wrotham Through the Ages* which I purchased for £3 in the hope that it might provide me with some useful information. The booklet mentions that: *there is reputed to have been a Roman vineyard on the slopes of the North Downs above Wrotham* (funny how reports about Roman vineyards always start with 'reputed' or 'thought to have been') but about the location of the cottage it gives little help. The booklet merely says that the cottage wall was *on the main road in the village* which is hardly giving the game away. The text in the booklet is more or less the same as can be found on Peterson's website: www.richardgrant-wine.com.

Today in Britain there are to my knowledge no commercial plantings of Wrotham Pinot and the Wine Standards vine variety database does not list any. Sara Bell of Sunnybank Vines, a vine nursery that holds the National Collection of outdoor vines, can supply cuttings that she says are Wrotham Pinot, but as she inherited the stock from the previous owner of the nursery, Brian Edwards, who was unsure of where his original vines came from, the chances are that it's a bog-standard Meunier. Sarah Bell says its 'indistinguishable from Meunier'. In 2017 I visited a grower in Novia Scotia who claimed to have some (he wouldn't tell me where he got the vines from) and when he realised I knew something about the supposed variety, he changed the subject.

Wrotham Pinot was always a chimera, a unicorn, a phoenix, a rare beast that was never going to re-appear. It would be good to find a *vinifera* vine that was resistant to Powdery Mildew because vine researchers all over the world would be beating a path to your door.

Appendix VII – The Complete Story of Sparkling Wine production in Great Britain – 1662-2020

The production of sparkling wine in Britain – although not from home-grown grapes – is verifiably over 350 years old, and we know from the two papers read at the newly-founded Royal Society in December 1662 that sugar added to a fermented product and sealed in a bottle with a tightly bound stopper produced a 'brisk and sparkling' product. The Reverend John Beale's 'Aphorisms on Cider', read to the Royal Society on 10 December 1662 says that 'bottling is the next best improver' for cider and that 'two to three raisins into every bottle' plus 'a walnut of sugar' – a recipe guaranteed to produce a secondary fermentation – works wonders on the cider. A week later, on 17 December, it was the turn of the now famous Dr Christopher Merrett to read his paper, 'Some Observations Concerning the Ordering of Wines' and describe how Britain's seventeenth-century 'wine coopers' were making their wines 'brisk and sparkling' by the addition of sugar. This practice was certainly happening before 1662 and followed the development of the strong *verre Anglais* bottles which Sir Kenelm Digby had (probably) been involved in perfecting since around 1628-1630.[1]

Exactly when the first sparkling wine made from English grapes was produced is open to debate. Certainly, wines being made in England in the 1750s were considered comparable to Champagne and as has been documented by me in other books, the wines produced at Painshill Place between 1741 and 1779 were often described as such. Of course, Champagne in those days was not always the sparkling wine that we know today. I have a wine list from the *Magaszin de Vins Fins Chez Terral* from Pontac, a village just outside Bordeaux, dated 1760, which lists *Champagne mousseux* and *Champagne non-mousseux* both at the same price.

The first recorded production of bottle-fermented sparkling wines – made from British-grown grapes – is probably that carried out by Raymond Barrington Brock at his Oxted Viticultural Research Station in the 1950s. The *Daily Mirror* of 17 August 1950 carried an article entitled 'A bottle of Maidstone '49' which praised the work of Brock and that other viticultural pioneer, Edward Hyams and ended by saying: 'perhaps ten years hence you'll be raising a glass of sparkling Canterbury in honour of the men who made an English wine industry possible'. In September 1959 Brock welcomed members of the wine trade to a tasting and offered them a number of different wines, including sparkling wines. I have a letter dated 11 September 1959 sent to Brock from John Clevely, then a young Master of

Wine, in which he thanks Brock for the visit and tasting and ends with a postscript saying: 'Moët must look to their laurels if you really start going "commercial" with that sparkling wine. I thought it was wonderful.' Praise indeed. Some of these sparkling wines survived undisturbed in the Station's cellars until the 1980s.

Sir Guy Salisbury-Jones at Hambledon, whose initial (1953) plantings included 20 Chardonnay vines, experimented with the production of a bottle-fermented sparkling wine, and in 1969 Bill Carcary, his vineyard manager, produced a batch of 60 bottles. Salisbury-Jones expanded the plantings of Chardonnay in 1970 with a further 1,000 vines but whether to make still or sparkling wine is not known. In 1979 his winemaking consultant Anton Massel helped produce a batch with apparently favourable results and as Salisbury-Jones also grew Auxerrois and Meunier, which ripened more easily, these became the basis of their sparkling wine *cuvée*. However, Sir Guy considered that the production costs were too high and the length of time the wine needed to mature was too long to make the product commercially viable and production ceased. Little did Bill Carcary know it at the time, but his experiments were to be the foundation stone of Britain's highest profile sparking wine producer, Nyetimber.

The first producers to make commercial quantities of bottle-fermented sparkling wines were Nigel (de Marsac) Godden at Pilton Manor Vineyard in Somerset – first planted in 1966 – and Graham Barrett at Felsted (or Felstar[2] as it was originally known) Vineyard in Essex – first planted in 1967. As was quite usual at that time, the main varieties grown were Müller-Thurgau and Seyval blanc and it is probable that it was these that were used. Their wines – never produced in large volumes – were certainly interesting, maybe even worth drinking and in the 1979 English Wine of the Year Competition (EWOTYC) the *1976 Felstar Méthode Champenoise* won a silver medal and the following year, 1978, the *NV Pilton Manor De Marsac Brut Méthode Champenoise* won a bronze. These early successes, however, didn't seem to help sales much and their production faded out.

The next appearance of a bottle-fermented sparkling wine in the EWOTYC (ignoring the carbonated *1983 Barton Manor Sparkling Rosé* that won a gold medal in the 1984 competition – delicious though it was) was in 1987,

---

[1] My next book is all about Digby and his bottle.

[2] The story was that he was not allowed to call it after the name of the village where the vineyard was situated, Felsted, as the village didn't have its own appellation!

when the first Carr Taylor sparkling wine won a medal. David and Linda Carr Taylor first planted vines at their vineyard in Westfield, near Hastings, East Sussex, in 1973 and until 1983 their grapes were sent to Lamberhurst Vineyards for winemaking. From their huge 1983 vintage however, when Reichensteiner cropped at 15 tonnes-acre and their total output came to 186,000 bottles, they decided to start making bottle-fermented sparkling wines. They engaged Clement Nowak, a Champagne-based Polish-French consultant winemaker, whose name at one stage actually appeared on the neck-label. For a few years Carr Taylor became the major producer – in fact almost the only producer – of bottle-fermented sparkling wines in Britain and achieved considerable success. Their *Vintage Sparkling* won a gold medal in the 1988 EWOTYC and their *Non-Vintage Sparkling* won the Jack Ward Trophy (best large volume wine) in the 1989 EWOTYC. In 1993 they won the IWSC English Wine Trophy with their *1987 Vintage Sparkling*. They also entered their wines into overseas competitions – a rarity in those days – and did surprisingly well. Their *1988 Vintage Sparkling* was awarded a gold medal at the prestigious Concours European des Grands Vins beating 1,800 Champagnes and other bottle-fermented sparkling wines from around the world, and in 1999, in the same competition, their 1996 vintage was awarded a gold medal, this time out of 4,300 entrants. A fact that tends to get forgotten in these days of Britain's mega-vineyards planted with Champagne varieties is that the Carr Taylors were certainly the first to make serious commercial quantities of bottle-fermented sparkling wines. They did, however, only ever use what might be termed 'native' varieties for Britain: Reichensteiner, Schönburger, Kerner and Huxelrebe being the most important ones. This reliance on non-classic varieties, whilst it gave their wines a point of difference from other Chardonnay and Pinot based wines, also gave the wines a character more akin to Sekt or Asti than Champagne, something not all critics and commentators liked.

In 1985, Karl-Heinz Johner, then winemaker at Kenneth McAlpine's Lamberhurst Vineyards, started making sparkling wine for Piers Greenwood at New Hall Vineyards in Essex, a major grape supplier to Lamberhurst. New Hall had started growing Pinots in 1971 when they planted 0.47-ha (1.16-acres) of Ruländer, the name under which Pinot gris used to be known in southern Germany, especially Baden, and in 1973, 1974 and 1975 they continued and planted 0.62-ha (1.53-acres) of Pinot noir in each of those years. Unknown to them at the time of planting, 50 Chardonnay vines had been included by mistake with the Pinot gris, and in 1983 (or it could have been 1984), 1,300 kgs of Pinot noir and 150 kgs of Chardonnay, 'the very last grapes picked from the harvest' according to Greenwood, were sent to Lamberhurst for Johner to make into sparkling wine. This wine was therefore almost certainly the first 'Champagne variety', bottle-fermented sparkling wine to be produced in Britain. Greenwood says the wine 'was released in December 1986 after ten months

on the yeast' (which makes it more likely to have been a 1984 wine, bottled in March 1985) and is the wine which Chris Trembath, Piers' brother-in-law and now part-owner of New Hall Vineyards, remembers having at his wedding in 1987. More sparkling wine was made in 1985 (verified by a New Hall stocktaking report from 1992 which lists '1985 sparkling' on it). These wines were probably those that I listed as my 'Vineyard Choice' in my 1989 book *The Vineyards of England*, which said '1984/5 New Hall Sparkling Wine Medium Dry £9.95'.

McAlpine was so impressed with this wine that he asked Greenwood to plant more Pinots, exclusively for Lamberhurst, and this he did in 1988 with 3.98-ha (9.83-acres) of Pinot noir and 1.80-ha (4.45-acres) of Pinot blanc. At that time, most people considered Chardonnay as being too difficult to ripen. When I took over as winemaker at Lamberhurst in May 1988 (after Johner was dismissed for dubious blending practices) there was a stock of bottle-fermented sparkling wine in the cellars and I was told that this was made from Pinot noir, Pinot blanc and Chardonnay, all grown at New Hall. After getting Tom Stevenson to taste the wine and give it his seal of approval, it was released as *Lamberhurst Brut* in 1988 or 1989.[3] Whether this was from the same 1984/1985 grapes as New Hall's or from a subsequent vintage, I am unsure.

The next on Britain's sparkling wine scene was David Cowderoy, son of Norman Cowderoy who had established a vineyard at Rock Lodge near Haywards Heath in 1963 (the vineyard is now leased by Plumpton College). David had been to Wye College to study agriculture (and met Jo, now his wife and currently WineGB Operations Manager who was also studying there) and after getting his degree went to Roseworthy College in Australia to study winemaking. Returning to Britain in 1986, and with the knowledge that marriage to Jo was on the horizon, he made a batch of sparkling wine using Müller-Thurgau and Reichensteiner from that vintage. Thus, when he and Jo got married in 1988, they and their friends and family were able to celebrate with what was to be the first of many batches of English sparkling wine. David continued with his experiments in sparkling wine and, working at his father's winery at Rock Lodge, produced the 1989 Rock Lodge Impresario[4] which won the IWSC English Wine Trophy in 1991.

In 1992 David joined forces with others to create Chapel Down Wines and one of their first wines, the non-vintage Epoch Brut, made from a blend of Müller-Thurgau, Reichensteiner and Seyval blanc, was in fact a re-badged Rock Lodge wine. The fact that Chapel Down was not using the classic Champagne varieties (which, with the exception of New Hall Vineyards, were not being grown in enough quantity for them to buy), gave them something

---

3 My main memory of Tom and his wife's visit to taste the wine was that he locked his car keys in his Renault and had to wait several hours for the AA to arrive and break into it, which they did in seconds. My offers of smashing the window were very soundly refused.

4 It had originally been called 'Rock Lodge Imperial' until Moët & Chandon complained as Impérial is one of their brands.

of a marketing advantage and enabled their prices to remain reasonable – under £10 – although at the time this was at least twice that of still wines. In the end though, once Chardonnay and Pinot-based wines started to appear in 1997–98, this marketing edge disappeared and their Müller-Thurgau, Reichensteiner and Seyval blanc based wines, although very good and well-priced, were always playing second fiddle to the Champagne lookalikes in quality (and quality perception) terms, as well as in price. At much the same time, John Worontschak, winemaker at Thames Valley Vineyard (today's Stanlake Park) made a sparkling wine using Pinot noir from Ascot Vineyard, a 1-hectare (2.47-acre) vineyard planted in 1979 on Crown land near Sunninghill Park and owned by Colonel Robby Robertson. Called Ascot Brut NV, it was released in 1992 and won a silver medal in the 1994 EWOTYC. Worontschak produced a number of Ascot sparkling wines from both Pinot noir and (unusually) Gamay noir, winning several silvers and bronzes between 1994 and 2004.

The production of sparkling wines using the three classic varieties – Chardonnay, Pinot noir and Meunier – started in the mid-1980s when growers like Piers Greenwood (see above), Martin Oldaker at Surrenden Vineyard, near Ashford, (planted between 1984 and 1986) and Karen Ostborn and Alan Smalley at Throwley, near Faversham (planted in 1986),[5] both in Kent, all started growing Chardonnay and Pinot noir with the encouragement of Christopher (Kit) Lindlar. After leaving the Merrydown Wine Company,[6] based in Horam, East Sussex, where he had been one of the winemakers since 1976, Lindlar set up as a contract winemaker, firstly at Biddenden Vineyards, and then, in 1986, at his own High Weald Winery at Grafty Green, near Ashford, Kent. Lindlar, who also supplied vines, persuaded the two Kent vineyards above to experiment with these varieties, which had until then been very unsuccessful in Britain. Brock had grown Chardonnay in his collection at Oxted but could never get it to ripen properly. Ian and Andrew Paget at Chilsdown Vineyard planted Chardonnay and also had no luck getting it to ripen. In 1981, a very dismal year for British vinegrowers, the acidity (in grams per litre) in their Chardonnay was higher than the degrees Oechsle. Ouch. Extreme unripeness was a common finding among those early growers who persevered with it, although most decided to give up and removed the offending variety. Only in really hot years would Chardonnay produce anything like ripe grapes and tolerable wine. Pinot noir, like many of the black varieties then being grown, suffered from terrible botrytis and was very difficult to ripen without huge losses. It is only since the arrival of better anti-botrytis sprays – initially Rovral

and Ronilan, but more recently Scala, Switch and Teldor – that growing fungus-sensitive varieties like Pinot noir has been possible. Meunier, in the guise of Wrotham Pinot, had always been grown in small amounts, but never used for anything other than blending with other, riper, reds. Lindlar's biggest, and subsequently best-known clients, were Stuart and Sandy Moss who decided, in 1988, to plant a vineyard at Nyetimber near Pulborough in West Sussex.

The Mosses had, by all accounts, been looking at various locations to plant a vineyard – California was at one time the front runner – but it was Sandy's love of (and business in) early English oak furniture, that persuaded them that England was the place. In 1985 Hambledon Vineyards was up for sale and the Mosses viewed it and made a bid for it, but lost out to another bidder, John Patterson, who owned it until 1994. Bill Carcary, who had been at Hambledon since 1966, got to know the Mosses quite well at the time and when they then bought the 49-hectare Nyetimber estate in 1986, they asked Carcary to come and work for them as estate manager and eventually as winemaker and got so far with this idea as to refurbish a cottage for him and offer him a contract of employment. In his discussions with them about planting a vineyard on the land at Nyetimber, Carcary remembers it being his idea that they should plant the Champagne varieties for sparkling wine production, something he had long wanted to do at Hambledon, but which, as has been stated above, Salisbury-Jones had ruled out on cost grounds. In the end, Carcary decided for family reasons not to leave Hambledon and stayed, working for the new owner. Whoever actually came up with the idea to produce bottle-fermented sparkling wines on this (for the time) very large scale, the Mosses went ahead and planted the classic Champagne varieties, something which at the time was revolutionary – some said bonkers.

The vines for the Nyetimber plantings between 1988 and 1991 were sourced from France and it was to Lindlar's High Weald Winery that the first commercial vintage, the 1992, was taken for processing under the watchful eye of consultant Jean-Manuel Jacquinot. As Lindlar modestly says, 'while they did hire Jacquinot, the winemaking buck stopped with me; that is to say, had those early vintages flopped it would definitely have been down to me.' Given the importance of the Mosses' enterprise, which when all was said and done was still something of an experiment, one has to give praise to Lindlar where it is due.

The Mosses released their first wine in 1997, the 100 per cent Chardonnay *1992 Nyetimber Première Cuvée Blanc de Blancs,* which won an IWSC[7] Gold Medal and that competition's English Wine Trophy. Then, in 1998 they released a second wine, the *1993 Nyetimber Classic Cuvée* (a true Champagne-variety blend) which went one better, winning not only an IWSC Gold Medal and the English Wine Trophy, but also the Bottle Fermented Sparkling Wine Trophy. Suddenly, everyone woke up to the fact that good wine, even stunningly good wine, could be made from

---

5 The 1989 Throwley Chardonnay Sparkling won the IWSC English Wine Trophy in 1992.

6 Merrydown was partially owned by Jack Ward who was its MD, as well as being Chairman of the English Vineyards Association. Merrydown owned a small vineyard, Horam Manor, but more importantly operated a co-operative winemaking scheme for British vineyards, where in exchange for a percentage of the grapes, they made your wine.

7 International Wine and Spirit Competition.

hitherto seemingly unworkable varieties – Chardonnay, Pinot noir and Meunier – and what was more, the wine could be sold at a premium price. Some of had been wondering if things were about to change, but after these successes a few of us realised that the game was up for German-variety based still wines. With these two wines and these two significant trophies, English Sparkling Wine had definitely arrived. Nyetimber went from strength to strength, winning the Gore-Browne Trophy in 2001, 2003, 2004, 2005 and 2006 and the IWSC International Sparkling Wine Trophy in 2006, 2008 and 2009.[8]

A few years after the Mosses planted, another Lindlar client, Mike Roberts, also decided to establish a dedicated classic-variety, bottle-fermented sparkling wine business at Ditchling in East Sussex. Ridgeview Winery was established in 1995 with thirteen clones of Chardonnay, Pinot noir and Meunier and today it covers 6.48 hectares (16 acres), although it has access to grapes from a much larger area. A modern winery, with underground storage cellar, was built and equipped with the contents of the High Weald Winery, which was acquired when Lindlar closed the winery. In order to kick-start Ridgeview's production line, Chardonnay and Pinot noir grapes were bought from other growers, including Surrenden and Throwley, and the 1996 *Cuveé Merret Bloomsbury* was produced. This wine won the 2000 EWOTYC Gore-Browne Trophy, awarded for wine of the year. Since that first release, Ridgeview has produced a range of wines, all named after London squares or areas – Belgravia, Bloomsbury, Cavendish, Fitzrovia, Grosvenor, Knightsbridge and Pimlico – and the tally of awards has been impressive. They won the Gore-Browne Trophy in 2000, 2002, 2009, 2010 and 2011 and regularly win gold and silver medals in the major wine competitions. Their most notable success was probably winning the Decanter World Wine Awards International Sparkling Wine Trophy (beating four very prestigious Champagnes in the process) with their 2006 *Grosvenor Blanc de Blancs*.

When first Nyetimber and later Ridgeview started selling wines and achieving the sort of prices that many in the wine business in Britain had thought impossible, the way forward for home-grown sparkling wines started to look a lot different. Following their significant commercial and competition success, plantings of the three classic Champagne varieties in Britain increased year on year and since the very warm year of 2003, several significant vineyards have been planted. Nyetimber changed hands twice and under its current ownership has expanded on various sites from its original 15.8 hectares to a whopping 322-hectares (796-acres) spread over 12 sites from Stockbridge in the west, to Chartham in the east, 177 km (110-miles) away. Owner Eric Heerema told me he was always looking for more land. Other large sparkling wine producers include the largest, Chapel Down, who own or lease 220-ha (544-acres), plus take grapes under contract from another 101-ha (250-acres) and

have a further 64-ha 158-acres) of land to plant at Boarley on the North Downs. The third largest by vineyard size, although with only a small percentage of the planted land yet cropping, is Mark Dixon's MDCV Ltd with vineyards in East and West Sussex, Kent and Essex currently totalling 284-ha (700-acres). The next three largest sparkling wine producers are: Hambledon with 90 hectares; Rathfinny with 93.5-ha; Gusbourne with 93 hectares; and Ridgeview with 90 hectares (both owned and under contract). Other major sparkling players include Bluebell Estates, Bolney, Camel Valley, Coates and Seely, Laithwaites, Langham, Exton Park, Furleigh, Greyfriars, Hundred Hills, Hush Heath, Roebuck, Simpson's, Squerryes, Tinwood, and Wiston. There are also several quite sizeable growers who mainly grow grapes under contract to the major producers. Together, there are around 25 producers who control 50 per cent of Britain's vineyard area and probably nearer 65 per cent of its sparkling wine production. And then there are the French.

The French connection

French producers, both from Champagne and elsewhere in France, have been interested in wine production in Britain for almost twenty years. In 2003, following the hottest and earliest ever harvest in Champagne, a grower and winemaker from Avizes, Didier Pierson, and his then (life) partner, British-born Imogen Whitaker, entered into a joint venture with a Hampshire farmer, and in 2004 and 2005 planted 4 hectares of the three classic Champagne varieties on a somewhat isolated, exposed and high (up to 203 metres above sea level) site. The vines were planted very much *à la méthode Champenoise* with narrow rows and low trellising (ideal height for pheasants), and farmed using the only *enjambeur* tractor in Britain. The vineyard, known as Meonhill, was subsequently bought by nearby Hambledon Vineyards following the break-up of the Pierson-Whitaker relationship. The vineyard has now been abandoned.

In around 2004–5, Champagne producer Duval-Leroy instructed the Canterbury office of Strutt & Parker (well-known Land Agents with strong farming links) to seek out suitable land for planting. Their brief was short, if slightly eccentric: the land must be within one hour of the Channel Tunnel, the soil must be chalky and there must have been no livestock on the land in recent years. This last request baffled most of those involved, who were told that there might be residues in the soil which would interfere with the fermentation process. (I suspect there was something lost in translation.) Although several sites were found and one selected (Squerryes Estate) and lengthy negotiations entered into, nothing came of Duval-Leroy's quest and they retreated back to Épernay.[9] In 2006 (on an exceptionally wet and windy June day) I took Louis Roederer boss Frédéric Rouzaud, his winemaker

---

8 Since then, they have declined to enter most wine competitions.

9 Squerryes Estate subsequently planted up and today have around 21 hectares of vineyards.

Jean-Baptiste Lecaillon and Mark Bingley MW, their British agent, on a tour of English vineyards simply to show them what we were up to. Whilst this visit fuelled a lot of speculation about their intentions, it had only ever been a fact-finding trip and I never expected it to be anything else.

In 2007 Champagne producer Billecart-Salmon, in a joint venture with London wine merchants Berry Brothers and Rudd, found land in Dorset for the creation of an English sparkling wine brand. They were all set to go, but the financial crisis of 2008, when Champagne sales took a sudden dive, put them off the whole idea and it was shelved. When well-known wine writer and taster Steven Spurrier wanted to plant a vineyard on his wife's farm in Dorset, he first approached Duval-Leroy, who (again) proved difficult to negotiate with; then he tried teaming up with the Burgundy producer Boisset (whose sparkling wine house Louis Bouillot produce around 10 million bottles of Crémant de Bourgogne a year). However, Boisset were convinced that home-grown English sparkling wine couldn't sell for more than £12.99 a bottle (this was in around 2006/7) and wouldn't take things any further. Spurrier subsequently went ahead on his own, planting 10 hectares under the Bride Valley label.

The first Champagne house to actually take the plunge was Taittinger who, after over two years of discussions and negotiations, put their mouth and their money behind Domaine Evremond. On 26 February 2014, Patrick McGrath, a fellow Master of Wine and MD and part-owner of Hatch Mansfield Agencies Ltd (Taittinger's British agents amongst other things) who wanted to discuss English sparkling wine. As a successful wine wholesaler, Hatch Mansfield had decided that they needed an English Sparkling Wine on their books. How should they go about this? Should they look at buying an existing producer? Should they go the Champagne route – buy grapes and create a blend which they would market exclusively? What were the options? Buying an existing producer I said was out. The best were not for sale and you wouldn't want any of the rest. Buy grapes? Forget it – too many of the large wineries were already chasing them and supplies were too erratic and – more importantly – too expensive (at that time). Why not go the whole hog I inquired? Buy land, plant vines, harvest grapes, make wine, bottle wine, wait three years and – hey presto – wine for sale? Is that a practical option asked McGrath? That's what I help people do I said. When do we start looking for land?

The next eighteen months were spent finding a suitable site. The south-east was chosen as the target area and Kent the favourite county. Why? I knew it well, it was near the Channel Tunnel and I considered it to have the best sites and best growing conditions for vines in Britain. As Hatch Mansfield were agents for Taittinger, they had to be consulted and Pierre-Emmanuel Taittinger, grandson of the founder and most definitely an Anglophile, agreed that Taittinger would become an investor. There several fraught Taittinger board meetings, at which the board decided eventually not to get involved. This put something of a damper on the project. However, after McGrath gave a passionate defence and promotion of the project, the board decided that they would get involved, but only if they could be majority shareholders. This was agreed and today, Taittinger hold around 55 per cent of the equity with the other 45 per cent owned by some significant minor shareholders, plus around thirty very small shareholders, mostly Hatch Mansfield employees, plus myself.

Various sites were found, looked at and rejected, but eventually what I considered to be a perfect site, south-facing, less than 100 metres above sea level, well sheltered and with mainly chalk soil was found between the villages of Selling and Chilham, eight miles south-west of Canterbury. After lengthy negotiations with the landowners, buying land they didn't want to sell and not buying blocks they wanted to sell, a deal was done and on 16 November 2015 the purchase was completed (it should have been the previous Friday but that, being the thirteenth, was rejected by the French as being inauspicious). The press launch was held on 9 December 2015 in the magnificent surroundings of Westminster Abbey, final resting place (in Poet's Corner) of Charles de Saint-Evremond. He was another Anglophile and Champagne lover who did much to introduce late-seventeenth-century Londoners to sparkling Champagne, which was becoming more popular than the predominantly still wine that was the norm at that time. The next 15 months were spent getting the site prepared, removing the apple, pear and plum trees that covered most of the site. The first 20 hectares (50 acres) of vines were planted in May 2017, another 8 hectares (20 acres) in 2019 and 17 ha (42 acres) in 2020. More will be planted as land becomes available, some of the land they bought being under a short farming lease (known as an FBT) to the original owners. The first (very small) harvest was taken in October 2018, with 50+ tonnes in 2019. The plan is to make a three or four vintage NV blend before bottling and ageing. Therefore, the first bottling will probably be in April-May 2022, with the first commercial release not scheduled until 2025 at the earliest. In June 2020, planning permission was granted for a £10 million winery (and modest visitor centre) and groundwork started on it almost immediately. The winery will be finished in time for the 2021 harvest.

Following Taittinger, the floodgates hardly opened, but certainly, tongues were set wagging in Reims and Épernay and it wasn't too long before the patter of *vigneron*-sized French feet were heard in England again, this time in Hampshire. Champagne Vranken-Pommery bought (from Malcolm Isaacs, owner of nearby Exton Park Estate, for, it is said, £20,000 an acre) a 40-hectare site near Old Arlesford and planted 19 hectares in 2018, with a further 14 hectares in 2019. In the meantime, they have collaborated with near neighbours Hattingley Valley to produce some English sparkling wines under the *Louis Pommery* brand name, the same name as they use for their Californian *méthode Champenoise* wine (no doubt they wish they could call it that in Britain). Boutinot, a major British-based wine wholesaler, bought Henners Vineyard in 2015 and the South African Benguela Cove Winery established vineyards in West Sussex in 2016 and 2017.

The latest large-scale overseas-based investor to set up shop in England is Mark Dixon who made his fortune with Regus, the global workspace and serviced offices company, and now owns the 121-ha Château de Berne, a Provencal Rosé producer. In around 2016 he bought Sedlescombe Organic Vineyards, followed shortly afterwards by buying Kingscote Vineyards, East Grinstead. Following the purchase of these, new vines were planted at both places, plus at Bodiam Vineyard, which Sedlescombe had leased since 1994. Further land was leased from the owners of the Bodiam vineyard, the Sternberg family, at Old Place Farm, Sandhurst, just below Sandhurst Church. Here, around 40 ha (100 acres) of vines were planted in 2019, somewhat perversely (and baffling to local vineyard owners) at an angle across a perfectly fine south-facing slope. Not content with this number and area of vineyards, the already-organic Court Lodge and Brookers farms in the hamlet of Luddesdown near Gravesend were purchased and in 2019 some of the land was planted with 177-ha of Champagne varieties. This was followed in 2020 by another new vineyard, this time a mere 38-ha in Essex in the village of Althorne facing onto the River Crouch in the by now much-bevined CM3 postcode. In total, Dixon now has control over 283-ha (700-acres) of vineyards, most of which is either organic, biodynamic or in conversion. Quite what the long-term plan is for all these vineyards is not clear. Both Sedlescombe and Kingscote were put on the market in early 2020 as they 'didn't fit in with the business plan going forward'. I heard it on good authority that Dixon's aim was to produce an entry-level (£12.99 was mentioned) organic, Charmat method sparkling wine. It will be interesting to see how this empire of wine develops. See www.mdcvuk.com for more details. Other southern French, Italian and even Californian producers have also been on reconnaissance missions to look at vineyards in Britain and although they are all surprised how cheap vineyard land is compared to where they come from, their main discovery is how low yields are. With still wines selling in the £10–15 range and sparkling wines retailing between £16 and £35, can you really make money when you only harvest 5 tonnes-ha (2 tonnes-acre)? Those yields might be OK if you are selling high-priced Champagne, Bordeaux or Burgundy, or New World wines with more than 95 Parker points, but they don't work for these relatively modestly priced wines.

The future for English sparkling wines is as uncertain as most things these days. There is a current (2020) imbalance between demand and supply with stocks large (very large) in relation to current sales, and with around 50 per cent of the land currently planted not yet producing wine for sale. Yields in the last two years (2018 and 2019) have been well above average and 2020 is looking good. With 300-350 ha of new plantings coming on-stream every year, and with vineyard plantings showing little sign of stopping, at some point, demand and supply have to reach an equilibrium. Many think that substantial price reductions of both wine and grapes cannot be far away.

Appendix VIII – Jack Ward, Horam Manor and the Merrydown Wine Company

No history of the revival of viticulture in the British Isles would be complete without recognising the part played by Jack Ward, his vineyards at Horam Manor in East Sussex and the company he co-founded, the Merrydown Wine Company.

Merrydown was founded in 1946 by two friends, Ian Howie and Jack Ward, to make cider and fruit wines from local Sussex produce. Jack, who had graduated from Trinity College, Cambridge with a degree in English, had become interested in wine while studying his first love, music, at the *Frankfurt Conservatorium*. In the early 1950s, he became interested in the revival of viticulture and in what Ray Brock, whom he visited, was doing at Oxted. In 1953 he planted six vines, obtained from Oxted, in the grounds of Horam Manor: two Riesling Sylvaner (Müller-Thurgau), two Madeleine Royale and two Gamay Hâtif des Vosges.

No sooner had these trial vines been planted than a property known as The Grange, a house with several acres of garden attached, came up for sale across the road from Horam Manor and was bought by the company to provide accommodation for some of its employees. Ward promptly claimed the gardens and in 1954 planted a 2-acre vineyard. This was at a time when Hambledon was the only other vineyard in the country, planted two years earlier in 1952 and yet to produce its first harvest. The varieties that Ward chose do not seem to be recorded, although one of them was certainly Müller-Thurgau, as a wine from this variety from the Grange vineyard was served to Prince Philip at a dinner held at The Vintners Hall in London. In 1962, however, with the vineyard only just reaching maturity, the company decided that the offer they had received from a local builder to develop the site was too tempting and it was sold.

Ward had perhaps foreseen this and had already thought about where else he might establish a vineyard. The company owned, in addition to the Horam Manor site, a property a mile up the road behind a row of houses known as Kingston Villas. This had been the site of an old brickworks and the heavy clay soil, although suitable for making bricks, could not be considered ideal for vines.

Unperturbed by this, Ward planted 800 grafted vines in the spring of 1963 but the heavy soil conditions were too much for them and not a single one survived! He reckoned that they had all been planted too deeply and had been killed by waterlogging. Not one to be put off easily, Ward planted further vines, this time using some rooted cuttings that had been raised on the site, as well as some others that had been grafted by Merrydown as an experiment. However, the latter were not a success and home-grafting

was soon abandoned. At the Brickyard, as the vineyard came to be known, the vines slowly established themselves, not helped by the extreme winter frosts of 1967 that cut a lot of the vines back down to the ground. The main business of Merrydown, the making of cider, perry and fruit wines, produced a by-product in the form of vast heaps of pomace (fruit skins, pulp and pips) left over from the pressing process. Not knowing quite what to do with it, they experimented with feeding it to pigs but soon realised that even pigs could only eat so much of the stuff and looked for other outlets. They had the idea of using it as a compost and soil conditioner and contacted a recently opened broiler-chicken producer at nearby Buxted (Buxted Chickens) who were only too happy to supply chicken manure for nothing. This, when mixed and turned with the press waste and then thoroughly composted, was bagged and marketed as 'Pompost'. Initially it was a great success and although eventually forced off the market by an entirely unfounded *salmonella* scare, it was not before large quantities of it had been spread about the Brickyard vineyard. This greatly improved the clay soil and by 1969 12 acres had been planted. Ward was constantly trying to improve the soil at the Brickyard by growing 'green manure' crops in the rows and I well remember seeing most of the vineyard down to sunflowers which, at the end of the summer, were cut down and then rotovated into the soil. The Brickyard vineyard was eventually expanded by cutting down a three-acre coppice that adjoined it and grubbing-up all the tree roots that remained. A proper land drainage system was then installed before it was planted with vines.

In addition to the Brickyard, Ward had found a spare patch of ground at the Horam Manor site and in 1968 planted almost half an acre of vines. That year saw the first harvest from the Brickyard site, reported as: *12 cwt of grapes, which yielded 100 gallons of wine*. The next year was marginally better and a yield of *106 gallons* was recorded, together with the juice from some Baco noir vines on a wall at the Manor. 1970 was a real bumper year, and some 42 tons were picked from both vineyards. In 1971 another five varieties were planted at Horam Manor: Reichensteiner, Ortega, Faberrebe, Kerner and Augusta Luise, the last an early table grape. The harvest that year was 1 ton 6½ cwt from the Horam vineyard and 1 ton 3½ cwt from Horam – a grand total of 2½ tons. Some plastic tunnels were put up at Horam and the crops under these helped improve the harvests. Somebody who shared the same enthusiasm for the vines as Ward was Reg Parsons who had been gardener at The Grange, the site of the first vineyard. Together, he and Ward managed both the Brickyard and the Horam Manor vineyards.

In 1979, Ward retired as Managing Director of Merrydown and his successor, Richard Purdey, was forced by commercial considerations to take a hard look at the future of the vineyard and (grape) winemaking operations. Reluctantly it was decided that both the vineyard and the Merrydown Co-operative Scheme (see below) would have to go as the company could no longer continue to subsidise it. The 1980 vintage had not been large and the vineyard would have to be sold.

By good fortune, Ward happened to share a railway compartment back from an English Vineyards Association Board Meeting with Kenneth McAlpine, owner of Lamberhurst Vineyards.[10] Ward explained the predicament that the vineyard was in and by the time they got to the end of the train journey, it was decided that the vineyard would be bought by McAlpine and become an outpost of the Lamberhurst operation. Reg Parsons had also been facing an uncertain future and although well past retirement age, he was asked to continue to help look after the vineyard. The vineyard at Horam Manor was needed for the expansion of storage tanks and bottling facilities and this was grubbed-up. Reg continued to live in a cottage overlooking the vineyard until he died in 1988.

Under Lamberhurst's management, the Brickyard vineyard was almost entirely replanted with new grafted stock. The original vines were planted on a 5-foot (1.52-metre) row width, far too narrow for the tractors and implements used by Lamberhurst, and the new vines were planted with 7-foot (2.13-metre) wide rows and 4 feet (1.22 metres) between the vines. Varieties planted were Kerner, Müller-Thurgau, Reichensteiner and Schönburger. During one stage of the replanting it was realised that parts of the soil consisted almost entirely of broken bricks, probably the waste heap from the brick kilns and the vines had to be pickaxed into their new homes. Despite this, they seem to grow happily and full production was resumed in 1986. The vineyard remained in McAlpine's ownership until October 1994 when it was bought by John Worontschak, the flying winemaker then based at Valley Vineyards (now called Stanlake Park), who had plans to build a winery, reduce the vineyard in size and graft the vines over to Pinot noir. Sadly, these plans never came to fruition and John sold the vineyard to some travellers who parked their caravans on it (and eventually got planning permission to build a house on it). Curiously though, it still appears on the WSB's Vineyard Register!

The establishment of the Merrydown Co-operative Scheme in 1969 was another important milestone in English viticulture. Ward had seen a growing number of small vineyards with owners who, having managed to grow a good crop of grapes, had problems turning them

into commercially acceptable wine. Their vineyards were either far too small to warrant the equipment needed or they lacked the expertise required to make good wine: very often it was both. Ward's ambition was to see the revival of English winegrowing well established and anything that he or his company could do to help it would be done. It is interesting to read in one of the first contracts issued by Merrydown, the reason given for the scheme having been instituted: *Because it* [Merrydown] *considers that the enterprise* [that of growing grapes in the United Kingdom] *presents a unique challenge to the British people and is therefore worthy of such support as we are able to give.*

The scheme was essentially one of contract winemaking whereby grapes would be individually processed for their owners and returned to them as finished wine. The difference between this and other contract winemakers was that the service could either be paid for in the normal way or a proportion of the wine made would be retained to cover all costs. The proportion of a grower's wine retained varied, depending on a number of factors, but to start with it was 70 per cent for the grower and 30 per cent for the co-op. As the years went by and the true costs or running and administering this scheme became apparent, the amount retained rose to 44 per cent. The scheme was run as a co-operative non-profit making enterprise and the charges were consequently very reasonable. The wine retained by the scheme was blended together and sold under the 'Anderida' label, either through the Merrydown wine-shop or through their trade sales division. A Merrydown wine list from 1974 shows that the 1971 Anderida was selling for £1.26 a bottle, while the Brede Riesling Sylvaner, which the year before had won the first Gore-Browne Trophy, was selling for £1.83!

In 1976, Christopher (known as Kit) Lindlar, who had recently returned from some practical winemaking experience on the Mosel in Germany, joined Greg Williams in the winery just before the harvest and together they handled the largest crop ever pressed at Horam – over 180 tons – which severely strained the facilities. In 1977, the whole of the English Wine interests of Merrydown were put into a new company 'Merrydown Vineyards Limited'. Over the 10 years that it was in existence, the Merrydown Co-operative Scheme made wines for almost all the vineyards then cropping. At a time when good equipment and technical knowledge were both in short supply, it had certainly enabled many vineyards to get a properly made and presented commercial product on the shelf. It also had Royal patronage. In *Merrydown, 40 Vintage Years* by Graeme Wright, it is recalled that Princess Margaret once sent some grapes that had been grown on a wall at Kensington Palace to be made into wine, although only enough for 6 bottles!

In one sense the co-op scheme was a victim of its own success. As vineyards sending grapes to Horam became bigger and as the vines became more mature, their owners gained in confidence and started to set up their own wineries, leaving only the smaller, newer vineyards to use the scheme. Eventually it became obvious that the costs

10 Ward was extremely lucky in this as Kenneth McAlpine told me that he never took the train having a car and driver at his disposal at all times (as well as flying his own helicopter for longer journeys).

of running the scheme were too high and in 1980 it was wound up. Lindlar and Williams decided that the time had come to find a new home and in August 1980 they both moved to Biddenden Vineyards, Lindlar to become a contract winemaker on his own account and Williams to concentrate on vine sales. Most of the old Merrydown customers moved with Lindlar to Biddenden. Lindlar later left Biddenden to set up his own winery, High Weald Wines and made the first few vintages of an experimental sparkling wine called – wait for it – Nyetimber. Lindlar eventually gave up winemaking, sold most of his winery equipment to RidgeView and took up the cloth, becoming a minister in the Church of England. Williams continued to sell vines for a number of years until tragically killed in an accident in 1988.

Apart from putting the services of Merrydown at the disposal of English winegrowers through the Co-operative Scheme, Ward was also very involved with the English Vineyards Association. He was its first Chairman from the initial Annual General Meeting on 18 January 1967 until April 1981. During his time as Chairman, the Customs and Excise were persuaded to grant growers their 'Domestic Use Allowance', an annual duty-free award of up to 1,100 litres of wine; the Wye College experimental vineyard was planted and produced some good data for growers; and the EVA Certification Trade Mark or Seal of Quality as it became to be known, was eventually agreed to by the Board of Trade and put into operation. He remained as a Director until 1983 but continued to attend Board Meetings until 1985. He was awarded the OBE in the 1979 New Year's Honours List for 'Services to English Wine' and was presented with his award by the Queen at Buckingham Palace on 27 March (and missed – for the first and last time – an EVA Board meeting). After the presentation he was given a celebration lunch at the Farmers Club by the whole Board together with Sir Guy and Lady Salisbury-Jones. Ward also wrote *The Complete Book of Vine Growing in the British Isles* (1984, Faber and Faber) and although now out of print, is remains the only book on viticulture written specifically for this climate.

Jack Ward died on 10 August 1986. He was responsible in no small way for the development of the industry at a time when few believed it had a future. He was a kind, sensitive man, who gave time to anyone who wanted to speak to him about growing vines or making wine and must be judged as one of the founding fathers of English wine.

Bibliography for *Wine Growing in Great Britain* 2nd Edition

Basler P, Scherz R, *PIWI Rebsorten* Stutz Druck, Wädenswil, Switzerland, 2011.

Bishop J, McKay H, Parrott, D and Allan, J *Review of international research literature regarding the effectiveness of auditory bird scaring techniques and potential alternatives.* DEFRA, 2003.

Campbell, Christie *Phylloxera.* HarperCollins, London, 2004.

Coombe, B. G., and Dry, P. R., (eds), *Viticulture Volume 1, Resources,* Adelaide, Winetitles, 1992 and 2nd edition, 2005. *Viticulture, Volume 2, Practices.* 1995.

Crossen, Tom *Venture into Viticulture.* Country Wide Press, Australia. 2003.

DEFRA *Code of practice for using plant protection products* (PB 11090). DEFRA, London, 2006

Dry, Nick *Grapevine Rootstocks,* Lythrum Press. Adelaide in association with the Phylloxera and Grape Industry Board of South Australia 2007.

Galet, Pierre, *A Practical Ampelography – Grapevine Identification,* translated by Lucie T. Morton, London, Cornell University Press, 1979.

Goode, Jamie. *Wine Science. The Application of Science in Winemaking.* London, Mitchell Beazley, 2005.

Iland P, Dry P, Proffitt T, Tyerman S, *The Grapevine, from the science to the practice of growing grapes for wine* Patrick Iland Wine Promotions Pty Ltd, Adelaide, 2011.

Jackson, David, *Monographs in Cool Climate Viticulure – No.1. Pruning and Training,* Christchurch, New Zealand, Lincoln University Press, 1997.

Monographs in Cool Climate Viticulture – No.2. Climate, Wellington, New Zealand, Daphne Brazell Associates Ltd with Gypsum Press, 2001.

Kliewer, W. M. *Grapevine Physiology. How does a grapevine make sugar?* University of California, Division of Agricultural Sciences, Berkeley, California, USA. 1981.

Landers, Andrew *Effective Vineyard Spraying* www.effectivespraying.com New York, 2010.

Lott, Heinz and Pfaff, Franz, *Taschenbuch der Rebsorten,* 13th edition, Mainz, Germany, Fachverlag Dr Fraund GmbH, 2003.

Lytle, S. E., *Vines Under Glass and in the Open,* Liverpool, Horticultural Utilities Ltd, ca 1951.

– *Successful Growing of Grape Vines,* ca 1954.

MacGregor, C.A. *Cool climate crop size estimation: Site specific.* Fifth International Symposium on Cool Climate Viticulture and Oenology. Melbourne, 2000.

Maltman, Alex. *Vineyards, Rocks and Soils, a Wine Lover's Guide to Geology* Oxford University Press, Oxford 2018.

May, Peter *Flowering and Fruitset in Grapevines.* Phylloxera and Grape Industry Board of South Australia with Lythrum Press. Adelaide. 2004.

McGrath, M. T. What are Fungicides? The Plant Health Instructor. The American Phytopathological Society 2004

Ministry of Agriculture, Fisheries and Food (MAFF)
– *Cane Fruits, Bulletin 156,* 1975
– *Soils and Manures for Fruit, Bulletin 107,* 1975
– *Outdoor Grape Production, HSG 22,* 1978
– *Grapes for Wine, Reference Book 322,* 1980
– *Lime and Fertliser Recommendations No: 3, Fruit and Hops* 1978
– *Weed control in bush and cane fruit HGG 25,* 1979, London, Ministry of Agriculture, Fisheries and Food.

Mollah, Mahabubur, *Practical Aspects of Grapevine Trellising.* Winetitles. Adelaide. 1997.

Nicholas P, Margery P, and Wachtel M, *Diseases and Pests,* Winetitles, Ashford, South Australia, 2011.

Nix, John, *Farm Management Pocketbook 44th Edition* Agro Business Consultants Ltd, Melton Mowbray, 2014

Ordish, George, *The Great Wine Blight,* London, J. M. Dent and Sons Ltd, 1972, 1986.

Pearkes, Gillian, *Vinegrowing in Britain,* London, J. M. Dent and Sons Ltd, 1982, 1989.

Pearson R, and Goheen A, *Compendium of Grape Diseases,* APS Press Minnesota, USA 1988.

Robinson, Jancis, *Oxford Companion to Wine* (ed.), Oxford University Press, 1994, 1999, 2006.

Robinson J, Harding J and Vouillamoz J, *Wine Grapes* Allen Lane, London, 2012.

Rousseau J, Chanfreau S, Bontemps E, *Les cépages résistants aux maladies crptogamiques. Panorama européen,* Groupe ICV, 34970 Lattes, France, 2012.

Skelton, Stephen P., *Viticulture – an introduction to commercial grape growing for wine production,* London, S. P. Skelton, 2009 and 2020.

Smart, Richard and Robinson, Mike, *Sunlight into Wine – A Handbook for Winegrape Canopy Management,* Adelaide, Winetitles, 1991.

Trought, M.C.T, Howell, G.S, Cherry, N. *Practical Considerations for Reducing Frost Damage in Vineyards* Report to New Zealand Winegrowers 1999, Lincoln University.

Trout, R.C. and Pepper, H.W. *Forest Fencing. Forestry Commission Technical Guide,* Forestry Commission, Edinburgh, 2006.

van Leeuwen, C, Trégoat O., Choné X, Gaudillère J-P, and Pernet, D. *Different environmental conditions, different results: the role of controlled environmental stress on grape quality potential and the way to monitor it.* Proceedings.

13th Australian Wine Industry Technical Conference 2007, AWITC inc, Glen Osmond,   Australia.

Waldin, Monty *Biodynamic Wines*. Mitchell Beazley. London. 2004.

Ward, Jack, *The Complete book of Vine Growing in the British Isles*, London, Faber and Faber, 1984.

White, Robert E., *Understanding Vineyard Soils*, New York, Oxford University Press, 2009.

Wilson, James E. Terroir. *The Role of Geology, climate and Culture in the Making of French Wines*. Mitchell Beazley. London. 1998.

Wrotham Historical Society, *Farming in Wrotham Through the Ages – The Wrotham Grape*, Wrotham Historical Society, 2004, pp. 56–8.

Biography – Stephen Skelton MW

Stephen Skelton has been involved with growing vines and making wine since 1975. He spent two years in Germany, working at Schloss Schönborn in the Rheingau and studying at Geisenheim, the world-renowned college of winegrowing and wine-making, under the late Professor Helmut Becker. In 1977 he returned to Britain to establish the vineyards at Tenterden in Kent (now home to Britain's largest wine producer, Chapel Down Wines), and made wine there for 22 consecutive vintages. From 1988 to 1991 he was also winemaker and general manager at Lamberhurst Vineyards, at that time the largest winery in the country. During his time as a winemaker he won the Gore-Browne Trophy in 1981, 1990 and 1991. Stephen now works as a consultant to vineyards and wineries in Britain, setting up new vineyards for the production of both still and sparkling wine and helping existing growers expand. He was instrumental in finding the site for and setting up Domaine Evremond, the Champagne Taittinger vineyards and winery in Kent and has planted all their vineyards to date.

In 1986 Stephen started writing about wine and lecturing for the WSET and has contributed articles to many different publications. He has written several books on the wines of Great Britain: *The Vineyards of England* in 1989, *The Wines of Britain and Ireland* for Faber and Faber in 2001 (which won the André Simon Award for Drinks Book of the Year), the *UK Vineyards Guide* in 2008, 2010 and 2016 and *The Wines of Great Britain*, part of the Classic Wine Library, in 2019. In 2014 he wrote and published *Wine Growing in Great Britain* and published the 2nd Edition of this book in 2020. His best-selling book is *Viticulture – An introduction to commercial grape growing for wine production* which was first published in 2007, with an updated 2nd edition published in 2020, and has sold well over 10,000 copies. This book is aimed at WSET Diploma and Master of Wine candidates and is now being translated into Japanese and Chinese. He was for many years the English and Welsh vineyards contributor to the annual wine guides written by Hugh Johnson and Oz Clarke and currently writes the section on English and Welsh wine in both Jancis Robinson's *Oxford Companion to Wine* and Hugh Johnson and Jancis Robinson's *World Atlas of Wine.*

Stephen was a director of the English Vineyards Association (EVA) from 1982–1995 and of its successor organisation, the United Kingdom Vineyards Association (UKVA) from 1995–2003. He was Chairman of the UKVA from 1999–2003. He was also at various times between 1982 and 1986 Treasurer, Secretary and Chairman of the South East Vineyards Association, Secretary of the Circle of Wine Writers between 1990 and 1997 and has served on various EU committees in Brussels representing British winegrowers. Since 2018 he has chaired WineGB's Viticultural Working Group.

In 2000 he completed a BSc in Multimedia Technology and Design at Brunel University. While at Brunel, Stephen was awarded the Ede and Ravenscroft Prize for his final year project, a touch-screen 'retail wine selector'. In 2011 he was awarded an Honorary Doctor of Business Administration from Anglia Ruskin University and in 2012 he completed an MA in Creative Writing (Biography) at the University of East Anglia.

In 2003 Stephen became a Master of Wine, winning the prestigious *Robert Mondavi Trophy* for gaining the highest marks in the Theory section of the examination and in 2005 won the *AXA Millésimes Communicator of the Year Award* for services to the MW education programme. Between 2003 and 2009 he served on the MW Education Committee and was Course Wine Coordinator, served on the Council of the Institute of Masters of Wine between 2009 and 2015 and was Chair of the Research Paper Examination panel from 2013 to 2021.

Stephen has been the panel Chairman for English and Welsh wines for the Decanter World Wine Awards since 2008 and has judged in the past for the International Wine Challenge, the International Wine and Spirit Competition, the Japan Wine Challenge and the Veritas Wine Awards.

In 2020, Stephen was awarded the Wines of Great Britain *Lifetime Achievement Award* even though, at the age of seventy-two, he believes he still has a few years remaining.

Index

Note, reference in italics indicate items that are not concomitantly mentioned in the text on the same page. Footnotes are indicated by 'n'. TDs are trunk diseases.

www.ingramcontent.com/pod-product-compliance
Lightning Source LLC
Chambersburg PA
CBHW061104210326
41597CB00021B/3974